The Huxleys

The Huxleys

An Intimate History of Evolution

ALISON BASHFORD

THE UNIVERSITY OF CHICAGO PRESS

The University of Chicago Press, Chicago 60637
© 2022 by Alison Bashford
All rights reserved. No part of this book may be used or reproduced in any
manner whatsoever without written permission, except in the case of brief
quotations in critical articles and reviews. For more information, contact the
University of Chicago Press, 1427 E. 60th St., Chicago, IL 60637.
Published 2022
Printed in the United States of America

31 30 29 28 27 26 25 24 23 22   1 2 3 4 5

ISBN-13: 978-0-226-72011-1 (cloth)
ISBN-13: 978-0-226-82412-3 (e-book)
DOI: https://doi.org/10.7208/chicago/9780226824123.001.0001

First published in the United Kingdom as *An Intimate History of Evolution:
The Story of the Huxley Family* by Allen Lane, an imprint of
Penguin Random House UK, 2022.

Library of Congress Cataloging-in-Publication Data

Names: Bashford, Alison, 1963– author.
Title: The Huxleys : an intimate history of evolution / Alison Bashford.
Description: Chicago : The University of Chicago Press, 2022. | Includes
    bibliographical references and index.
Identifiers: LCCN 2022009756 | ISBN 9780226720111 (cloth) | ISBN
    9780226824123 (ebook)
Subjects: LCSH: Huxley, Thomas Henry, 1825–1895. | Huxley, Julian,
    1887–1975. | Evolution (Biology)—History.
Classification: LCC QH361 .B375 2022 | DDC 576.8—dc23/eng/20220331
LC record available at https://lccn.loc.gov/2022009756

♾ This paper meets the requirements of ANSI/NISO Z39.48-1992
(Permanence of Paper).

For
Keith Harvey Bashford, 1934–2022

Contents

CONTENTS

List of Illustrations

ix

Author's Note on Names

There are many Huxleys in this book. To distinguish between the two on whom I focus most closely, Thomas Henry Huxley and Julian Huxley, I use 'Huxley' for the grandfather and 'Julian' for the grandson.

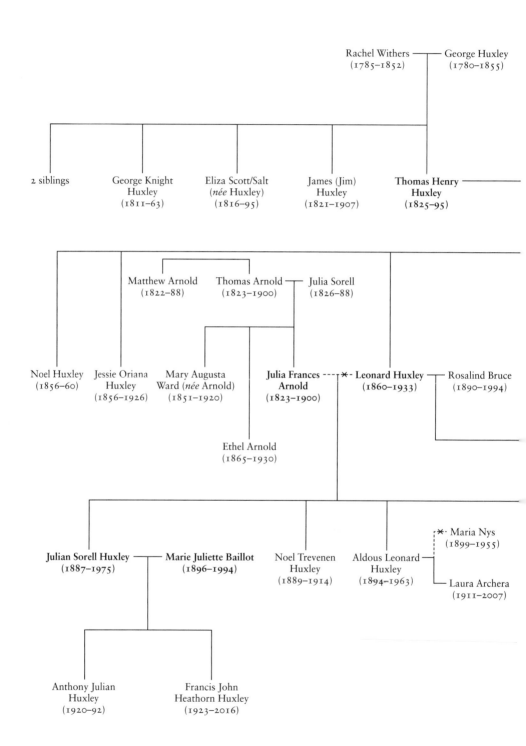

Rachel Withers ——— George Huxley
(1785–1852)   (1780–1855)

2 siblings   George Knight   Eliza Scott/Salt   James (Jim)   **Thomas Henry**
              Huxley   (*née* Huxley)   Huxley   **Huxley**
              (1811–63)   (1816–95)   (1821–1907)   **(1825–95)**

Matthew Arnold   Thomas Arnold —— Julia Sorell
(1822–88)   (1823–1900)   (1826–88)

Noel Huxley   Jessie Oriana   Mary Augusta   **Julia Frances** --- ╳- **Leonard Huxley** —— Rosalind Bruce
(1856–60)   Huxley   Ward (*née* Arnold) **Arnold**   **(1860–1933)**   (1890–1994)
               Huxley   (1851–1920)   **(1823–1900)**
               (1856–1926)

                              Ethel Arnold
                              (1865–1930)

                                                                                        ╳- Maria Nys
                                                                                           (1899–1955)
Julian Sorell Huxley —— **Marie Juliette Baillot**   Noel Trevenen   Aldous Leonard —
(1887–1975)   **(1896–1994)**   Huxley   Huxley
                                                           (1889–1914)   (1894–1963)
                                                                                        └ Laura Archera
                                                                                           (1911–2007)

Anthony Julian   Francis John
Huxley   Heathorn Huxley
(1920–92)   (1923–2016)

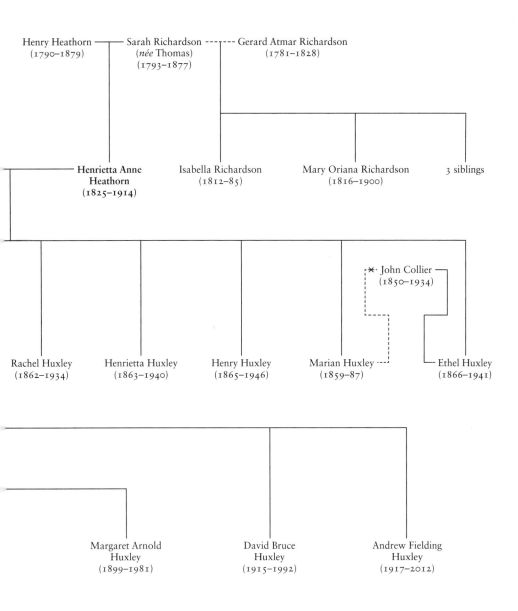

Henry Heathorn (1790–1879) — Sarah Richardson (*née* Thomas) (1793–1877) ------ Gerard Atmar Richardson (1781–1828)

Henrietta Anne Heathorn (1825–1914) Isabella Richardson (1812–85) Mary Oriana Richardson (1816–1900) 3 siblings

John Collier (1850–1934)

Rachel Huxley (1862–1934) Henrietta Huxley (1863–1940) Henry Huxley (1865–1946) Marian Huxley (1859–87) Ethel Huxley (1866–1941)

Margaret Arnold Huxley (1899–1981) David Bruce Huxley (1915–1992) Andrew Fielding Huxley (1917–2012)

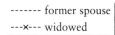

------- former spouse
---✕--- widowed

Introduction

How are humans animal and how are we not? What is the nature of time and how old is the Earth itself? What might the planet look like – with or without humans – 10,000 years hence? More modestly, who asks such questions for a living, and who thinks they have the answers?

Thomas Henry Huxley and his grandson Julian Huxley did. In 1947 they featured in *Life* magazine, one dead, one living. At the photo shoot in his Hampstead library, Julian self-consciously arranged himself in front of a portrait of his grandfather (figure 0.1). In the foreground, a well-known mid-twentieth-century science writer, zoologist, conservationist. In the background, a mid-nineteenth-century scientific naturalist – Charles Darwin's most outspoken spokesman. Julian was displaying his scholarly bloodline over one shoulder and the authority and expertise inherited from the library over the other, many of them his grandfather's own books. Well might he do so. Over the nineteenth and twentieth centuries, these two men of science, these self-appointed cosmologists, communicated to the world the great story of evolution, including the birth, death and rebirth of the idea of natural selection. They were 'trustees of evolution', a phrase that Julian often used to describe all of humankind, but which in truth applies more neatly to the Huxleys themselves.[1]

The younger man constantly fashioned himself after his Victorian grandfather, pursuing those signature Huxley knowledge-quests, some profound, others simply grandiose. They were both remarkable and both, on occasion, tortured. Writing these natural scientists together permits a kind of time-lapse over the nineteenth and twentieth centuries, precisely because they were so similar. We might even think of

Figure 0.1: Julian and Thomas Henry Huxley,
trustees of evolution: *Life* magazine photo shoot, 1947.

them as one very long-lived man, 1825–1975, whose vital dates bookended the colossal shifts in world history from the age of sail to the space age; from colonial wars to world wars to the Cold War; from a time when the Earth was 6,000 years old according to Genesis, to a time when it was 4.5 billion years old, according to rock samples returned from the Apollo missions.

Two other portraits help us understand these trustees of evolution, and how they were keen players in the great modern effort to comprehend and fix 'nature' and 'culture'. The likenesses are windows into their very different scientific souls, but we need not look into their eyes, rather at what they hold.

Julian holds 'culture', an Igbo wooden sculpture from Nigeria, bought on the cheap in his West African travels in 1943 (figure 0.2). It was, he tells us, a sacred object, a 'small idol . . . destined for household shrines'. Instead it ended up on a Hampstead shrine of a very different order.[2] He was self-fashioning his curiosity and his cosmopolitanism, as well as his scientific inheritance. Student and lover of nature, inheritor of agnostics, brother of Aldous, believer in evolution by natural selection, and (his own invention) evolutionary humanism, Julian was from the beginning inclined to culture. Even as a zoologist, he was foundational to 'ethology', the study of animals' behaviour, how they and we act, think and even feel. Yet culture was itself part of evolution and technically so, not just loosely or illustratively. Human minds are unique in evolutionary terms, Julian always insisted, precisely because we can comprehend just that.

T. H. Huxley, by contrast, holds nature (figure 0.3). Like a sovereign holding an orb, Huxley claims the skull-of-everyman. The representative of God's dominion on earth is transformed into secular trustee of all humankind. His most famous likeness was captured in heavy Victorian oils by the celebrated portraitist John Collier, as it happens Thomas Henry Huxley's son-in-law, Julian's uncle. The patriarch of one idiosyncratic family announces himself not just as patriarch of science, but also of what he called the Man Family, *Homo sapiens*, including its pre-human ancestors. This cranium was highly symbolic, as the family artist well knew, a sign of Huxley's ascendancy over British natural sciences, and a secular talisman of his great intervention: 'evidence as to man's place in nature'.[3]

Figure 0.2: Julian Huxley and Culture: *Life* magazine photo shoot, 1947.

Figure 0.3: Thomas Henry Huxley and Nature:
oil by John Collier, 1883.

One winter's evening in 1867, T. H. Huxley delivered a lecture on ethnology, the science of man, to working men in the Mechanics' Institute in Chancery Lane and was enjoined to take a collection to relieve distress in the East End. Huxley said he 'would simply place his own subscription in one of the skulls on the table', and others could follow if they wished. A human skull served as a receptacle for the collection; the strangest secular offertory box. The entire lecture hall fell in with this mixed choreography of ethnology, self-help, poor relief and almsgiving.[4] Many would then, as now, rightly consider human remains to be somehow sacred, certainly as sacred as the effigy Julian held. Yet Huxley the grandfather comprehended nothing of the sort. Skulls were workaday material objects, fascinating, but in no sense sacrosanct. Neither in death nor in life was the human body the receptacle of a soul, at least we can never show that to be the case, Huxley the agnostic insisted. He battled his whole life for nature, and for a strictly scientific method to understand it.

The modernization of natural sciences of which Huxley was a key player, even dramaturge, challenged the magical and the supernatural. It heralded the death of the divine, 'disenchantment' as the sociologist Weber had it, the death of God according to the philosopher Nietzsche. In different ways they were both responding to the ascendance of scientific naturalism that T. H. Huxley spearheaded, and in a German scholarly tradition in which Huxley was deeply learned – self-taught, of course. There are no spirits, no Creator, no divine, at least none of which we can be sure. Wonderful! Huxley would have exclaimed had he lived long enough to read the great scholarly debates that unfolded on secularization and disenchantment over the twentieth century. And yet he might also have questioned why, then, his biologist grandson Julian was so drawn to sacred idols, studied parapsychology and wondered occasionally about ghosts; why his literary grandson Aldous sought radically other perceptions and doors into them; not to mention his anthropologist great-grandson, Francis Huxley, who became so immersed in the magical and the mystical, in voodoo and in rituals of sacralization, as to become himself a kind of animist.

One trajectory of this intimate history of evolution, according to the thesis of disenchantment, *should* be a history of secularization, a victory of T. H. Huxley's rational agnostics in a dynasty of scientists. In fact, it is a story of Huxley re-enchantment. Over three and four generations of modernity, the Huxleys searched proactively for something to put in the place of religion, compatible with Grandfather's scientific naturalism but somehow exceeding it. Julian ended up a kind of neo-Romantic naturalist, complete with visions and poetry, closer in some ways to the early nineteenth-century *Naturphilosophie* that his grandfather had railed against. In short, these Huxleys travelled over the modern period, symbolically *from* the disenchanted skull-object of nature *to* the enchanted, carved, sacred object of culture, not the reverse. Julian sometimes even imagined himself as a kind of Creator who bestowed his own Word on humankind: evolutionary humanism. Or perhaps, like Moses, he authored this new scientific religion for the world, Truth certainly not acquired from heaven, but instead from Earth, from nature itself.

The cranium that Thomas Henry Huxley holds with such propriety, almost defiance, stands for the species *Homo sapiens*. But we know that a particular human used to be 'in' it, used to *be* it. It once held a brain, a mind, a consciousness, a human (or perhaps a pre-human) *being*, even if, for Huxley, a soul-less one. But agnosticism about a soul did not mean that Huxley wasn't interested in 'feeling'. On the contrary, consciousness and perception, sense and sensation were all fascinating to him as a physiologist and a learned philosopher of mind, not just a comparative anatomist. He was a relentless inquirer into how the brain and mind inside a human skull senses, thinks, perceives, feels and does so similarly to or differently from other animals, radically extending ancient debates about human reason and animal instinct. Well known as a founder of British anthropology, he was part of the beginning of psychology too. By his grandson Julian's early adulthood, this particular Huxley family inquiry had evolved from studies of consciousness and unconsciousness – mesmerism, hypnotism, anaesthesia, reflex – to the analysis of a twentieth-century phenomenon called the unconscious. Julian consumed Freud. And by the 1950s and 1960s, the Huxley brothers, Julian and Aldous, were actively wondering about altering consciousness, through telepathy, through *Vedanta*, through mescalin and LSD. The brother who lived in safe and cosy Hampstead was a *philosopher* of new kinds of human perception, at one remove. The brother who lived in edgy California was a *practitioner*, fully immersed in new experiential perception.

The material relationship between a skull, a brain, a mind and individual human consciousness turned out to be intergenerational Huxley business. It turned out to be intergenerational Huxley suffering as well. Here, the history of evolutionary ideas becomes as intimate as it is possible to be. The Huxleys were a family deeply afflicted by mental illness *and* they were a brilliant and inquiring family. They both embodied and thought through genius and melancholy as these states had been imagined and put together over the nineteenth century in science and letters, and in the twentieth century by psychoanalysis, behavioural psychology and psychopharmacology. A Huxley collective depression was only matched by the family capacity to think about it in highly curious ways, to be intrigued by the mysterious relationship between mind and body via the brain that sat inside those skulls.

The grandfather and grandson looked often to outer worlds and sometimes to other worlds. They were cosmologists of sorts. Over their lifespans, 1825–1975, time got longer and longer, the past deeper and deeper, not just the prehistory of humans, but also the geohistory of the Earth. This was a phenomenal conceptual revolution. Yet this one long-lived man was Janus-faced with regard to the time revolution, the Victorian Huxley looking backwards to a receding past, while Julian the modern was enchanted by the temporal beyond. Julian's compelling non-fiction essays about the future of humankind, and even of the planet, were perhaps offset by his second-rate science fiction attempts, shadows of his brother's brilliance. Yet in the twentieth century, Julian and Aldous Huxley shared visions of all kinds of possible new worlds, an intellectual and fraternal intimacy of a profound order.[5]

As often as they looked to outer or future or nether worlds, the Huxleys looked inwards, into their very selves. Such introspections were sometimes public in the form of poetry books and unrestrained memoirs, or else were entwined in essays on religion, on ethics, on philosophies of knowing and ways of being. For neither Thomas Henry nor Julian Huxley were their depressions ever fully hidden or sequestered. And this introspection rarely manifested as withdrawal from the world. Thomas Henry Huxley was hardly retiring. He was both self-made and self-satisfied, especially over his fights with the world. He was a big personality, bold and fractious. Yet he nurtured honest, emotional and not infrequently confessional intimacies with his many close friends, men of science and men of God. Victorian masculinity was so much more forgiving and enabling than its carica-ture of repression. Thomas Henry Huxley needed it to be; his struggle with melancholy was relieved in meaningful ways by his tight scien-tific circle.

Thomas Henry Huxley's character booms through the records, but what was Julian Huxley like? There is no lack of description, since his circle included as many poets and writers as biologists, observers good with their words and sharp with their insight. It puts him at an unfair disadvantage, biographically speaking. One author who mar-ried into the wider family, Elspeth Huxley, found him rude, vain and selfish.[6] Another writer – who knew him well – thought him 'brilliant, charming, idealistic, gregarious', but also 'unstable, full of self-doubt, and a raging egotist'. After Julian's death in 1975, the distinguished American poet May Sarton wrote to his wife Juliette, diagnosing without mercy; 'a very spoiled baby person'.[7] This 'intimate history' is more sympathetic to Julian. Partially hidden to such poet-observers but known too painfully by his long-suffering wife, who had her own elegant and honest words to put to her life-partner, was Julian's para-lysing depression. On occasion his mind and body slowed almost to a standstill, his soul all but too heavy to bear, his affliction dreaded and terrible. An accomplished poet himself, when Julian was unable to gather words coherently to describe anything let alone such pain, he reached for those of others, confessing publicly how his mind and soul were tortured, how he lived Walt Whitman's 'terrible words': 'Hell under the skull-bones; Death under the breast-bones'.[8] The fam-ily's most famous wordsmith, Aldous, reintroduced to the twentieth century another term, 'accidie'. Chaucer's Middle English described the condition that 'forsloweth and forsluggeth' a man.[9] This was what was passed from Thomas Henry Huxley to his grandson, and what was inherited here and there across the wider Huxley family.

In the psychoanalytically intense 1920s, Julian disclosed to the world the crippling melancholy he shared with his grandfather – that is to say, crucially and technically, that he knew he had *inherited*. His public honesty in early adulthood makes the child's hand in this early studio portrait softly covering – claiming, seeking – his grandfather's poignant and painful (figure 0.4). It holds a truth. These men of science were privately tied together by their suffering. Publicly and professionally they were tied together by Julian's own hand, by his work and will, as he actively grafted himself onto his grandfather's lineage.

T. H. Huxley (1825–1895) was one of the Victorian world's best-known zoologists and comparative anatomists, doggedly promoting and defending Charles Darwin in the earliest public dispute over the theory of evolution by natural selection. Huxley was up for the fight. He sought it, enjoyed it, won it. In the process he challenged Britain's established power structures, most especially the Church of England. Raised outside London and then in the Chartist Midlands in an educated but struggling and socially declining family, Huxley was not simply granted his position as professor of palaeontology and natural history at the Royal School of Mines, what had started as the Museum of Economic Geology and became Imperial College London. Rather, he fought for it, retaining a lifelong belief in merit-based rewards and effort-based achievement over inherited authority and position. This stance was explicit and political, sometimes bitter and always antagonistic. It rocket-fuelled his battles against conventions, against authorities and against orthodoxies. The explosive idea of evolution by natural selection was as suited to Huxley's character as it was a paralysing ill match for Charles Darwin. Together, and with a close circle of like-minded friends, they drove a new scientific naturalism in the middle of the nineteenth century, contesting any explanation of nature, old or new, that relied on a beyond-natural or supernatural force or origin.

Victorian science and letters were conjoined when Huxley's son Leonard married Julia Arnold of the literary family – Thomas Arnold, Matthew Arnold, Mary Arnold (Mrs Humphry Ward). Julian was their first surviving child, born in 1887 and crossing vital dates with his famous grandfather for eight important years. The eldest of four siblings, with Noel Trevenen (1889–1914), Aldous Leonard (1894–1963) and Margaret Arnold (1899–1981), his pastoral Surrey childhood was spent in birdwatching, verse lessons and German induction. Julian was

Figure 0.4: Thomas Henry and Julian Huxley, 1895.

a beneficiary of his grandfather's fights in scientific and educational fields. A great rise in family fortunes saw him a student at Eton and then Balliol College, Oxford, from 1906. There, natural sciences had completely transformed from his grandfather's era. Julian learned evolutionary theory as a matter of course, not as a matter of controversy. But he also studied the then-strong critique among scientific discontents about natural selection specifically. A technical problem with Darwinian natural selection was something that Julian helped communicate and resolve in the 1930s and 1940s, what he called 'the modern synthesis' of Mendelism and Darwinism, genetics and evolution.

In a remarkable turnaround – one that we can put down to larger-than-life H. G. Wells – Julian traded in an academic career in zoology for what we would now call science communication. He went on to become an adventurous communicator in media unimaginable to his grandfather; in broadcasting, in film and eventually in early television. He even won an Oscar. In so many ways, Julian Huxley was David Attenborough's antecedent. More than that, in the early 1950s, Julian Huxley was in front of the camera, with a young David Attenborough, producer, behind it. Along the way – and in between paralysing depressions – Julian was controversial secretary of the London Zoo in the late 1930s and equally controversial director-general of Unesco in the late 1940s. He was an A-list international speaker on population control in the 1950s and 1960s, the catastrophic era when the end of the world seemed imminent. Ours is not the first generation to think so.

The grandfather invented the term 'agnostic'. The grandson followed through, inventing the term 'transhumanism', the idea that both individual humans and the species as a whole can be radically enhanced and improved, mentally and physically. Highly political in a remarkable era when many natural scientists were just that, Julian offered key public statements about the human species and its variations. He was an anti-fascist signatory to the so-called Geneticists' Manifesto (1939) that opposed Nazi racial dogma and early driver of Unesco's anti-racist 'Statements on Race' for the new world after fascism and after the Holocaust. He was also post-war president of the British Eugenics Society, an apparent contradiction that in fact was perfectly reconcilable in his own intellectual ecosystem. We shall see how and why. As an elderly man Julian put his name to the Humanist Manifesto (1973), carrying signature Huxley rationalism into the age of Aquarius.

What agonized Julian Huxley personally fascinated him intellectually: what do we inherit and what inheres in us? What happens over and between generations of living beings? How are similarity and variation carried forwards over time, measured in the generational years of a family tree; in the tens of thousands of years in 'the tree of man'; or even more expansively in the great geological epochs in the 'tree of life'? What do we really know about breeding, whether of peas, pigeons, horses or humans? The family history of the Huxleys doubles as an account of evolving ideas about generations and genealogy, genes and eugenics.

Evolution on the one hand and inheritance on the other are two very different natural phenomena, both revolutionized by and through Charles Darwin and Alfred Russel Wallace's idea: natural selection. Evolution was T. H. Huxley's intellectual business in the nineteenth century, while genetic inheritance was Julian's in the twentieth. Combinations and recombinations of Huxleys and Darwins routinely drew down on their shared intellectual legacy, sometimes looking to their own ancestry, eventually to consider how selective breeding might be brought to humans, as individuals, as families, as societies, as a species. They began to connect the idea of improved inheritance and the possibility of social interventions into the reproduction of mental and physical states. Generations of twentieth-century Darwins and Huxleys crossed paths in and around the Eugenics Society in London, rebooting an idea born from Charles Darwin's interest in artificial selection and folding it into Mendelian genetics on the one hand and modernist futurism, even progressivism, on the other. Thomas Henry Huxley was not so sure about this ambition to improve future humans. Julian Huxley was 100 per cent sure.

For decades now, historical and political analysis of eugenics has rightly bedded down a wide-ranging critique of its ambitions as a race-oriented programme, linked to genocide. It is difficult to comprehend how, for some post-war actors like Julian Huxley, eugenics and *opposition* to biological racism were compatible. For Julian, the Nazi version of eugenics was just that – a dangerous, misplaced, expedient programme, egregious in its racist implementation. Strange to say, for him, eugenics needed rescuing, not abandoning. At one level, this is explained by the fact that British eugenics was in the first instance about mental health and ill-health, fitness and unfitness in early twentieth-century language, rather more than an intervention into race difference, race policy or racial reproduction of various kinds. Implicitly and occasionally explicitly, British eugenicists discussed 'the race', by which they sometimes meant the nation, sometimes the species, *Homo sapiens*. Occasionally they did mean 'white' or 'Caucasian', 'Negro' or 'Asian' or 'Aryan', or a hundred other spurious divisions of humankind. Herein lies a history of the very idea of 'race', the actual term, as its meaning changed from species differentiation to intra-human difference. Thomas Henry Huxley generally disavowed the term 'race' with regard to humans, preferring 'stocks', while his grandson influentially argued from population genetics that there was no such biological thing as 'race', installing instead the idea of cultural 'ethnicity' as part of his active mid-twentieth-century anti-racism.

The age of evolution recast key components of the age of revolution, not least the fundamental question: 'Are men born free and equal?' This was an Enlightenment proposition over which battles were fought and lives were lost, over which republics were born and liberal democracies reformed. It turned into a biological question in the nineteenth century, not only, but not least, by the hand of Thomas Henry Huxley.[10] He penned the essay 'Emancipation – Black and White' in the 1860s, in the context of the Civil War on one side of the Atlantic and the women's movement for civic and educational equality on the other. Julian, by contrast, considered political in/equality in an age when human rights were being codified. A high internationalist, he himself was on the outer edge of the new United Nations that rendered the recognition of equality between all humans aspirationally universal.

Confusingly, even shockingly, both Huxleys' biological answer to the modern political question 'Are men born free and equal?' was categorically no. Over and again, just when the Huxleys seem to redeem themselves by our measures, they incriminate themselves again. The reverse is also true. Just when their bigotry, racism and sexism were at their most blatant, cracks appear in our judgement from the present, usually because someone else from *their* present steps in to present a more complex picture. T. H. Huxley's insistence on a singular human species, for example – not multiple kinds of biological humans as some others thought – was adopted by anti-slavery emancipationists to further their cause. Julian, to take another example, wrote an egregious suite of articles on 'The Negro Problem' and a ridiculously colonial and superior book on his African travels, and yet this prompted the sociologist W. E. B. Du Bois to respectfully invite a Huxley contribution to his *Encyclopedia Africana*, the great African-American ambition to offset the *Encyclopedia Britannica*. In 1946 Julian made *Man – One Family* with the Jewish communist filmmaker Ivor Montagu, pressing for an anti-racist and (what we would later call) multicultural post-war world.[11] In the very same year, he wrote eugenics into his manifesto for Unesco and declared therein that humans are not born equal. Herein, we try to understand these flawed Huxleys less through our own imperfect measures, and more through the measures and responses of their own changing times. In many ways, it is harder to do so. The positioned and politicized Huxleys are a good measure of the complexities of these matters over two centuries.

Determining, fixing, substantiating 'man's place in nature' was the most enduring Huxley project of all. And yet that required as much focus on non-human animals as on *Homo sapiens* and earlier versions thereof. Both the grandfather and the grandson were zoologists before they were human scientists of various nineteenth- and twentieth-century orders, enchanted by accumulating knowledge about creatures of the sea, of the land, of the sky.

   T. H. Huxley's journey to the Pacific in the 1840s was his one major experience as a zoologist 'in the field', even though he was commissioned to the Royal Navy's HMS *Rattlesnake* as assistant surgeon, not as a naturalist. From Rio to New Guinea, up and down the east coast of Australia, the *Rattlesnake*'s voyage ranged for Huxley from becalmed boredom and deep melancholy, to an almost unbearable love-sickness for his Sydney-based beloved, Henrietta Heathorn, who became his long-lived wife. Along the way he saw dugongs and starfish, marsupials and tropical birds of colours and sizes never to be adequately caught by his watercolour on parchment. It was analysis of the unlikely jellyfish and sea anemone that earned him a precocious fellowship of the Royal Society. He scooped them out of the warm waters, cut them apart and studied them minutely – often with great maritime difficulty – under his microscope. Thereafter, his zoological expertise turned to vertebrates, famously primates. And while as a physiologist he was interested in living responses, he learned most from dead animals and bits of them: from those skulls; from the fossils that gradually revealed the link between dinosaurs and birds; and from so-called 'soft anatomy', nerves or brains or spinal cords that could be dissected. In his museum and laboratory in Jermyn Street and later in South Kensington, Huxley collected, researched and taught for decades from fossils and skeletons and pickled embryos. The most recently animate things around him were the pinned bugs and beetles in the entomology collections and the preserved mammalian, reptilian and avian trophies suspended in taxidermic life.

Julian's world as a zoologist was vital and animate, not lifeless. For him, animals were best observed alive in wonderful interaction with each other. Far from his grandfather's zoological method that relied on dead beasts, some of them so old as to be stone, 'fossilized', Julian observed behaviour and even 'emotional' relationships between living creatures, an ethnographer of the animal world. His first and enduring love was with birds. Julian's best-known zoological work remains, to this day, 'The Courtship Habits of the Great Crested Grebe' (1914), drawn from fieldwork observations in Hertfordshire undertaken in 1912 with his brother Trevenen.[12] It is a much-cited classic in ethology, the foundation for his study of rituals, those enchanting 'dances' between the grebes. Importantly, also, it was a challenge to Charles Darwin's other theory, that of sexual selection. These water birds, unlike Darwin's peacocks, were minimally sexually dimorphous. It was difficult to distinguish between the female and the male. How then did sexual selection work?

Julian's animals were not always wild or even faux wild, but he definitely preferred them alive. If Thomas Henry Huxley inhabited a museum, Julian Huxley inhabited a zoo, for much of the late 1930s and the war years living in an apartment inside London Zoological Gardens. In his later life, however, it was not Regent's Park but Africa that provided him the opportunity to serve as ecologically responsible trustee to animals. The conservation of African wildlife in situ became his mission. For his grandfather, 'Africa' was only to be found in the Athenaeum Club or the British Museum libraries, and that special African primate, the gorilla, he only ever saw as a drawing or as a skeleton. Julian, by contrast, was captivated by, we might well say in love with, London Zoo's most famous gorilla, Guy. He was, in this regard, more the intellectual grandson of Charles Darwin than of Thomas Henry Huxley, the great evolutionist a century earlier climbing into some of those very same primate cages in Regent's Park to observe the expression of animal emotion.[13]

Part of the great modern ambition to distinguish culture from nature *was* the project of distinguishing humans from animals, a drive which has run through philosophy, science and art for most of the 200 years of late modernity and which drew down on the entire classical, ecclesiastical and humanist canon before that. What does it mean to be human? Where does humanity end and animal begin? Historian Joanna Bourke suggests that 'Darwinian arguments may have contributed to the deconstruction of the radical differences imagined between humans and animals, but humanism survived this attack.'[14] This is precisely where the Huxleys sat intellectually, performing the dual work of modernity that both affirmed us as animal, part of evolution, and that in the end found ways clearly to separate humans from brutes.

The Victorian grandfather insisted on natural humanity, eroding a theologically endorsed human uniqueness offered by that other 'Origin' book Genesis. Most natural scientists by the mid-nineteenth century did not need Thomas Henry Huxley's mocking of the special creation of man then woman on the sixth day, after the creation of fowls, creeping things, fishes and beasts of burden, to question Genesis. Not for several generations had most naturalists read Genesis literally. And yet as a rule they stuck one way or another to the idea of humans as special, by reason, by language, by capacity to conceptualize an afterlife and, more modestly, by an opposable thumb. Julian knew far more profoundly, technically and fully than his grandfather, even than Charles Darwin, that evolution by natural selection acted on everybody, everything and always, yet he was the one who maintained, even extended, this tradition of insisting that humans *were* special: we evolved that way. One extraordinary effect of the long evolutionary history of *Homo sapiens*, of 'man's place in nature', was that culture and knowledge itself could be inherited, a story of evolution from a happenstantial opposable thumb to a particular kind of mind. For Julian, this set humans apart entirely.

These Huxleys dealt with life on Earth on every scale imaginable, from the solar system to the soul, in orders massive and minute. They were masters at communicating how the smallest matter of nature fitted into a whole, and the bigger the whole the better. Evolution, and human knowledge of it, was everything, Julian proposing in the mid-1960s that evolutionary science 'has its roots in cosmology and its flowering in human history'.[15] Yet between the atomic bomb and the population bomb, modern humanity was poised to obliterate itself. As Julian would often put it, this threatened to extinguish all the evolutionary work that had produced humankind itself over thousands of previous generations. The response was a political one: 'The threat of the atomic bomb is simple – unite or perish.'[16] In a post-war catastrophic register, species unity recommended global organizational unity, and more to his point, global organizational unity would save the species *Homo sapiens* from itself.

The grandfather had compiled *Evidence as to Man's Place in Nature* (1863), looking backwards to an evolutionary past that was becoming older before his eyes with every palaeontological discovery. One century later, his grandson considered 'The Future of Man – Evolutionary Aspects' (1963), and looked forwards from his present, 'a crucial moment in the cosmic story'.[17] Julian was so far 'in' nature as to be in charge of it, to have dominion over it: 'Man's destiny is to be the sole agent for the future evolution of this planet'.[18] They – we – stand alone as trustees of evolution since we alone can comprehend it, whether we like it or not carrying the responsibility of the 'business of evolution on behalf of the earth and the rest of its inhabitants'.[19] On the one hand, this was a new kind of dominion 'over the fish of the sea, and over the fowl of the air, and over every living thing that moveth upon the earth'.[20] On the other, this is Julian Huxley as commentator and theorist of an imperilled global ecology. The resonance with our Anthropocene is a true one.

For both of them, evolution was cosmological because it linked all matter in all time. Even non-organic and organic matter are connected: 'Evolution, from cosmic star-dust to human society,' wrote Julian, 'is a comprehensive and continuous process.'[21] His grandfather had said more or less the same thing in 1894:

> not merely the world of plants, but that of animals; not merely living things, but the whole fabric of the earth; not merely our planet, but the whole solar system; not merely our star and its satellites, but the millions of similar bodies which bear witness to the order which pervades boundless space, and has endured through boundless times; are all working out their predestined courses of evolution.[22]

These evolutionists, 1825–1975, not only showed us, but also actively gave us, models to pursue and to avoid: for thinking humanity and nature together; for placing the species into deep pasts and distant futures; for tracking a repetitive human tendency to dominion and for considering the Earth with and without *Homo sapiens*, the apparent trustees of evolution.

PART I

Genealogies

I

Generations: The Huxleys, 1825–1975

The Inheritor

BABY mine, how strange to see
Other faces blent in thine,
Other greatness touching thee,
Baby mine.

Something in a curve or line
Here revives thine ancestry:
Each on thee has laid his sign.

And thyself? Ah! thou for me
Shall this heritage enshrine;
All I was not, thou shalt be
Baby mine!

Leonard Huxley, 'The Inheritor', in *Anniversaries and
Other Poems* (London: John Murray, 1920), 54.[1]

The night that Leonard Huxley became a father, he transformed the
wonder into poetry. For no one but himself, he penned 'The Inheritor'
to mark the birth of Julian Sorell Huxley in 1887. The awe and the
worry of reproduction and family lines, of descent, breeding, inheritance and heritage were on Leonard Huxley's mind. In the baby's face
ancestors were reborn: perhaps Leonard's own father Thomas Henry;
perhaps the other grandfather, the doctrinally troubled Tom Arnold;
perhaps the baby's great-grandfather, headmaster, historian, Thomas

Figure 1.1: Thomas Henry, Leonard, Julian Huxley, 1895.

Arnold. 'Other greatness' from the Huxleys and the Arnolds had been revived with the miracle that was grown by Julia Arnold Huxley and birthed that summer night, 1887. None of it was yet thought of as 'genetics', let alone 'Mendelian inheritance', but it soon would be: the newborn would be one of the key actors in that reworking of 'breeding', and of 'pedigree'.

Over these three generations – the Victorian, the Edwardian and the Modern – the Huxley fortunes rose, and their legacies flourished, publicly and privately. Yet Leonard Huxley was long subject to Thomas Henry Huxley's oppressive greatness. So too his beautiful newborn Julian would grow up, then step up and into the spectacular scientific position that Thomas Henry Huxley had single-handedly conjured, if not out of thin air, certainly out of not much. Julian did inherit a good visage from his Huxley and Arnold families, just as he inherited their inclination to reading and writing poetry (figure 1.1). But Julian the Inheritor also absorbed and carried forward his father's self-recrimination vis-à-vis Thomas Henry: 'all I was not'. On the night of his birth, even as he came out of his mother and into the world, Julian Sorell Huxley had the expectations of his line bestowed upon him. Leonard's private longing was to become his own weight to bear.

VICTORIANS: THOMAS HENRY HUXLEY AND HENRIETTA HEATHORN

Thomas Henry Huxley, unlike his son Leonard, was not inclined towards his familial forebears. Rather he tended to distance himself from them, even cut himself off. He found little satisfaction in either looking or going backwards in most matters. At the height of his career, his interest in the past lay more in the pedigree of *Homo sapiens* than in the Huxleys. Looking forward, however, he willed himself to be patriarch of a new dynasty, as well as a new science. And he was.[2]

When Huxley did look back, it was not far, and it was to his mother, not his father. 'Physically and mentally I am the son of my mother so completely . . . that I can hardly find any trace of my father in myself.' Rachel was London-born and had five surviving children before Thomas Henry was born in May 1825, in Ealing. She was forty, unusually old to be bearing children in the 1820s. He wrote of her wit and her quickness. From his mother, by Huxley's own account, he acquired 'piercing black eyes', 'rapidity of thought', 'excellent mental capacity' and very little modesty.[3] He was devoted to her. Less so to his father George Huxley, who was a teacher. The Ealing School was on the decline even as George Huxley taught there between 1807 and 1835 and certainly in the brief time that his son was a student. Thomas Henry enrolled when he was eight, in 1833, and left two years later, the worst society he had ever known, he recalled.[4] Remarkably – especially given how entwined the Huxleys were to become with education – this was his sole experience of school. Huxley finances spiralled down along with Ealing School's, and eventually George resigned and took his wife and children to Coventry, where his family had originated.

That was in 1835, a Midlands time and place foundational to Huxley's political and philosophical inclinations.[5] We might imagine an industrializing town, nestled up to Birmingham. Yet this was not a factory town, even though, or really because, Coventry was the home of silk weavers and manufacturers of ribbons. Several steam looms were introduced in the 1830s, but the town's masters, journeymen and apprentices were cautious, independent craftsmen more troubled than enabled by the great industrial changes sweeping across the Midlands. Coventry battled poverty and inequity and it did so partly through a long tradition of dissenting politics. T. H. Huxley's signature opposition to establishment authority started here. It was a town where privilege and convention were questioned, almost as a matter of course. In Coventry over the 1830s and 1840s, there were Chartists, there were freethinkers and secularists, and there were phrenologists. There was the socialism of second-generation Owenism, especially championed by Charles Bray, a Coventry ribbon manufacturer, busy publishing on new mechanisms for working-class education and improvement, on a new kind of social science, and on *The Education of the Body: An Address to the Working Classes* (1836). For young Thomas Henry Huxley, this Coventry culture was a primer in the critique of orthodoxy. He set himself determined rituals of self-instruction, monumental by any measure, consuming a whole range of political philosophies that grounded his later actions, and indeed formed his very self.

For a boy born into both class and financial decline, reading was worthy and, as it turned out, a good investment over many long nights, but it brought no income. Not working-class, but financially struggling to fit into a lower middle class, Huxley pursued his own 'education of the body' as a young medical apprentice.[6] His two sisters happened to each marry medical men in 1839, and this set a course both for himself and for his elder brother James (Jim). His long-favoured sister, Lizzie, married John Salt, who apprenticed Jim. And Huxley was apprenticed to the other brother-in-law, John Cooke, at just thirteen. In 1841 Huxley followed this half of the family to London – the two sisters, his medical brothers-in-law and brother Jim. There, he was re-apprenticed to Thomas Chandler in the miserable Docklands. Over his lifetime and through dogged hard work, Huxley was to extract himself and his wider suite of semi-hopeless siblings from family financial and social decline. But at this point, he was indebted to his brothers-in-law and his elder brother George, who funded much of his medical education in London. With University College initially out of reach, Huxley first attended Sydenham College, surrounded by medical radicals. With Jim he later gained enrolment at Charing Cross medical school in September 1842, both as 'free scholars'. Huxley had something to prove. By 1845 he took University College's Part I Bachelor of Medicine examination and won a gold medal for anatomy and physiology. He was ambitious and he was all-in.

It heralded a bright future, but what income was to be had? He needed to work – and quickly – to repay debts for his medical education and training. In 1846 Huxley was commissioned to the Royal Navy's HMS *Rattlesnake* as assistant surgeon. Under Captain Owen Stanley's command, the instructions were to explore New Guinea, and the Torres Strait, north of Australia, and to chart the Louisiade Archipelago as yet unexplored by the British Navy. For the twenty-one-year-old, this was to be life-changing, his own Melanesian coming-of-age. Even before departure, Huxley was catapulted into a whole new company: erratic Stanley introduced Huxley to Richard Owen, the great comparative anatomist of the era who a decade later was to become Huxley's intellectual nemesis. He met John MacGillivray, the *Rattlesnake*'s naturalist, Oswald Brierly, the expedition's artist, and the surgeon John Thompson. Thomas Henry Huxley was poised between medicine and natural science, and for years he could have gone either way.

The ship departed after a long preparation for what ended up a four-year voyage. Sailing in chilly December 1846 from Portsmouth, HMS *Rattlesnake* returned to Chatham in November 1850. Sydney was the base for the expedition's numerous cruises north to Cape York, to the Torres Strait and to the islands of Melanesia. And while poor Captain Stanley fell ill and died in Sydney – in young Huxley's arms – Huxley himself fell into other arms. He met the woman with whom he was to reproduce a scientific dynasty, and through whom he could become the patriarch he aspired to be, looking forwards not backwards in time.

Long-lived Henrietta Heathorn (1825–1914) came from an altogether more complex family than Huxley, daughter of Kent brewer and some-time merchant seaman Henry Heathorn and Caribbean-born Sarah Richardson. Sarah was in fact married to someone else at the time, a man with whom she had (at least) two children already, Oriana and Isabel (or Isabella). Henrietta's place of birth is unclear but may have been in the West Indies, possibly Jamaica given Sarah Richardson's pre-existing connections. But there are no birth records for Henrietta Heathorn because she was illegitimate, Sarah having eloped with Henry Heathorn, a difficult fact that Henrietta was shocked to discover late in life.[7]

Born in 1825, Henrietta was the same age as Huxley, but as a young person was far more travelled than he, and more formally educated. If Huxley moved from London to Coventry and back to London in his childhood and youth, Henrietta Heathorn travelled much further in her early years, following her father's multiple and mostly unsuccessful business ventures from the West Indies to England and then to the Australian colonies. She was fortunate to spend late-childhood years in Kent with an aunt, Catherine Heathorn, through whose initiative she received thorough if unconventional schooling. Catherine Heathorn happened to be a subscriber to the London Association for Moravian Missions, and through contacts and family funding secured Henrietta's enrolment in a boarding school, the Moravian Settlement, in Neuwied on the Rhine near Koblenz. Henrietta was educated there between the ages of fifteen and seventeen, over 1840–42, the years when Thomas Henry was apprenticed in Coventry and then London. Their contexts could not have been more different. While Huxley was miserably dissecting and dispensing by day and teaching himself German and Ancient Greek by night, Henrietta was being taught basic natural sciences, arithmetic, Latin, German and French by the Moravian sisters, the as-yet-unmarried women of the community. The Moravians endorsed significant mental as well as physical education for girls as well as boys; in 1841 there were seventy boys and around fifty girls.[8] Henrietta was also absorbing a particular form of Protestantism, even as young Huxley was being led away from God altogether by Coventry secularists, pan-theists and a-theists, and by Sydenham radicals. Yet the Moravians were tolerant of denominational differences – ecumenical – and stressed personal piety and the intimate experience of faith, all significant for Henrietta's future life with Huxley. Tolerance for his agnosticism was perhaps seeded in her German boarding school.

In 1843 Henrietta moved to New South Wales, where her father had just taken on the Woodstock Mills in Jamberoo, a remote hamlet, barely that, in the coastal hinterlands south of Sydney. He was milling rainforest timber in what was still – but only just – a penal colony built on convict labour. Jamberoo was a difficult place to live – snake-infested and frontier – and when her half-sister Oriana married merchant William Fanning, Henrietta moved into their comfortable house in Newtown, Sydney. But she kept the house as much as lived in it, sometimes imagining herself as Cinderella, the obedient servant-relative.

> She was the last of sisters three,
> Fair, golden-haired, a merry maid . . .
> She journeyed to a distant land,
> With waters blue, and sunshine bright,
> Where lofty columned palm-trees stand,
> And moonbeams make a day of night.

But Cinderella – as she titled this poem – had a higher destiny. And there he was on the deck of the *Rattlesnake*, the brooding and melancholy prince.

Twelve moons had waned, there came a youth
Fate wafted to this self-same shore
A princely soul who sought for truth
And honour too – but truth still more.[9]

The prince (whom she called 'Hal') and Henrietta (whom he cast less as Cinderella and more as Jane Eyre)[10] met at a Sydney government-house ball, where they fell hopelessly and instantly in love. They danced on the deck of the *Rattlesnake*, they picnicked on sparkling Sydney harbour foreshores, and Huxley became the favoured visitor at the Fannings' house. For Huxley – perhaps for both of them – there followed a love-sickness, for just as they discovered one another, the *Rattlesnake* departed for yet another long northern cruise. Young and stricken with love, he gifted a portrait locket of himself to Henrietta. She was rapturous in gratitude at the token, but little did she know that this image was to substitute for his own vital self, and for too long.[11] As letters slowed from the *Rattlesnake*, the treasured portrait-locket was all that was left of him, like a relic. Theirs was a long and tortured engagement, intimacy deferred by his naval obligations, the interminable *Rattlesnake* cruises, and his return journey to England.

Back in London, Huxley was as committed to the Royal Navy as he was to Henrietta. He was also committed to the Malthusian-derived cultural convention that prospective husbands defer marriage until they are able to provide adequately. As things stood, he could not, and it took him five further years over the early 1850s – he in London, she in Sydney – to secure an income adequate to his own expectations.

In London, Huxley took leave from the navy to write up various findings, committed to pursuing not the medical practice of his naval appointment, but the by-product, natural science. His work in marine biology was fairly swiftly published in the *Transactions of the Royal Society*, and he wrote to Henrietta in February 1851 that 'FRS . . . is nearer than one might think – to my no small surprise'.[12] He was indeed elected into a precocious Fellowship of the Royal Society that year, followed by its medal in physiology (1852). His star was rising, but no job with it. Despite his publishing success and affirmation by the Royal Society, Huxley was passed over for a sequence of positions any one of which would have offered a decent salary. In a sequence of minor humiliations, he failed to secure a position at the universities of Toronto, Aberdeen, Cork and Kings College, London (where his grandson Julian was to pick up a professorship seventy years later, with notable ease). The prospect of a professorship in natural history at the new University of Sydney also came to naught. Henrietta was getting word from other members of the family that her Hal was 'anxious and depressed'.[13] He was. But finally, in a rush, it all fell into place. In July 1854, he secured a lectureship in natural history at the Government (later Royal) School of Mines associated with the Geological Survey in London. His offer was 'Paleontologist and Lecturer on Natural History', and although he later said he refused the first element of the title, imagining himself uninterested in fossils, in fact he held this post for thirty-one years.[14] At long last, Huxley had a good salary intact, strongly supplemented by external lecturing and by writing. His state of indebtedness was over, and he gradually became the Huxley to whom others turned for financial support, even rescue. He was ready to marry, and Henrietta sailed from Sydney at the end of 1854, arriving in May 1855.

Thomas Henry and Henrietta Ann wed on 21 July 1855 in the Anglican All Saints Church, Finchley Road. For all her own complex heritage, the match provided a fresh starting point for a new dynasty. Malthus had instructed young men to delay marriage – which Huxley had – and thereby to reproduce fewer children – which he did not. Henrietta bore eight. The firstborn came on New Year's Eve 1856, at midnight recalled Henrietta: 'coming at Christmas I named him Noel'.[15] Then came Jessie Oriana, artist Marian – Mady – who was to marry the celebrated portraitist John Collier, Leonard, father of Julian, Rachel, singer Henrietta, doctor Henry and the youngest, Ethel, who – scandalously and illegally so far as English law and conventions were concerned – also married John Collier after the untimely death of her older sister.

For Huxley and Henrietta, Marian's death in 1887 (the year that Julian was born) was the second great family tragedy, since firstborn Noel had long before died from scarlet fever in December 1860, four years old. Mortality might have been commonplace for the Victorians, but Henrietta mourned deeply and for years. A great friend, the Rev. Charles Kingsley, gave her a book of his sermons for contemplation, and it is clear that Henrietta's faith did console. At the same time,

Huxley's soon-to-be-named 'agnosticism' was hardened in conversation with Kingsley, and literally at the graveside as the tiny coffin was being lowered. He was honest with his grief – there was no 'Victorian' suppression of emotion – yet Huxley managed to get on with things. Henrietta's grief, however, was disabling. One stint at recovery was at the home of new friends, the Charles Darwins. 'Ah! life's music fled with him!' Henrietta mourned in her poem 'Now and Then, 1860'.[16]

This mid-Victorian generation of Huxley children grew up in St John's Wood. After Noel's death they moved to 26 Abbey Place and then in 1872 purchased 4 Marlborough Place. Their neighbourhood boasted any number of mid-Victorian intellectual and literary celebrities, especially importantly for Huxley the philosopher Herbert Spencer.[17] They were neighbours also with 'George Eliot' and George Lewes. Huxley visited the unmarried couple and the exciting philosophical and literary circle around their home, but disapproving Henrietta did not.[18] In St John's Wood, as the Huxley children grew up, the Victorian couple hosted well-known Sunday 'tall teas', welcoming all kinds of literary and scientific Londoners and international visitors to their own home. But Huxley's health was not good, and in 1891, when their children were all married – or had died in untimely ways – Henrietta and Huxley retired to Eastbourne, to a house that son-in-law Frederick Waller had designed. Huxley died there, not an old man, in 1895, Henrietta in 1914.

EDWARDIANS: LEONARD HUXLEY AND JULIA ARNOLD

Thomas Henry left a line of Huxleys with more opportunity, more money and more institutional education than he had ever received from his own socially and financially declining forebears. Huxley had almost no schooling, and had presented himself for medical examinations and qualifications through an apprenticeship and sheer hard work. His children, by contrast, were highly and formally educated. The sons, and as it turned out grandsons, studied at Oxford. The girls studied at the Slade School of Fine Art. It all signalled a stunning middle-class leap from Huxley's Coventry and London-based medical apprenticeship and nocturnal self-instruction.

Leonard Huxley was the middle child and the eldest surviving son, granted his name because it 'held our lost Noel's'.[19] Poor Leonard had the parents' grief written into his very newborn self. Psychoanalytically literate Julian was later horrified. Leonard's own childhood – and especially his educational path – was as different from his father's self-instruction as it was possible to be. His was the generation which benefited from, rather than duplicated, the patriarch's rise, but which also had to endure expectations of new standards and of more conventional advancement. Leonard attended University College School, Gower Street, part of the nonconformist educational structure and culture of University College London, which aligned with Huxley's anti-authoritarian politics and his opposition to the Established Church. Leonard then studied at St Andrew's University and finally graduated from Balliol College, Oxford, at a time when its Broad Church affiliation was affirmed by the Master, Benjamin Jowett.[20] While there was no scientist within Leonard, there was a man of letters – of classics, of poetry and of literature. He became assistant master at the grand Charterhouse School in Godalming, Surrey, teaching classics. Later he worked in publishing as assistant editor of *Cornhill* magazine from 1901, and as its editor from 1916. *Cornhill* was a literary magazine whose initial success was the serialization of distinguished Victorian authors: Elizabeth Gaskell, Anthony Trollope and Henry James, for example. It had also published *Culture and Anarchy* in the late 1860s, a series of essays by the renowned poet Matthew Arnold.

Thomas Henry Huxley and Matthew Arnold were well acquainted. Over the 1870s, they exchanged books – their own and others'. Matthew Arnold expressed sympathy with a range of Huxley's views, predicting an imminent religious revolution as significant as the Reformation.[21] They discussed Spinoza. They discussed 'the character of Christ'. They discussed Hume, as well as Huxley's *Hume*.[22] Yet through all this, little did they know that the Huxley and Arnold families would be joined in matrimony, a match that grafted two rising dynasties of Victorian science and letters.

When Leonard was at Oxford he met smart Julia Arnold (1862–1908) one of the first students at the new Somerville College for women.

Granddaughter of Thomas Arnold of Rugby, daughter of Tom Arnold, sometime chair of English literature in Dublin, niece of Matthew Arnold, she was, naturally, reading English literature. Her sister was the learned novelist Mary Arnold, later Mary Augusta Ward, who published with stratospheric success under the unlikely name Mrs Humphry Ward.

Leonard and Julia married in 1885, and after she had borne all their children – Julian Sorell (1887), Noel Trevenen (Trev) (1889), Aldous Leonard (1894), Margaret Arnold (1899) – Julia Huxley continued the educational endeavours of her father and grandfather, but in a completely different manner, method and style. In 1902, she took an alternative educationist's path and started her own progressive school for girls in Surrey, calling it Prior's Field. This was no Rugby and no Charterhouse, from which Leonard resigned in 1901. It was certainly no Moravian boarding school. It was liberal and non-sectarian. It was progressive and adventurous, both for Julia and her boarding girls. Julian, Trev and especially the younger Aldous and Margaret grew up in rural Surrey alongside the Prior's Field girls, in the splendid modern Arts & Crafts house and school designed by C. F. A. Voysey. The young Huxleys shared their mother, and much of their familial and domestic space, with multiplying boarders and a few day girls, some of them Huxley cousins (figure 1.2).

Leonard and Julia Huxley led active and pastoral lives with their four children and their many domestic animals. By all their accounts, it was easy and pleasant: they walked, swam, cycled, rode and read. Leonard, Julia and their four children took long summer holidays with both sets of extended families. They visited Huxley and Henrietta – Pater and Gran'moo – in Eastbourne. Leonard took Julian and Trev mountaineering in France and Switzerland – they were skilled and athletic. And they summered with Julia's sisters, especially Mary Augusta Ward in her grand Hertfordshire home, Stocks. And from those summers, the English novel-reading public enjoyed versions of the Huxley children, the Julian-inspired character 'Sandy', for example, in Mrs Humphry Ward's *The History of David Grieve*; 'a most delightful imp', grandfather Huxley wrote to her.[23] The three Huxley boys were wide-eyed and pretty, impossibly so. There was nothing yet inherited of Thomas Henry Huxley's signature scowl (figure 1.3).

Figure 1.2: Prior's Field School, *c.* 1909.

Figure 1.3: Trevenen, Aldous and Julian Huxley, *c.* 1897.

Leonard and Julia were comfortable enough to employ a governess-tutor to look after them, who was almost always German and charged with their fluency. 'I shall expect you to say everything in German soon,' Leonard wrote to his young boys Julian and Trev in 1896.[24] The littlest, Aldous, had some way to go. 'Liebe Mutter,' the four-year old wrote to his mother in 1898, on holiday again at Aunt Mary's. 'I like Laleham's house better than Stock, liebe mutter ich liebe Frl Ella, ich liebe mein Laleham's presents, mein solders, my horse and my clown. I am very good indeed, I kiss you, Aldous'[25] (figure 1.4).

German literacy, literature, philosophy and science remained inter-generationally important to the Huxleys, as writers and as scientists. But the manner in which this generation acquired it was just one marker of change and upward class mobility. No governess or tutor for the struggling Midlands Huxley of the 1830s. He had had to teach himself German, and did so diligently and effectively. He read German science as well as German philosophy throughout his life, beginning with Goethe, another poet-scientist. This made him an important conduit between English and German ideas and a translator of major biological and zoological texts into English. Henrietta's fluency was probably superior, from her Moravian schooling. They early shared German philosophy, he courting her with a precious twelve-volume set of Schiller's works, her favoured poet. Sometimes Huxley would make a show of criticism of her translations yet relied often enough on her skill. It was she who translated lines from Goethe for his much-discussed and unlikely first article in the original issue of *Nature*, which was more poetry than science.[26] Julian was to continue that tradition of engagement with German science and literature. As a young biologist he spent long periods in German laboratories getting up to speed with the latest in Mendelian genetics, the next-generation knowledge that was momentarily to throw Darwin's ideas asunder.

Despite Thomas Henry Huxley's capacity to turn around a familial decline in the mid-nineteenth century, and despite the inheritance of his stellar scientific name, Leonard's Edwardian generation was certainly not standing to inherit any notable wealth. Leonard and Julia's family were middle-class, without a doubt more comfortable than Huxley's family of origin. But they nonetheless struggled to maintain the display of social and domestic standing expected of them by others

Figure 1.4: Mary Augusta Ward, née Arnold, with Aldous Huxley,
at Ward's Hertfordshire house, Stocks.

and by themselves. Certainly, they had to work and constantly to reckon income and expenses. Little could be taken for granted in Leonard and Julia's household. Their sons were launched into a privileged educational realm – Eton – but only just, and only via scholarships. With back-up Rugby plans, first Julian and Trev successfully gained scholarships, and then Aldous, who 'just scraped in'.[27] 'What should we do if the boys were stupid, & had no chance of getting scholarships?' Julia wrote to sister Mary in 1899, worrying again and hovering over their results: 'Julian began the term by being 3rd then 2nd now has got top again & keeps there steadily … Trev has been top & 2nd in his new form & is really doing splendidly in mathematics.'[28] When Julian was undecided on science *versus* letters, Julia signalled his dual heritage, classics from his father, biology from his grandfather: 'It would have been very nice if you could have stuck to classics, & got it. However Biology must come out I suppose, like measles or chickenpox.'[29] In 1902 she was pleased that Julian's grades and form had jumped, but he should be able to do even better. 'I don't see why Huxley cum Arnolds shouldn't manage it! There's good blood in you on both sides, & that means staying power, wh. is everything. All the same don't over work yourself to please an ambitious mother.'[30] It all turned out well enough, especially when it came to Aldous. 'Congratulations on being top,' big brother Trev encouraged little Aldous. 'I expected you would pick up well when you had got over the first week of loneliness.'[31] The brothers exchanged strategies for coping at Eton (figures 1.5 and 1.6).

Julia Arnold Huxley relied on Mary Ward to keep the household together, including servants, as they struggled on Leonard's salary and occasional income from writing. She drew on her sister's novel-derived as well as matrimonially derived affluence, money freely given by one sister to another. Julia worried a good deal, and clearly it was her responsibility, not Leonard's, to manage the household books. She shared her anxieties as well as her budgets with Mary Ward regularly, withholding both advice and loans from Leonard, especially regarding the money borrowed to keep Prior's Field buoyant. She turned to Mary for far more mundane matters as well. Might she borrow £25 for the doctors' bills, she asked her sister in 1898, especially those for sickly Aldous: 'he is so thin and weak, poor mite'.[32] These next-generation

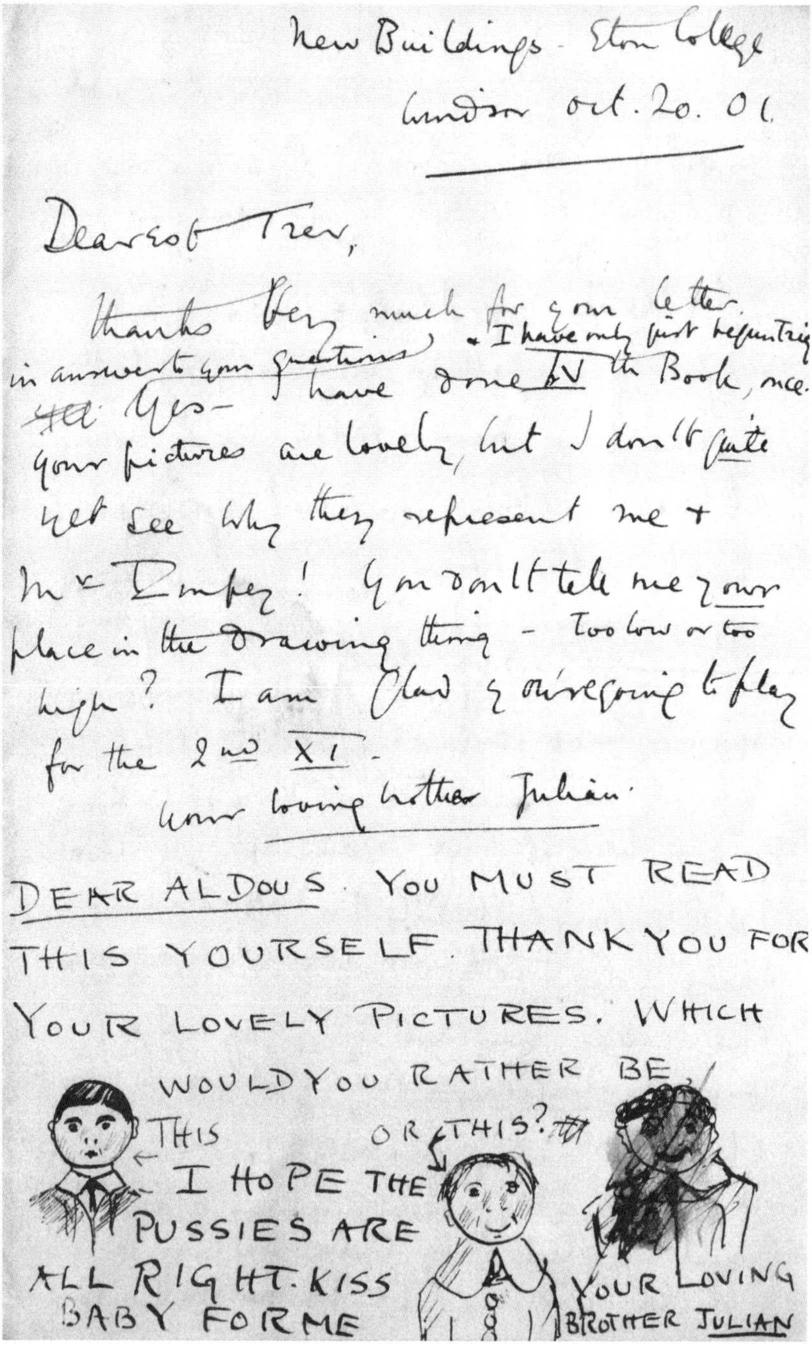

New Buildings - Eton College
Windsor Oct. 20. 06.

Dearest Trev,

Thanks very much for your letter. I have only just begun trig in answer to your questions Yes - I have done XV th Book, nce.
Your pictures are lovely, but I don't quite yet see why they represent me + Mr Dempey! You don't tell me your place in the Drawing thing - Too low or too high? I am glad you're going to play for the 2nd XI.

Your loving brother Julian.

DEAR ALDOUS. YOU MUST READ THIS YOURSELF. THANK YOU FOR YOUR LOVELY PICTURES. WHICH WOULD YOU RATHER BE, THIS OR THIS? I HOPE THE PUSSIES ARE ALL RIGHT. KISS BABY FOR ME

YOUR LOVING BROTHER JULIAN

Figure 1.5: Julian to Trev and Aldous Huxley, 20 October 1906.

Figure 1.6: Julian Sorell Huxley at Eton.

Huxleys, seemingly well-to-do, were in debt, and not just to the doctor, but shamefully even to the butcher and the grocer. Sensibly, they should not be having another baby, but she longed for a girl, and Margaret was soon added to the household (figure 1. 7). 'Of course we are living beyond our income.'[33] So much for Malthus.

Scholarship pressures continued into the brothers' university years. There was no other way that they could follow Leonard's (and Great Uncle Matthew Arnold's) footsteps to Balliol. Julian secured the Brackenbury Scholarship to study zoology (1906–9), and Trev expanded his splendid talent for mathematics at Oxford from 1908. They were the elder pair, tall, fit and wiry, climbing Oxford's spires together as they had the Alps with their father. Aldous followed at a distance, first at Eton, then at Balliol, where he read English Literature from 1913. 'Little' brother of the three, Aldous grew to six feet four and half inches, the tallest of these long-limbed Huxley brothers.

There seems to have been wide affection, both within this immediate family as with the wider family, but also difficult separations. In the years that Sigmund Freud was thinking through the connection and individuation of mothers, fathers, infants and children, the Huxley boys were routinely separated from Julia and Leonard, and from each other. First Julian, then Julian and Trev, then Julian, Trev and Aldous, were sent away to various aunts while their mother was confined to give birth at Goldaming, often to Aunt Mary Ward's. Then, when she had completed her own biological family, Julia took in and taught the girls of others. Her considerable energy and attention were consumed by her school and 'her girls', impressively so, sometimes poignantly so. Underneath the Huxley children's memories of ease, pleasure and leisure there was more going on. At Eton, Julian missed his mother painfully and sensed her focus on her girls. He confessed to hating Eton, but routinely she was too busy with Prior's Field matters to make visits, describing all the new girls in detailed letters and then apologizing for not coming up to Eton.[34] Indeed Prior's Field went from strength to strength, and she delighted in announcing its success. Writing to sister Mary in 1908, she was proud that Julian was about to go up to Balliol, and in the next breath detailed her own achievements: 'The school is absolutely packed 59 boarders, 6 day girls.'[35]

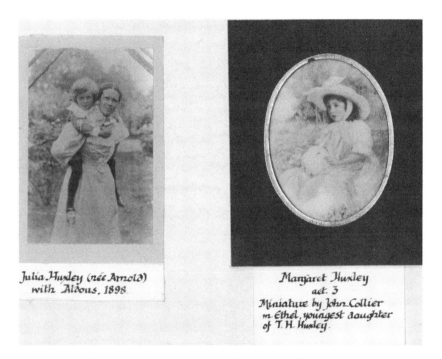

Julia Huxley (née Arnold) with Aldous, 1898

Margaret Huxley act. 3
Miniature by John Collier
m. Ethel, youngest daughter
of T. H. Huxley.

Figure 1.7: The first page of Prior's Field's albums of historic photographs. Julia and Aldous Huxley, 1898, and Margaret Huxley, age three, miniature by John Collier, *c.* 1902.

Figure 1.8: Julia Arnold Huxley, the New Woman, *c.* 1902.

Edwardian Julia Arnold was no Victorian Henrietta Huxley. If Henrietta had imagined herself as Cinderella, clingingly dependent on her own 'prince', and Huxley had imagined her as Jane Eyre, the governess figure seeking the legitimacy and purpose of the patriarch husband, Julia Arnold of Somerville College held her own authority, learning and confidence. She was very sure of herself. Bespectacled in one stunning portrait, her glasses were not just for reading, but a symbol to be read: Julia Arnold Huxley was a New Woman (figure 1.8).

And then she died. It was painful and it was too early. Julia noted a small lump in her left breast in early 1906. It was treated with belladonna by her local doctor, then homeopathically by Charles Thomas Knox Shaw, recommended to her as the 'best Homeopath in Town'. By 1908 she was having electrical treatment, the new röntgenotherapy named for Wilhelm Röntgen, who had discovered X-rays just a decade before Julia's illness.[36] Well might we shudder. Leonard had to write to poor Julian at Oxford that his mother was declining: 'very weak & sometimes not clearly conscious ... You must attune your mind, dear boy, to the sad expectation of hearing with the next 48 hours that the end has come, tranquilly I trust & believe, in sleep & peace.'[37] As it was, Julia held on for another eleven days, and died on 29 November 1908, aged forty-six. She was with Leonard and her much-loved sister as she passed into the next life, or as she herself said to Mary, into 'the next stage of this – because after all it is only one life right on thro' eternity'.[38]

MODERNS: JULIAN HUXLEY AND JULIETTE BAILLOT

Julian Huxley was a student at Oxford when his vital mother passed away in November 1908. He completed his studies in 1909, and his early career in biology fell as easily into place as his grandfather's attempts to enter natural sciences professionally had been blocked and difficult. It was not solely because of his grandfather's name and fame that Julian's zoological path was clear, but certainly partly so. All manner of opportunities and connections came his way. In June 1909, fresh from Oxford, he was invited to attend the Charles Darwin Centenary in Cambridge, an anniversary of his birth in 1809 as well as of the publication of *On the Origin of Species* in 1859. It was a family affair, as much as a scientific one. Long-lived survivors of Charles Darwin's own generation attended, including Henrietta Huxley.

While Darwin the man was memorialized in 1909, his actual ideas were just then at a low point, as later chapters will explain. The next generation of zoologists, only too aware of some problems with Darwin's thesis on natural selection, was making major breakthroughs. Julian was party to much of this new work, in which evolution and genetics were rubbing up against one other. First, from September 1909, he went to Naples for a year to work at the Stazione Zoologica, where his grandfather had also spent time. Then, from October 1910, he gained a lectureship at Balliol and a demonstratorship in Oxford's Department of Zoology. While his grandfather had craved such appointments as a young man, it all fell into Julian's lap, or so it would seem. By 1912 he was offered the Chair of Biology at the new Rice Institute, Houston, Texas, lured by a salary far beyond meagre Oxford offerings. It had been under discussion at least since late 1910, when the biologist Edwin Conklin from Princeton was asked for his opinion on Julian Huxley by Rice Institute's first president, Edgar Odell Lovett. 'Charming,' Conklin confirmed, 'with many of the qualities of his grandfather.' Princeton University, via Conklin, in fact also pursued Julian over 1911, to the point of an informal offer – Julian asked him just what 2,500 dollars per year was in pounds. Conklin presumed that Princeton would win out since Julian was accustomed to 'an ancient academic center'. But Julian indicated that 'the very newness & remoteness' of Rice, in Houston, was precisely what attracted him.[39] Lovett won the day and the Huxley prize. It was a coup. Julian indicated that he wanted to spend time in German laboratories first, but that he would attend opening ceremonies in the autumn of 1912.

All seemed promising, even glowing, for the Inheritor. But the following few years were very difficult indeed, and almost nothing fell into place for the too-sensitive Julian. Indeed, everything fell apart. While his father found love again after Julia's passing – Leonard married the much-younger Rosalind Bruce in 1912 – Julian failed in first love. And he became sick over it. He was engaged to Kathleen Fordham, a former student from Prior's Field School, but over the years he felt far more obligation than desire; suffocatingly so, and the long engagement was broken off in 1913. Impossibly difficult for her, for him this was a trauma that precipitated a first hospitalization for depression.

As low as it was possible to be, struck down by circumstance and the melancholy inherited from his grandfather, Julian finally re-emerged and rallied, and headed back to Houston and Rice Institute duties in 1914, by way of Thomas Hunt Morgan's already famous fruit fly laboratory in New York: genetics was being born there. Julian's commitment to Rice was on-again and off-again, coming and going between 1912 and 1916. It was a messy and short-lived appointment, not just because of the Great War, but partly so. In no small measure through Aunt Mary Ward's patriotic urging, even shaming, by September 1916 he resigned, enlisted back in England and began war work at the Censor's Office.[40]

Julian had an easy war compared to most. His personal battles preceded it, though he reflected later that having 'breathed the international atmosphere of Science', and having learned important new biology in German and Italian, as well as British and American, laboratories, he felt intensely compromised by wartime allegiances. It all turned him into an anti-nationalist. Nationalism was an offence to him, and he was shocked when he first heard an American declaring, 'My country, right or wrong'.[41] Nonetheless, he served, unlike some in his biology circle who conscientiously objected. If anything, the war for Julian Huxley was social and productive. He met all kinds of people at Aldershot and had remarkably good leaves from duty. The summer of 1917, for example, was spent climbing at Lake Ogwen, north Wales. There, while US troops were arriving in France and George V changed the family name to Windsor, Julian read *Abelard and Heloïse* in the evenings to improve his French, though the combination of romance and tragedy can't possibly have elevated his mental health.[42]

Later that summer Julian was a guest of Lady Ottoline Morrell at Garsington Manor, near Oxford. Garsington was a home, a centre, for English cultural modernism, and Ottoline Morrell a highly unconventional patron of new-generation writers and artists. D. H. Lawrence, Virginia Woolf, T. S. Eliot and Katherine Mansfield were all hosted and nurtured there. So was Aldous Huxley, who perhaps ungenerously satirized both Morrell and Garsington in his debut novel, *Crome Yellow* (1921). Morrell's household and circle were young, culturally adventurous, avant-garde, experimentalist. By introducing his elder brother Julian, Aldous brought him into a sexually and emotionally charged time and place. Aldous met the Belgian refugee Maria Nys at Garsington, whom he married in 1919. Julian was taken by Swiss Juliette Baillot.

Marie Juliette Baillot was not at Garsington as an artist, however. Rather she was governess to Ottoline Morrell's daughter, also, as it happened, named Julian. Like any Victorian governess – real or literary – early twentieth-century governesses like Juliette were part of households, but precariously so. They were in between family and service, salaried but special. At Garsington, Juliette Baillot was neither a guest nor a host, and yet her artistic sensibilities (she later sculpted and wrote superbly) recommended her for such highly unusual sociability. She later recalled her naivety about the Garsington lifestyle and sexual arrangements, and she did not find it an easy place, though her devotion to, and long friendship with, Ottoline Morrell was grounded there.[43] She served as a kind of muse for Morrell and her camera. Like many others at Garsington Manor in those years, Juliette Baillot was captured nude, many photographs now held by the National Portrait Gallery, London. Juliette and Julian married in July 1919. Their friend H. G. Wells called them the 'Huxleyettes and Huxleys'.[44]

That Julian and Aldous were now married to these two modernist women – one a participant in early twentieth-century aesthetic nudity, the other a Belgian bisexual beauty – signals a massive generational shift in the Huxley world, as in the wider world. If Thomas Henry Huxley had lived through an era of political and religious dissent, his grandsons Aldous and Julian lived in an era of modernist cultural and artistic dissent. As young boys they may have learned from the theological morality tales that Aunt Mary wrote, but this Garsington milieu produced the erotic *Lady Chatterley's Lover* (Lawrence inspired by Ottoline Morrell), as well as the more stylistically adventurous modernist writing of Mansfield, Woolf and Eliot. Aldous was also experimenting with form and structure.

The great change was embodied in, and displayed by, the generational couples. Thomas Henry Huxley had chosen religious and earnest

Henrietta, who underwrote her husband's public sphere theological controversy with studied domestic sphere propriety, convention and personal faith. They were a patriarch and a matriarch occupying conventional gender roles. The outer extent of Henrietta's public unconventionality was her Sunday 'tall teas', which guests loved. Her private unconventionality – her late-discovered illegitimacy – was shameful and secret. Leonard, by contrast, had chosen a smart New Woman Julia Arnold, clever and determined, with a significant English literary pedigree. Julian and Aldous chose icons of early twentieth-century modernity and apparent sexual freedom – both continental women, Aunt Mary Ward frowned.[45]

On the surface, this line of marital and cultural descent over time shifted from the mid-Victorian Huxleys' traditional order to the 1920s partnering of the Garsington moderns. But it turned out that gendered conventions within marriages change far less easily than, say, trends in thought over natural selection. Over a long and difficult twentieth-century marriage, Juliette Huxley née Baillot fell into a domestic and public support-and-hostess role for Julian, as outwardly conventional as Henrietta's had been for Huxley in the nineteenth century, if more contextually exciting: less St John's Wood and more Nairobi and Paris. As Julian and Juliette's relationship unfolded, a freedom to occupy and satisfy different sexual attractions and expressions was by agreement afforded to both of them, but in practice taken up by him more than her. He had long affairs that were painful to all, including to his father Leonard and brother Aldous, who routinely witnessed, intervened in and tried to short-circuit the familial dramas. Juliette suffered the emotional pain and then looked after her husband's pain – Julian's repeated breakdowns – all the while accompanying him in his life's work as zoologist and writer, in the latter respect much as Henrietta Huxley had done. These women both outlived their men and managed their memories and memorials. Henrietta published Thomas Henry Huxley's occasional poetry and his 'Aphorisms and Reflections'.[46] Perhaps with more honesty and less default devotion, Juliette managed Julian's memory, reputation, representation, books and papers after his death in 1975. And she recalled their married life with brutal honesty in her own elegant autobiography. It was the middle-generational couple who were different; Leonard and Julia. In founding and running her own unconventional and highly successful girls' school, Julia pursued a career and an enterprise parallel to Leonard's own work in the educational then literary and publishing world. Together they modelled a new possibility for paired men and women, even as they reproduced and raised the next generation of Huxleys into both science and letters.

MEN AND WOMEN OF LETTERS

The Huxleys are remembered as a scientific dynasty, but they are at least as significant as a literary family. They wrote in multiple genres and for different audiences, in private and in public. That the Arnolds grafted onto the Huxleys in the middle generation, meant that this became a conjoined family of highly regarded literary figures – Matthew Arnold, Aldous Huxley, Mary Augusta Ward – and of essayists, memoirists and science writers – Thomas Henry, Leonard and Julian. More privately, though still intermittently in published form, all of the Huxleys were poets. This was a family of professional wordsmiths.

Thomas Henry and Julian were what we now call 'science communicators' and were unquestionably the founding masters. They both wrote in a highly skilled manner, but also in a manner that paid the bills, drawing income from their pens and typewriters. Grandfather and grandson wrote in obviously different cultural, political and technical contexts, and in diverse media: Julian worked in radio, television and film, for example, as well as in print. But in other respects, their public communication of zoology, evolutionary ideas, ethics, religion and science were similar. They both undertook private lecture tours. Julian's US lecture tours were exhausting, he thought, 'but well paid'.[47] They both turned their talks into pamphlets, magazine articles and short books for popular consumption. Their journalistic essays were published across all manner of outlets, from Huxley's assessment of the *Origin of Species* in *Westminster Review* in the 1860s to Julian's well-paid catastrophizing of world population growth for *Playboy* in the 1960s.[48] They both wrote what came to be called 'text books', as well as high-level and highly proficient philosophical studies about the human and natural world. Speaking on the BBC, Julian later called his grandfather, quite rightly, a 'master of English prose', although Aldous's assessment was more qualified.[49]

Thomas Henry Huxley was raised – or really raised himself – on a literary and philosophical as well as a natural history canon. For his polymath generation, these were indistinct genres, combining ways of knowing the world in which science had not yet fully separated from philosophy, classics, literature and theology. He inherited also an early nineteenth-century genre of commentary via the essay. A commercial shift in the publishing industry had taken place, from books to those great Victorian periodicals: *The Fortnightly Review*, *The Nineteenth Century*, *Westminster Review* and the literary *Cornhill* magazine that Leonard was eventually to edit. The looming Victorian historian and essayist Thomas Carlyle might have bemoaned this new marriage of literature and commerce,[50] but for entrepreneurial and hard-working Huxley it was ideal; it was his supplementary living.

He wrote constantly for periodicals to enhance a salary of £400 a year. Henrietta remembered that while they had to 'economize in every way', especially with more children, they would have been 'in a sore plight' had it not been for the writing.[51] As soon as Huxley took on the job at the School of Mines, he took on 'lectures to working men' as well. These were lecture courses – typically a suite of six – with an affordable entrance fee. 'Six Lectures on Fossils', for example, 'what they are and what conclusions may be drawn from them'. The Gilchrist Fund supported some of these lectures, which were enormously popular, typically over-filling the London Mechanics' Institute Hall in Chancery Lane.[52] And then he was able to capitalize on this and rework much of the material into books and pamphlets, *Lectures on the Elements of Comparative Anatomy* (1864), for example.[53]

While the elder Huxley drew a base salary from the Royal School of Mines, his grandson was far more dependent on publication, indeed for much of his life solely so. Julian drew salaries from his early but intermittent academic posts – at Oxford, at Rice University, at Kings College, London – then from the Zoological Society of London in the 1930s, and for a short post-war period as director of Unesco in Paris. But otherwise, he supported himself, Juliette and their two sons entirely as a science communicator: a lecturer, broadcaster, non-fiction writer and sometime filmmaker.

If they were ever short on ideas, the later generations found they could trade on Thomas Henry's fame and his notoriety surrounding the Darwinian controversies over the 1860s and 1870s. Leonard and then Julian could both rely on retelling and reselling a Huxley family story, and many times over. Leonard started the tradition, the first of the generations to draw a full income from the literary world, beginning, sensibly enough, with *Life and Letters of Thomas Henry Huxley* in two volumes. His household was literally counting on it. In 1899, when Leonard was still assistant master at Charterhouse, Julia Arnold looked at the account books and anticipated £200 for *Life & Letters*, even as she was asking Mary Ward for a further loan.[54] On its success, Leonard turned to his wife's family, editing *Thoughts on Education Drawn from the Writings of Matthew Arnold* (1912), and then others of his father's circle, including the *Life and Letters of Sir Joseph Dalton Hooker* in two volumes (1918).[55] A little more emboldened, he later presented himself in the public literary domain, with *Anniversaries* (1920), a collection of his own superior poems, unmediated by famous forebears, though sometimes about them. If Leonard published accounts of his father, Julian did the same in print and as broadcasts, all authorized by his own growing stature in science. Winding up his own father's study after Leonard's death in 1933, Thomas Henry's 1840s *Rattlesnake* diary was miraculously uncovered among dishevelled papers. It had masqueraded for all those years as an account book and is now treasured in the Imperial College archives.[56] Julian published the diary in 1935 and received royalties for years.

We might recall Aldous as the Huxley man of letters and Julian the man of science, yet both brothers were professional authors. As a schoolboy, Julian was recommended for scholarships on the grounds both of exemplary understanding of zoological principles and the fact that 'his power of writing good English is exceptional'. His only fault was 'a tendency to think himself infallible'.[57] (A curious assessment, since not much later, Julian confessed about Eton 'I used to weep on slight provocation; & was bullied.')[58] Julian was a writer in his father's image, and a science wordsmith in his grandfather's image. He had a marked talent as a non-fiction writer, but far less successfully experimented with science fiction.[59] In the 1930s he tried the novel form as a container for the signature Huxley thematic: 'man's place in nature', eighty years after Thomas Henry Huxley's best-known book of that title appeared. Julian's version came out as 'Man's Place in the Universe', and he tried too hard to shoehorn his rarified scholarly ideas into an unpublished novel set in Oxford parlours: the device was Aunt Mary's convention of polemical instruction set out for the reader through drawing-room conversation.[60] But he couldn't control it. Over-long drafts morphed into disorganized sets of his own reflections on the 'uniqueness of man', Julian's enduring refrain. He learned that he was a more proficient essayist and poet than novelist, and that Aldous was the right one to continue the Arnold fiction-writing family tradition, if in modernist form.

These Huxleys were all poets. They composed verse both to, and about, one another. Thomas Henry Huxley's 'From Shanklin' recalled his many years with Henrietta (figure 1.9):

> March 1, 1887
> DEAR wife, for more than thirty years
> Have you and I, hand clasped in hand,
> Sometimes all smiles, sometimes in bitter tears,
> Wended our way through the strange land
> Of living men; until with silvering hair,
> And graver mien and steps more slow,
> Adown the strand of age we fare
> To the still ocean, out beyond time's flow.
>
> True wife, housemother, worn with many cares,
> Love's afterglow shall brighten all the years
> That yet are ours; and closer still shall be our clasp
> Of hands, until they nerveless fall and cease to grasp.[61]

Henrietta responded with more verse than her husband; she had more time on those hands, as well as a devotion that had to be caught and put down. She had two devotions really: one to her husband and another to God. With regard to the first, she wrote poems of love and anguish as early as their long, separated engagement in the 1840s and early '50s. One was inspired by her passage from Australia to England in 1855. Finally, she was going home. As she approached the English coast, she banished the Southern Cross. 'I will not look South any longer, / But I will

Figure 1.9: Thomas Henry and Henrietta Huxley, 1882.

look to the North ... Be my beacon, the North Pole star / Oh my soul! take thy leave of the Cross.'[62] While Huxley's poem was a private token, Henrietta eventually published hers. First bound in her own notated typescript, then privately printed in 1899, she finally published her own, along with three of Huxley's, in 1913. It was a memorial to their life together, but her inclination towards writing predated life with Huxley. A journal Henrietta kept in her twenties indicates a passing ambition 'to be famous as a poetess'.[63] She wrote stanzas of grief and ecstasy, and of Moravian-inspired religious rapture. Floods of emotion transformed into sometimes saccharine verse, marking family joys and crises, as well as a nature-oriented piety. Mary Ward loved Henrietta Huxley and affectionately called her, like all the kith and kin, 'Moo'. But this didn't mean she loved her poetry: 'of little or no technical merit, but raised occasionally by sheer intensity of feeling ... into something very near the real thing'.[64] Mercifully, it was a posthumous assessment. Leonard inherited a capacity and inclination to verse from both of his parents and passed on the craft of poetry and the expectation that it be mastered, in the form of verse itself. 'Poets to Be' was about, and directed to, his precious sons.[65]

This eloquent family traded their compositions within and between generations, sometimes for critique, sometimes as sentimental tokens or special-occasion iambic pentameter gifts. In 1910 Julian wrote 'Eastbourne Days' about his grandmother, grandfather and their retirement house that the grandchildren visited so often. Charming, Henrietta complimented him. 'It moved me deeply in its expression of love for me.'[66] Julian had recently excelled himself, winning the Newdigate prize for English verse at Oxford, following Great Uncle Matthew Arnold, awarded in 1843. The prize was a significant recognition of Julian Huxley's mastery, and 'Holyrood' was published in 1908, an Edinburgh-inspired reflection on the regal and common spirits inhabiting the abbey-turned-palace-turned-ruin.[67] Grandmother Henrietta wrote from Eastbourne to congratulate him. 'If only Pater could have lived on and know of your success this day.' Would he like all Pater's books, she was motivated to ask?[68] Her eight children meant that there was a good swathe of grandchildren to choose from, but Henrietta already had her eye on Julian as the Inheritor.

Leonard, Julia and their offspring sometimes made light of their 'pomes'. But in truth they took the craft very seriously. Julia – educationally and chromosomally steeped in English literature – offered frank commentary on another of Julian's. 'Leave out verse 2', she recommended, 'it gives too much detail to your simile.'[69] Poetry was a technical enterprise, as the younger generation learned not just from their Huxley father and Arnold mother, but also from the great living literary figure in the extended family, Mary Augusta Ward. Accounts of the Huxleys – perhaps especially, of Aldous – tend to presume the special significance of Uncle Matthew Arnold. But he had died in 1888 and was unknown to Julian and Aldous, even if they read his work. It was Mary Augusta Ward née Arnold who was the major life-time influence. She was a famous, learned and professional storyteller, and a bestselling one in Britain and even more so in the United States. Even her now lesser-known novels were fantastically successful. *Lady Rose's Daughter* (1903), for example, had a first print run of 52,000 in the US alone.[70] *Robert Elsmere* (1888), *Marcella* (1899), *The Marriage of William Ashe* (1905), right through to wartime novels such as *Eltham House* (1915) and *The War and Elizabeth* (1918) signalled one of the most prolific novel-writing careers of the period. She was a literary critic – of Austen, of the Brontës, of Keats – and but for the happenstance of her gender she could and should have been a professor of literature like her father. She would have made a far better one.

Julia sent her children's juvenilia to Mary for comment.[71] Aunt Mary routinely passed on high-level, technical advice and uncompromising critique to her nephews Julian, Trev and Aldous. And Julian continued to seek her literary advice as a young adult. They travelled together to Florence in 1910, with Julian newly minted from Oxford. She acculturated him in things Italian, and from Naples he enclosed his latest 'pome', his magnum opus he called it.[72] She was forthright in her response. 'A great deal of it is charming verse. Flexible & graceful, some passages particularly good . . . But if I am to be frank I think its deficit is its subject, or rather its point wh. seems to me really not a point.' She instructed him philosophically. 'I have been thinking over it in connection with William James's "Pragmatism" wh. I have just finished . . . the strangest attempt at a philosophic unification I ever read.' She sent him back to the German canon: 'In your spare moments if you haven't already read them – get and read Eckermann's Conversations with Goethe. They contain all wisdom and they are particularly wise on subject and composition.'[73] Mary Ward gave Julian advice on his prose as well. 'There are a few sentences that seem to want combing out a little,' she wrote of an early article in *Cornhill* magazine, 'but it certainly gives one great hope of your future as a writer.' She even imagined her young nephew Julian to be a philosophical writer: 'We want an English Bergson – do you feel inclined to fill the niche!'[74] The exclamatory punctuation – not a question mark – was hardly immaterial to the novelist. Mary Ward might well have been urging her thoughtful nephew on quite sincerely. Indeed, Julian dedicated his early book *The Individual in the Animal Kingdom* (1912) to Henri Bergson, whose *Creative Evolution* (1907) and notion of *élan vital*, the idea of a vital force, or impulse for life inherent in all organisms, then held strong attachments.[75] Far too magical an idea for his grandfather.

Mary Ward was important to her nephews Julian, Trev and Aldous, whom she certainly perceived and valued as literary Arnolds as much as naturalist Huxleys. In return, they read her novels diligently. Perhaps her most famous and enduring is *Robert Elsmere* (1888), drawn from her own father's – Julian and Aldous's other grandfather's – doctrinal crises and written in large part from Peper Harow, Surrey, adjacent to Godalming, where Mary Ward and her own family spent six consecutive summers.[76] Aldous consumed his aunt's novels too, despite his poor eyesight: he had developed keratitis at Eton. In 1911 Julian wrote to his Huxley grandmother that Aldous's eyes were looking better, 'on the upgrade', and that he was reading Aunt Mary's latest, *The Marriage of William Ashe*. It was a bestseller in the United States and had been serialized in *Harper's* magazine. Aldous was reading it in Braille.[77]

INTROSPECTION: HUXLEY SELF-PORTRAITURE

There is no question that generations of Huxleys presented themselves to the world through the printed word. They also did so in visual art, in countless drawings, paintings, photographs and sculptures. Portraits and self-portraits saturated the Huxley universe.

Thomas Henry Huxley was fully engaged with explorations of the self, of character, of sensation. Yet he hardly displayed that other Victorian virtue, studied selflessness. His mature home was filled with images of himself, not just in oil, but in marble and terracotta. He greeted himself every time he ascended his stairs (figure 1.10). Still, this Huxley *was* striking; a favourite for both sombre portraitists and piercing caricaturists. His black eyes, his signature sideburns, his confident carriage all transferred well into oil, ink, pencil and marble. He loved being public.

If Huxley liked to look at himself, taken with his own visage, Henrietta was a student of other faces: 'nothing delights me so much as dwelling on physiognomy, I mentally trace each feature with my invisible pencil.'[78] Like most educated or self-educated Victorians, they both sketched and painted. Huxley made quite a bit of his own talent, a piece of self-aggrandizement that various biographers have repeated, but in truth his sketches and watercolours are ordinary enough by the standards of his time and his milieu. Many not even that. And yet art and talent to observe and represent faces certainly passed to the next generation, including the tendency to represent their Huxley selves. Daughter Marian – Mady – could put strong emotion into and onto her highly accomplished portraits, especially those of her striking father, who sat for her many times (figure 1.11). They shared a melancholy.

Mady studied at the Slade School of Fine Art, where she met and then married John Collier in 1879. This was a marriage of the late Victorian era's best portraitist into the Huxley dynasty, twice over. Perhaps John Collier's most intimate Huxley portrait was of his wife Marian (figure 3.4). Her Huxley melancholy is caught, just as she caught her father's. In a reverse gesture, she depicted him painting a portrait of her.[79] The layers of Huxley introspection were endless.

Through Collier's own relentless interest in generations of Huxleys, they became, if anything, over-represented in oil. His style was pre-Raphaelite: rich and thick. In fact, John Collier couldn't stop painting the Huxleys. He painted Henrietta and offered a delicate miniature of Margaret Huxley to Leonard and Julia (figure 1.7). As an elderly man, long married to Ethel, he wanted to paint his nephew Julian and H. G. Wells together, caught in the process of writing *The Science of Life*, but he couldn't pin down the sought-after Wells.[80] So he considered

Figure 1.10: Huxley greeted himself every time he climbed the stairs.

Figure 1.11: *Thomas Henry Huxley* by Marian Collier (née Huxley).

one just of Julian, the writer: 'My crude idea of a portrait of you is to express you in writing at a table covered with books & documents & litter generally. I have a morbid hankering after a skeleton in the background.'[81] Any skeleton in the Huxley context was a reference to the long-dead patriarch. Collier also sought out Juliette Huxley to sit for him several times – he was not satisfied that he had caught 'her elusive charm'.[82]

Most famously, Collier painted perhaps the most recognizable images of the naturalist friends Darwin and Huxley: Darwin humbly holding his hat, Huxley grandiosely holding a skull. Etchings of the pair still grace a Huxley-Darwin home (figures 1.12 and 1.13). Thomas Henry Huxley loomed over all his descendants, especially Julian Huxley. Collier's oil of Grandfather hung for decades in his library at 31 Pond Street, Hampstead, the home renovated by architect Ernst Freud, Sigmund's son. Julian gladly announced his genealogy via this portrait, for *Life* magazine and for the world. He was the Inheritor.

Figure 1.12: Etching by Leopold Flameng of John Collier,
Charles Darwin, 1883.

Figure 1.13: Etching by Leopold Flameng of John Collier,
Thomas Henry Huxley, 1885.

2

Trustees of Evolution:
The Intellectual Inheritance

Evolution: At the Mind's Cinema

I TURN the handle and the story starts:
   Reel after reel is all astronomy,
   Till life, enkindled in a niche of sky,
Leaps on the stage to play a million parts.

Life leaves the slime and through all ocean darts;
   She conquers earth, and raises wings to fly;
   Then spirit blooms, and learns how not to die, –
Nesting beyond the grave in others' hearts.

I turn the handle; other men like me
   Have made the film: and now I sit and look
In quiet, privileged like Divinity
   To read the roaring world as in a book.
     If this thy past, where shall thy future climb,
     O Spirit, built of Elements and Time?

Julian S. Huxley, 'Evolution: At the Mind's Cinema',
The Captive Shrew and Other Poems of a Biologist
(Oxford: Blackwell, 1932), 55.

Thomas Henry Huxley eventually set himself up as patriarch of a new familial dynasty, carrying less fit siblings, expecting great work from his descendants, and long after his death getting it. He installed himself also as defender-patriarch of a whole new science, weightily confirmed by Edward Onslow Ford's classically inspired statue in the Natural History Museum, South Kensington (1900) (figure 2.1). There sits Huxley, open, aggressive, fist clenched and ready for a fight, even now. His friend Charles Darwin, by contrast, sits cross-legged, closed and speculative: no fight in him at all (figure 2.2). As men and as working natural scientists, Huxley and Darwin were as different as their two Kensington statues. They came from widely separated social strata: the Darwins with considerable upper-middle-class fortune; the Huxleys in serious lower-middle-class decline. They might have begun similarly as medical students, but Charles Darwin did so at the University of Edinburgh in the 1820s, enjoying hunting and shooting outside of lectures in anatomy theatres, while Huxley was a teen apprentice hard at work in grim Coventry and then the grimmer East End in the 1840s, studying late into the night after gruelling days. Natural sciences took over from medicine for each of them, but differently again, despite their formative global maritime journeys.[1] Charles Darwin was a collector, always, as well as a field naturalist and rigorous experimentalist, undertaking decades of breeding research on all kinds of species from his Kent home. Huxley was never a collector in the same way; rather he was paid as a young man reluctantly to organize other collections of specimens, and as an older scientist he compared the specimens collected by others. He was never an experimentalist of breeding or inheritance: not for him the intricate, slow observation of generations of pigeons or peas. And while Darwin was cautious, worried and private about speaking the truth of his new idea on natural selection and the formation of new species, Huxley cast himself forth as public defender of the scientific method, to be proclaimed as loudly and as often as he could.

The friendship between this odd couple, Huxley and Darwin, was as intellectually fertile as the marital pairs – Thomas Henry and Henrietta, Leonard and Julia, Julian and Juliette – were reproductively fertile. Huxley is widely known as 'Darwin's bulldog', nominating himself thus according to the American Museum of Natural History's Henry Fairfield Osborn, one of Huxley's many students.[2] He was certainly a strong, even brutal advocate for Darwin, usefully described by his friend as 'my General Agent'.[3] Yet both of these descriptors imply that Huxley supported and argued for 'Darwinism' – evolution

Statue of Thomas Henry Huxley
Science Museum
South Kensington

Figure 2.1: Thomas Henry Huxley, marble, Edward Onslow Ford (1900).
Photograph inscribed by Ford to Henrietta Huxley.

Figure 2.2: Charles Darwin, marble, Joseph Boehm (1885).

by means of natural selection – from the beginning. This was not the case. Huxley was not fully persuaded by natural selection as a specific mechanism of evolution and barely entertained sexual selection, Darwin's other key intervention.[4] And so, for decades now historians have asked: 'Was Huxley a genuine Darwinian?'[5] To understand this question, we need to distinguish between general ideas about evolution on the one hand and on the other the specific thesis of natural selection that Darwin and Alfred Russel Wallace proposed simultaneously in 1858. Evolution as an idea existed as early as the eighteenth century, and the possibility of species changing over time – the 'transmutation of species' – was put forward not least by Darwin's own grandfather Erasmus.[6] But the idea of *natural selection* as a specific mechanism for this transmutation was the sudden breakthrough, considered by Darwin privately from the 1830s, discussed with Huxley and his other great friend the botanist Joseph Dalton Hooker over the late 1850s, and finally published in 1859. In their many conversations, Darwin had to work hard to win Huxley over. Persuaded by evolution in general, Huxley was never as fully disposed to natural selection as his grandson Julian came to be. Reassessing his grandfather's work in the mid-1930s, Julian was adamant that almost up until the actual publication of Darwin's *On the Origin of Species* (1859) his grandfather was barely an evolutionist at all, let alone a precocious Darwinist.[7]

Over the mid-1890s when old Huxley was dying in Eastbourne and young Julian was growing in Surrey, 'Darwinism' was in decline. Some even spoke of the death of Darwinism after 1900, when Gregor Mendel's rediscovered experiments in Brno on inheritance seemed to render natural selection dubious. This was the era of biology within which Julian Huxley was educated and first made his mark: strangely for the Huxley story, at the low point of Darwinian ideas. Julian himself was a key author of the narrative of an 'eclipse of Darwinism', yet he turned out to be an honourable and loyal trustee of evolution: more so than many doubters. Surrounded by a brilliant generation of geneticists, mathematicians and experimentalists hard at work over the 1920s and 1930s, it was Julian Huxley who managed to best tell and sell a whole new story about the resurrection of Darwinism, and the viability of its 'modern synthesis' with Mendelism.[8]

The Huxleys are a good measure of the fortunes of evolutionary ideas over the nineteenth and twentieth centuries. Between them, they show how misplaced is the common perception that 'evolution' arose suddenly with Darwin, became embattled with theological orthodoxy and then ushered in a secular victory. While in chapter 10 we will engage with the Huxleys' unique connection to religion, agnosticism and belief, here we journey with them through the complexities of the *scientific* fortunes of what came to be called 'evolution'. They have a counterintuitive relationship to Darwinian thought: Thomas Henry Huxley, so closely associated with evolution by natural selection, was an active if respectful doubter; Julian Huxley, biologically raised just when the idea of 'natural selection' was questioned by some biologists, came to be one of Darwinism's most loyal twentieth-century defenders.

ORIGIN STORIES: CREATION AND TRANSFORMATION

How did naturalists think about the animal and vegetable kingdoms in 1825, the year that forty-year-old Rachel Huxley gave birth to her son? What ideas on natural history, the age of the Earth, the creation of life and of variation within and between species circulated in that late-Georgian world?

That year, young Charles Darwin was a medical student in Edinburgh, sent north from Shrewsbury to follow in his doctor-father's footsteps. Darwin was reluctant and unimpressed with medicine and soon turned with more interest towards the students and teachers of natural history, joining the naturalists' Plinian Society in 1826 and taking some instruction in taxidermy. Even if mounting and stuffing animals might have had more then to do with hunting than zoology, it is clear that his scholarly attention was diverted from humans to animals, even before he transferred to Christ's College, Cambridge. It was also set, intellectually, by his own family.[9] Darwin brought with him to Edinburgh important ideas about evolution inherited from his grandfather. Another physician, Erasmus Darwin, published *Zoonomia* in 1794, raising kernels of an idea of common descent and describing a process of change in species that anticipated the

French naturalist Lamarck's slightly later and better-known intervention. In *Zoonomia,* late eighteenth-century readers learned that

> from their first rudiment, or primordium, to the termination of their lives, all animals undergo perpetual transformations; which are in part produced by their own exertions in consequence of their desires and aversions, of their pleasures and their pains, or of irritations, or of associations; and many of these acquired forms or propensities are transmitted to their posterity.[10]

But Erasmus Darwin was a poet-dabbler compared to Charles Darwin's Edinburgh mentor in the mid-1820s, the comparative anatomist Robert Edmond Grant. Grant studied anatomical homologies – similar structures – between species, trying to explain continuities among vertebrates, and working towards the description and explanation of a 'unity of plan' of the kind that Huxley also eventually endorsed.[11]

Grant held important first-hand knowledge of the great French debates then raging on *transformisme* and he brought these back to Edinburgh and London. Did animals change their structure and form between generations? More importantly, did they transform from one species into another over time, and if so, how? Grant spent long periods in Paris at the Muséum national d'histoire naturelle, absorbing and eventually agreeing with the ideas of the museum's professor of zoology, Étienne Geoffroy Saint-Hilaire. For Geoffroy, animals changed over time as a response to environments that stimulated a kind of potential: environments induced change. This was a slightly different proposition on the mechanism of *transformisme* than Lamarck's, who saw change over time due to the inheritance of habit or behaviour in each animal, through the use or disuse of a particular structure.

Parisian research and debate was brimming with disagreement. The stakes were high as this generation of zoologists and comparative anatomists reassessed the Enlightenment legacy, both of their founding predecessor Buffon and of a cluster of 'materialists' who had been more determined than Buffon to divest nature of any kind of divine creation and design. In 1830, Geoffroy and the palaeontologist Georges Cuvier, his colleague at the Museum, held divergent views on transformation – what would soon enough become 'evolution'. All the classificatory work done in the previous generations – by Buffon,

Linnaeus, Blumenbach – continued in European centres, but it was overlaid by new questions and methods, especially the investigation of morphology, the analysis of form and structure of different species. For Cuvier such forms were fixed and determined by the animal's needs. For Geoffroy, morphology showed a constant modification from one unified type. Cuvier perceived abrupt appearances and then extinctions of many creatures, including those in the fossil record, not a gradual transformation between species, which Geoffroy proposed. It was a highly charged difference in what was then the world centre for zoological research, and the Geoffroy–Cuvier debate set a long nineteenth-century pattern and precedent for inclination and dis-inclination towards *transformisme*, and later Darwin's version of evolution by natural selection.

It was Robert Grant who did most to insert gradual transformation or transmutation into British discussion. And so, one day in Edinburgh, when Thomas Henry Huxley was still a newborn, young Charles Darwin and Robert Grant were walking together when the teacher 'burst forth in high admiration of Lamarck and his views on evolution'. Charles Darwin was astonished, even though he had read 'similar views' in his grandfather's *Zoonomia*.[12] His shock is a meas-ure of how unusual and even startling the proposition of a gradual transformation of species, one into another, could be in 1825.

Darwin took his astonishment with him onto HMS *Beagle*, sail-ing around the world between 1831 and 1836, noting species' distributions and differences, variations in different geologies and cli-mates and evidence of extinctions. How is variation within species and between species to be explained? How were the living and fossilized animals he encountered related to their environments? How is a strange animal like the marsupial platypus to be classified? On his return he started to take notes on transmutation, the first of his notebooks titled 'Zoonomia', with more than a nod to his grandfather. Transmutation gradually shifted from a shock idea to a possible, even sensible explan-ation for what he had observed. At the very least, transmutation was a possibility that drove years of careful breeding experimentation. In 1838 he put his learning together with an idea derived from the British economist Thomas Robert Malthus: 'being well prepared to appreciate the struggle for existence which everywhere goes on from long-continued observation of the habits of animals and plants, it at once struck me that under these circumstances favourable variations would tend to be preserved and unfavourable ones to be destroyed'. He recalled this as a beginning: 'Here then I had at last got a theory by which to work.'[13] This was the natural selection of variations, one generation to another.

Charles Darwin's and Thomas Henry Huxley's lives coincided with a great age of debate about origins and change. Three kinds of origin were at issue: that evidenced by Genesis and other Scriptures; that evidenced by rock strata and fossil remains; and, on another scale altogether, that evidenced by embryos, by continual new life. One way or another, Huxley was to intervene in each of these origin stories.

First: Genesis. All and any ideas about species had to be reckoned alongside the Christian account of the separate creation of the fish, the birds, the mammals and the special creation of man and then of woman. Any idea of common ancestry, of transmutation of one species into another, of a fin turning into a wing let alone a hand, cut across the core proposition of Genesis: separate and divine creation. But religious response was never a simple or simplistic reaction, and certainly not when Huxley was learning and then teaching. Theological debate was in many instances itself the business of science: deeply learned, finely morphological and physiological, and carefully argued natural history that began with Genesis. Theologies of life were as complex and hotly debated as natural histories of life, and importantly were not necessarily separate, even rarely so.

The previous century had seen, however, a rise in materialist philosophies that sought to explain nature without God, especially in the anti-clerical French context. And while this transferred into a kind of dissident science in Britain, it also sparked there a renewed appetite for divine design arguments. In the early nineteenth-century British context, this was 'natural theology': in nature and its complex design we can see fine and detailed evidence of God's Creation. Demonstrating this in meticulous detail was the purpose of William Paley's *Natural Theology, or Evidences of the Existence and Attributes of the Deity* (1802). This was a famous and enduring book argued partly from analysis of elaborate and minute morphologies – the eye, being his most important instance – and ranging across everything from blood vessels to joints to birds' bills. All these complex parts add up to a uniformity of God's plan and design, Paley explained. The more other ideas emerged – French *transformisme*, Lamarck's inheritance of acquired characters, Erasmus Darwin's transmutation of species – the more natural theology's explanation of animal variation was insisted upon and expertly refined. Paley's book was part of Charles Darwin's Cambridge education and it was one of Huxley's boyhood favourites, even if an early 'indoctrinating' one, as he later put it.[14] Huxley took detailed notes from Paley's work.[15] Natural theology received a great boost in the 1830s – right at the moment of the Cuvier–Geoffroy debate – through the so-called Bridgewater Treatises, eight detailed and influential publications appearing between 1833 and 1836 on 'the power, wisdom, and goodness of God, as manifested in the Creation'. These were commissioned by the Royal Society, honouring the will of the Earl of Bridgewater, who died in 1829. The Scottish surgeon and anatomist Charles Bell wrote his treatise on the hand, for example, and found 'prospective design' even in that one structure of the human body.[16] It was the kind of argument that Thomas Henry Huxley was soon vehemently to oppose.

The second field of knowledge for new ideas on origins and changes of life was found in apparently inorganic rocks. The age of the Earth was a great question also to be reconciled with Genesis. Natural historians had scripturally reckoned an Earth around 6,000 years old through a generational calculation from Adam to Jesus Christ and Christ to the present, yet fossil records found in strata of rocks increasingly suggested great changes over much longer periods of time. Palaeontological discoveries revealed forms of animals no longer in existence: strange creatures from a distant past now extinct. These fossils and the strata in which they were found were time-keepers, telling of an age of the Earth that receded into an increasingly deeper time.[17] When Huxley was a boy, a 'young Earth', 6,000 years old, had already been challenged, and was redated by the geologist Charles Lyell to a previously unimaginable several hundred million years. New epochs were being invented, and paleontological evidence for change over time was accumulating. For transformationists – early evolutionists – such a gradually changing Earth could now for the first time support ideas about incrementally changing life forms.

For young Huxley and for Charles Darwin, Lyell was the most important Anglophone geologist. His thesis came to be assessed as depicting an Earth that was in a state of constant and more or less 'uniform' change. 'Uniformitarianism' countered 'catastrophism',[18] the latter a more biblically reconcilable theory of the Earth based on sudden events, including a great deluge. But Lyell's famous *Principles of Geology* holds a complicated place in the history of evolutionary ideas. Its original volumes published in the early 1830s included strong disputes with Lamarck, which Lyell extended to most versions of transmutation, deeply anxious about the implications of a logically bestial status for humankind. It is ironic, then, that it was Lyell's book which helped Charles Darwin think about biogeography, species differences and possible transmutation. Yet in conversation with Darwin, Lyell altered his *Principles of Geology* by its tenth, post-*Origin* edition, to align more or less with the transmutation of species. It was a striking change of mind.[19]

By the 1830s, few insisted on strictly literal readings of Genesis-creation, but none would fail to comprehend the theological significance of argument about gradual transformation (of life or of Earth) vis-à-vis sudden, 'catastrophic' or at least staged models of the past, more aligned with special creation and Genesis. Huxley received letters his whole life from clergymen trying to grapple with, accommodate or else doggedly resist the challenge to biblical accounts of Creation that 'transformism' and later 'evolution' posed. Most were agitated, some were intrigued, and a few were compromising and open-minded, genuinely striving to reconcile accounts and scales of time.

One such was Robert King, a senior clergyman in colonial Sydney, who had sailed on the *Rattlesnake* with Huxley: they had occasionally enjoyed companionable fishing from the stern. King wrote to his old shipmate in 1863, worried. He had been reading Charles Lyell's new book, *Antiquity of Man*, and was now at a loss as to how to put geological facts and biblical facts together. He didn't question the authority of an expert like Lyell, but why wasn't the record of the 'Creation of Man written by Him who created Man' also consulted? After all, God created 'the testimony of the rock as well as the work of Genesis'. If Man was created around 6,000 years ago, how does this possibly fit geologists' timescales, including the talk of humans on

Earth around 180,000 years before Christ? For Huxley's shipmate, now a learned theologian, one fact did not necessarily disprove another, but how was he to evaluate the various hypotheses? One possible reconciliation was to question whether this ancient man was in fact 'Adam's race'. Might the ancient *Homo* that Lyell was dating to 180,000 years BC be a different species altogether?[20] The Age of the Earth and the Antiquity of Man were all of a piece in the accumulating palaeontological evidence that was pointing to long periods and gradual processes of transformation of life on Earth.

The word 'evolution' was rarely used with regard to species' transformation until the 1850s. Robert Grant talked about animals having 'evolved' in 1826, and Charles Lyell used it in the context of species in 1836, although the latter explicitly distanced himself from transmutation and especially from Lamarck.[21] The term *was* familiar, however, in the context of embryology; the intricate and conceptually far-reaching research into the development or maturation of embryos of all kinds. Of a different order of time and scale altogether, then, embryology was a third and profoundly important site for the analysis of origins of, and changes in, life on Earth.

In the eighteenth century, the French naturalist Buffon had been deeply interested in embryological development, and postulated that life was created by an as-yet mysterious vital matter that held a capacity for like to create like. He originally considered the power of these non-specific organic molecules to have been divinely created but later shifted towards a spontaneous creation idea. From Buffon onwards, and in a German tradition as well, a core question of early nineteenth-century philosophies and sciences of life was this: is each individual's vital matter pre-formed ('preformationism') or is original material unformed, with structures that somehow formed over time? Both Huxleys would follow up this broad question, Julian via an entirely new twentieth-century understanding of genetics and inheritance, but the question remained more or less the same: how do individual embryos develop or 'evolve', including their mysterious capacity to carry forward the potential to reproduce both similarity and variation in the next generation? And the eventual question: when and how are new individuals reproduced with enough variation from their parents as to constitute a new species?

Over Thomas Henry Huxley's scientific career, the term 'evolution' shifted from its use with respect to embryos to the species-formation context in which we use it today. It did so not randomly, but precisely because the short-term 'development' of individual embryos came to be seen as a recapitulation of that species' long-term development: the embryo repeats over a matter of weeks or months the development of its ancestors over thousands of generations. 'Recapitulation', as this was called, was largely a German intervention and so one in which Huxley came to be especially well schooled, compared to many of his British colleagues. He was one of a few, along with Richard Owen, whose German was fluent enough to liaise and engage closely, as well as to translate authoritatively. This put him into close contact with the German zoologist Ernst Haeckel, who developed important recapitulation ideas, asserting an apparent match between ontogeny (an individual organism's development of form) and phylogeny (stages of development in remote ancestors). Contemporaries, Haeckel and Huxley were to become interlocutors and then friends from 1866, when they met in London, supporting Darwin's work in the German and British worlds respectively. Haeckel was a regular visitor to the Huxley home, appreciating both Henrietta's German fluency and her welcome, a long-lasting friendship which was meaningful right up to her death, and indeed beyond; the elderly Haeckel intermittently advised and instructed young Julian Huxley in his early ecological worldview. Julian's library included his grandfather's precious inscribed copy of Haeckel's 1869 *Über Entwicklungsgang und Aufgabe der Zoologie*.[22]

Notwithstanding a close scientific and familial attachment to Haeckel, in his early scientific years Huxley was rather more persuaded by the embryological work of another German, the older Karl Ernst von Baer, who researched the cells of embryos, so-called 'germ layers', that changed with fertilization and growth. Von Baer did not think that the whole adult form 'recapitulated', as did Haeckel, but that particular structures might. Huxley translated von Baer's work into English in 1853, engrossed with this particular embryology. And it was Huxley who provided the still-clearest four-point summary in English of von Baer's 'laws'. He considered the German's rigorous experiments and the principles drawn from them to have cut through the 'endless mere anatomical discussions' of Cuvier and Geoffroy and the 'hypothetical cobwebs' of German *Naturphilosophie*.[23]

Even as Huxley was translating von Baer and undertaking his own microscopic work on fertilization and 'germ layers', he was not necessarily connecting these interventions to transformationist ideas. Certainly he was not doing so as directly as others in the early 1850s, notably Darwin, who was still holding privately to his idea of common ancestors evolving through natural selection. Far more publicly, Huxley's close friend the polymath Herbert Spencer was applying embryological argument to the broad tenets of 'development', or what would soon enough become 'evolution'. In 1852, Spencer forcefully presented 'the development hypothesis', that complex forms can develop from simple forms over time and can do so in incremental or 'insensible' modifications. The simple and the complex, the immature and mature organism need not bear any resemblance, but nonetheless one can become the other. Thus a seed becomes a tree, a single cell becomes a man, and he linked this to vast amounts of generational time. 'Surely if a single cell may, when subjected to certain influences, become a man in the space of twenty years, there is nothing absurd in the hypothesis that under certain other influences, a cell may, in the course of millions of years, give origin to the human race.'[24] That was a blunt invitation for trouble. Huxley was only just forming his friendship with Darwin and not yet openly discussing transmutation, but already Huxley and Spencer were fuelling each other, especially within publisher John Chapman's London salon and the circle around the radical *Westminster Review* that was soon to support 'evolution' so strongly. Huxley and Spencer were as temperamentally attracted to trouble as Darwin was averse.

In the early 1850s, Huxley was still elbowing his way towards the centre of London zoology, and he recalled only two men then discussing transformism directly, and neither was Charles Darwin. The first was Robert Grant, who had taken up a position at University College London. The second was Spencer. 'Many and prolonged were the battles we fought on this topic,' Huxley recalled. Still, he early stood his ground on the question of evidence: 'even my friend's rare dialectic skill and copiousness of illustration could not drive me from my agnostic position ... up to that time, the evidence in favour of transmutation was wholly insufficient'.[25] Part of what eventually drew

Huxley towards Darwin's idea was that it was scientifically verifiable: the evidence might be acquired; the experiments might be repeated. Equally, it might be proven false.

EVOLUTION AND NATURAL SELECTION: HUXLEY AND DARWIN

It is insufficient to think of mid-nineteenth-century scientists being aligned 'for' or 'against' evolution, even after the publication of *The Origin*. Some were not even thinking of the transformation of species at all. Huxley himself recounted that he did not really consider the 'species question' until after 1850.[26] All his early work on the *Rattlesnake*, as we will see in chapter 4, was about the morphology of marine invertebrates, the then-chaotic classification of lower forms of life.[27] And when he did actively engage with the conversation on transformation over the 1850s, Huxley set himself unambiguously *against* two of the early 'stepping-stones' to Darwin's theory of evolution by natural selection. The first was Richard Owen, then perhaps the world's foremost palaeontologist. Second, and again counterintuitively for 'Darwin's bulldog', Huxley set himself very publicly against the bestselling *Vestiges of the Natural History of Creation*, the work that probably did more than anything to clear a path for the reception of 'transformation' and then 'evolution' within Britain.

If French debate had entered British discussion via Robert Grant in Edinburgh, contemporary German debate was imported into British discussion very much by Owen. In London, he was pre-eminent: Hunterian Chair in Comparative Anatomy from 1837 to 1856 and subsequently superintendent of Natural History at the British Museum. And yet he had not gained his position via Oxbridge privilege, even if he later maintained it through elite alliance. Like Huxley, Owen had been a young medical apprentice, first in Lancaster, and then at the University of Edinburgh and St Bartholomew's Hospital London. Like Huxley, he early abandoned medical practice for natural science work, initially with the extensive Hunterian Collection of the Royal College of Surgeons. This is where he began an extraordinary career as a comparative anatomist, a great expert on those 'terrible, powerful, wondrous lizards', which he named 'dinosaurs'.[28] Owen was not Lamarckian, like Grant, yet he was convinced that species changed over time, and that that change was apparent in the accumulating evidence of the fossil record. Some account needed to be made of all those extinct species and Owen developed the idea that species changed from a set of original and ideal types – 'archetype' was the word he used. Those archetypes themselves never changed, certainly not into one another, and so this was a kind of parallel descent, different to Darwin's model of a branching tree with common ancestors, an idea as yet privately held. And yet on occasion Owen tried out ideas that looked very much like 'evolution'. At the end of his 1849 lecture *On the Nature of Limbs*, for example, he gestured towards a natural law that governed the 'succession and progression' of all vertebrates, advancing 'with slow and stately steps . . . from the first embodiment of the Vertebrate idea until it became arrayed in the glorious garb of the Human form'.[29] Even though Owen cautiously prefaced this by recognizing the 'Divine mind', which planned all archetypes and always knew what modifications would unfold, his conclusion elicited instant criticism that made him suspend that highly controversial line of argument.[30]

While they disagreed, Owen was an enormously important figure for Darwin over the 1830s and 1840s, and for Huxley over the 1850s, offering considerable support and access to specimens and collections. Personally, though, Owen was no favourite to say the least, and he was to become Huxley's public and private enemy. But it is incorrect to imagine that Owen was therefore *anti*-evolutionist. Quite the opposite. Owen's archetype idea was a key stepping-stone in the emergence, then acceptance, of ideas about transformation. Indeed, at this point he was far more of a transformationist than was the younger Huxley.

In the early 1850s, when Huxley was desperately searching for a salaried position, Richard Owen was supportive. They shared a close knowledge of marine invertebrates, and often enough Huxley had to go through Owen to secure access to key London collections and specimens, in order to develop his own expertise in comparative anatomy. But they began to fall out. Owen was notoriously grasping and self-aggrandizing, successfully so with regard to his superiors (he was granted a house in Richmond by Queen Victoria), but mis-stepping badly with respect to his scientific colleagues. Owen was extraordinarily knowledgeable but occasionally passed others' work off as his own; he was slippery with his crediting of scientific knowledge and there were few greater sins in the gentlemanly circles of natural science. Richard Owen, far Huxley's senior and superior, was nonetheless taking liberties in the late 1850s, assuming for a lecture series Huxley's own hard-earned title 'Professor of Comparative Anatomy and Palaeontology, Government School of Mines, Jermyn Street'.[31] Unforgiven, there was no going back.

Intellectually, Huxley disagreed with Owen's unsatisfactory account of why and how animals changed from an original archetype. Owen's great strength was detailed comparative *description* of homologies, the similarity between species of structures inherited from a common ancestor, but not the biological *principles* to which all that detail might add up. Disposed to a Platonic idea of archetypes, Owen wrote of a non-specific organizing energy, a kind of life force that he eventually called 'ordained continuous becoming'. Huxley would have none of that.[32] Early and always, Huxley set aside any explanation that relied on an original creative or divine or metaphysical force. And yet Huxley also became interested in the idea of persistent types, and even used the word 'archetype' to describe a hypothetical structure representing general anatomical uniformities.[33] In 1855, for example, he prepared a course of lectures on the General Laws of Life. He detailed how an 'immense variety of form' was reducible to a few 'common-plans or archetypes'.[34] To some extent he accepted that there were large structurally related groups that changed, but they did not transform between groups, from one to another; this was not mutability or transformation of species. Huxley at this point did not perceive transitional form between types.[35] Indeed, he made some clear comments positively against the mutability of species. It was all an unlikely start for Darwin's bulldog.[36]

The second stepping-stone in the emergence of evolutionary theory that Huxley actively undermined was the anonymously published book *Vestiges of the Natural History of Creation*. Originally published in 1844, the republication of this important book popularized the idea of transformation or evolution at a key point. Its role in clearing ground for Charles Darwin's ideas lies both in its title, 'the *Natural*

History of Creation', and in its stellar record of editions. What a coup for publisher John Churchill. It was strictly anonymous – its author's identity kept secret for decades – which signals just how precarious, even dangerous the very suggestion of a natural history of Creation was at that point. The book presented a thesis on transmutation not just of species, but of all life and all matter, beginning with the origins of the solar system.

Huxley was invited to review the tenth edition in 1853. On the face of it, he might have liked the book. He was not strictly *opposed* to transformism in the way he *was* opposed to any idea of understanding nature through Divine Creation, or any other comprehension of a mysterious or metaphysical life force or vital energy. And politically, he should have approved of the book's appeal to progressive change. Instead he savaged it:

> We grudge no man either the glory or the profit to be obtained from charlatanerie, and we can hardly expect that those who are so ignorant of science as to be misled by the 'Vestiges,' will read what we have to say upon the subject; but a book may, like a weed, acquire an importance by neglect, which it could have attained in no other mode.

Mercilessly, he wrote of the author's 'blunders and mis-statements' and his 'foolish fancies', and the book itself as a 'notorious work of fiction'.[37] Poor Scottish Robert Chambers, who had devised elaborate mechanisms and go-betweens to keep his authorship a secret. Still, watching it reprinted into its tenth quick edition and into one of the most widely read books of the era, he could hardly feel too self-sorry.[38] Charles Darwin was more inclined towards the book than Huxley, but there were important points of disagreement. Chambers argued in part from embryology that generational differences were influenced by environmental change, but that this was in rapid and discrete stages. Thus, *Vestiges* did not support the slow and incremental change that first Charles Lyell and later Charles Darwin and Herbert Spencer advocated.[39] But what explains Huxley's cutting review? The answer speaks less to his intellectual position than to his person: his character and place in a new generation of fierce and bold professional – salaried – scientists. Here we see the young version of the man who became the clenched-fisted South Kensington statue.

First, Huxley exposed what he thought was Chambers's implicit retention of a theological interpretation of nature. Underneath a veneer idea of the 'natural history' of Creation, Huxley perceived a remnant of investment in a divine power that brought the natural world into being and then shaped it. Second, the review was a vehicle for a thinly disguised attack on Richard Owen.[40] Darwin and Huxley cemented their own friendship, perhaps a little meanly, through a mutual enjoyment of cutting the unpopular Owen down to size. If anything, Darwin urged the much younger Huxley on in this respect – who had more to lose – while never quite taking a public stand against Owen in the same terms himself. Third, and possibly most importantly, Huxley took issue with Robert Chambers's amateur and unschooled status: argument from geology and zoology was the highly complex business of professionals and serious laboratory and field practitioners, researchers and students, not gentleman dabblers and mere readers of books. Huxley's seat at the Jermyn Street School of Mines and Museum of Practical Geology was barely warm (he had been appointed in July 1854), as he declared professionalism against an older tradition of amateur inquiry. 'In the popular mind the foolish fancies of the "Vestiges" are confounded with science.'[41] This was Huxley precociously announcing himself – already! – as defender of that particular faith. Science belonged to him and his growing circle of true believers in provable hypotheses, and that alone. Science needed to be protected not just from presumptuous amateurs but also from established centres and holders of increasingly outdated orthodoxies, especially those linked to the established Church (Richard Owen). As for Chambers – the 'vestiginarian' – how dare a non-scientist venture into complex and intricate scholarship on the mutability of species: no wonder there were so many errors, Huxley snarled.

But, as we now know, and as Darwin if not Huxley correctly perceived, *Vestiges* did a great deal to get 'evolution' widely discussed, and in many circles and across many classes, more than a decade before *The Origin* appeared. Chambers had popularized and perhaps even de-exceptionalized a version of transformation, and Darwin was watching and learning, especially from the critique to which Chambers was subject. He was anxious about the likely responses to his own ideas about the transformation of species which he had already worked out in private. It was early enough in their intellectual acquaintance, but Darwin confessed to Huxley, 'I am almost as unorthodox about species as the Vestiges itself, though I hope not *quite* so unphilosophical.' Projecting, Darwin worried to Huxley about 'the poor author' and thought the book rather usefully 'spreads the taste for natural science'.[42] What if Darwin himself was the object of such a scathing review by peers? Even the thought sent him back to his day bed.

Huxley later regretted the review of *Vestiges* for its 'needless savagery'.[43] But it was an early learning exercise in other respects for Charles Darwin: this Huxley's tongue and pen were razor sharp. He quickly perceived the special place of this bold even reckless young controversialist, even though they were not necessarily aligned on evolution or uniformitarianism or embryological recapitulation. Over the 1850s, Darwin well understood Huxley's disinclination to the idea of transmutation of species but clearly valued his knowledge, especially of invertebrates, and was confident of likely persuasion, given the store of evidence in the Down House library and gardens. But as to this fierce character: perhaps Huxley was *too* bold. Darwin cautiously suspended the nomination of his friend to his club, the Athenaeum, after an especially adversarial Huxley lecture.[44]

The two corresponded initially in 1851, when Huxley sent Darwin his report on Johannes Peter Müller's anatomy and development of echinoderms: starfish and sea urchins.[45] The intellectual courtship and exchange of tokens had begun, and they bonded over marine creatures. By 1853, Darwin offered to send Huxley some of his collection of *Ascidiae*, invertebrate tunicates he had acquired in the waters off the Falkland Islands and in Tierra del Fuego, since Huxley was just then cataloguing the British Museum's collection. Darwin was loath to ask, but might Huxley like to review his recent publication on the barnacle *Cirripedia*?[46] Huxley sidestepped the invitation, but Darwin continued to press him on other ideas. He was taken with Huxley's particular deployment of the 'type', and flattered him. 'The discovery of the type or "idea" (in *your* sense, for I detest the word as used by Owen, Agassiz & Co) of each great class, I cannot doubt is one of the very highest ends of Natural History.'[47]

It fairly quickly became a familial friendship. Darwin extended invitations to both Huxley and Henrietta. 'A little change is good for man woman & beast.'[48] Over 1855 and 1856, between London meetings, letters and notes to and fro, and Darwin's hospitality in Kent, the two men began to join forces to influence professional honours and recognition, and thereby careers, in the scientific fraternity. They shared intelligence about who most deserved the Royal Society's various medals. And all the time Darwin was noting a change in Huxley's views. Darwin's ambition to persuade him over to transformation of species was working, slowly.

By 1857, Charles Darwin wrote to Huxley with a more authentic account of the implications of his idea of the continual and continuing relationship between all species, all life. Yet Huxley thought he had the past and future of zoology all mapped out and clearly did not quite perceive the significance of the idea of transforming descent from common ancestors, Darwin's 'pedigree business'.[49] Even in 1859, just before Darwin's book was published, Huxley was not convinced by the gradual-change thesis that geologists proposed, and that Darwin favoured. He wrote to Lyell that evidence showed that new forms sometimes appeared 'at once in full perfection'. 'I think *transmutation* may take place without transition.' It was a theory named 'saltationism', sudden leaps not gradual change. Nor did Huxley align himself at that point with Lyell's proposition that some fossil form 'intermediate between men and monkeys' will be found in the rocks. 'How do we know that Man is not a persistent type?'[50] For Huxley, the key word in this sentence was at least as much 'know' as 'type'. *How* do we know? *What* is the evidence? These were already Huxley's leading questions, and he brought them both to the *Origin of Species* when it finally appeared in 1859.

Huxley finished reading *The Origin* in November and paid Darwin the great honour of comparing him to von Baer. Parts of *The Origin* were persuasive, other parts he considered less so. The chapters on geology and the fossil record, followed by chapters on biogeography, classification, morphology and embryology (rather more Huxley's domain) he endorsed enthusiastically.[51] The most important part of the book, however, was where Darwin set out the new idea of natural selection, in Chapter III:

Owing to this struggle for life, any variation, however slight and from whatever cause proceeding, if it be in any degree profitable to an individual of any species, in its infinitely complex relations to other organic beings and to external nature, will tend to the preservation of that individual, and will generally be inherited by its offspring . . . I have called this principle, by which each slight variation, if useful, is preserved, by the term of Natural Selection, in order to mark its relation to man's power of selection.[52]

In fact, Huxley held reservations about this idea. At best Huxley was initially a qualified 'Darwinian'.

Huxley's friend had provided a theory of evolution that could be empirically tested, yet it had not been proven yet. The doubter had questions about the mechanism of natural selection and how it related to breeding and species. Natural selection required prolific breeding, but 'breeding in and in destroys fertility', as he put it to his friend Charles Kingsley, 'this is the weak point of Darwin's doctrine'.[53] He indicated very publicly that the physiological question of sterility often produced by selective breeding was the sticking point. Still, fertility and sterility were little understood, and so he would go so far as to adopt Mr Darwin's hypothesis subject to proof 'that physiological species may be produced by selective breeding'.[54] It was only from 1868 that evolutionary concepts were directly applied by Huxley to his own research, and it was less Darwin than Haeckel's application of Darwin's idea that finally convinced him, after he reviewed *Generelle Morphologie* in 1866.[55] In his own account, written close to his death, Huxley again signalled the pre-eminence of evidence – records – of evolution. It was not until 1878, he said, that the evidence of palaeontology had finally demonstrated that the many existing forms of animal life had evolved from their predecessors. At last, evolution by natural selection was no longer hypothetical but 'an historical fact'.[56]

Despite this qualified initial commitment to Darwin's thesis, Huxley nonetheless 'sharpened his beak and claws' in readiness to defend *The Origin*. In large part this was because of the delicious opportunity to talk up an idea that the Establishment definitely did not appreciate. He wrote several positive reviews immediately that offset Richard Owen's damning piece in the *Edinburgh Review*. And by the following year, 1860, Huxley was drawn into debate with Bishop Wilberforce at the Oxford meeting of the British Association for the Advancement of Science. There, his position was fixed as an 'evolutionist' and a 'Darwinist'.[57] He pressed the case professionally, institutionally, popularly, and ultimately successfully. For the sceptic who was interested in pursuing only facts that could be proven, his support for Darwin and for evolution, if not for natural selection, was epistemological and political as much as it was scientific. As for epistemology – ways of knowing – Huxley was most opposed to any kind of science that relied on a metaphysical causation. Darwin's theory categorically did not.[58] As for politics, Huxley was deeply invested in challenging a British scientific orthodoxy that was supported by the established Church and by elite class powers. Darwin was too, to some extent, though it was Huxley who took the personal risks.

There was certainly a marked difference between the Huxley of the 1850s and of the 1860s. He spent the first half of the 1850s trying to gain a professional scientific post; the second half of the 1850s securing his position and experimenting with just how abrasive he was prepared to be in his battles. 1860 was a turning point. Huxley willed it to be so, conjuring up his coming 'master years' in an end-of-year journal entry: 'In 1860 I may fairly look forward to fifteen or twenty years "Meisterjahre," and with the comprehensive views my training will have given me, I think it will be possible in that time to give a new and healthier direction to all Biological Science.' He itemized his plans:

> To smite all humbug, however big; to give a nobler tone to science; to set an example of abstinence from petty personal controversies, and of toleration for everything but lying; to be indifferent as to whether the work is recognised as mine or not, so long as it is done:– are these my aims? 1860 will show.[59]

Huxley was trying to set a 'new and healthier direction' for his own hot and adversarial tendencies as much as for science. Perhaps he achieved the latter, but it was not by tempering himself.

Over the 1860s Thomas Henry Huxley was in full and majestic and, for some, terrible flight. For the unlucky he swooped in for the kill in reviews and lectures and meetings. The lucky were brought under his ever-expanding institutional wing. But day to day this growing scientific influence was not just about remarkable character and capacity to think and fight with sharpness and precision. It was about a schedule and an amount of work that was relentless and, on occasion, sick-making. Julian later wrote of his grandfather that his life's work was 'almost superhuman in its arduousness'.[60] But Huxley was no superhuman, and the work took its toll, even when he was in his prime. A letter to Darwin in July 1863 shows both their familiarity and their affection, as well as Huxley's commitments that week and how it all felt: 'I wake up in the morning with somebody saying in my ear, A, is not done, & B, is not done & C, is not done & D, is not done &c &c.' What were all these pressures and deadlines, few of which Darwin had to endure? Huxley itemized them, a form of therapy:

A. Editing Lectures on Vertebrate skull & bringing them out in the Medical Times.

B. Editing & rewriting Lectures on Elementary Physiology – just delivered here & reported as I went along –

C. Thinking of my course of 24 Lectures on the Mammalia at Coll. Surgeons in next spring & making investigations bearing on the same.

D. Thinking of & working at a Manual of Comparative Anatomy (may it be d——d) which I have had in hand these seven years . . .

To top it all off, he finished to his friend, he is now 'pestered to death in public & private because I am supposed to be what they call a "Darwinian"!'[61]

Huxley also complained about the many societies, clubs, dinners, parties and 'all the apparatus for wasting time called "Society"'. Yet it was in and through those societies, dinners and especially clubs that he knew his 1860 ambition to 'smite all humbug' might be realized. He created one for himself in 1864 that did its fair share of smiting: the X Club.[62] This was a group of like-minded liberal evolutionists and scientific naturalists linked by an ambition to fully transform the country's key scientific institutions, especially the Royal Society, and to set a new and to them truer path for the next generation of scientific education and research. There were nine in the X Club, including Huxley: palaeontologist George Busk; chemist Edward Frankland, who taught alongside Huxley at the Royal School of Mines; mathematician Thomas Archer Hirst, linked to University College London;

botanist Joseph Dalton Hooker, Darwin's close friend and director of Kew Botanical Gardens; archaeologist and antiquarian John Lubbock; philosopher Herbert Spencer; physicist and publisher William Spottiswoode; and physicist, glaciologist and great mountaineering friend of Huxley's, John Tyndall. It was a dining club of scholarly and scientific friends, part of the intimate history of evolution. Many were already close. Huxley, Tyndall and Lubbock enjoyed a mountaineering holiday in Switzerland – 'capital fun', according to Lubbock.[63] Together the X Club was an unconventional force to be reckoned with, and one that succeeded in overturning Owen's pre-eminence, largely by casting him as anti-Darwin and anti-progressive. Over the 1860s, 1870s and 1880s Owen continued to be a player in London science, controlling the British Museum's collection and instrumentally pushing for the new Natural History Museum in South Kensington, where his statue still stands. But he was increasingly marginalized, and it was the X Club that managed to install its own sequence of Royal Society presidents and secretaries, to greatly influence the British Association for the Advancement of Science, and to bring Darwin and Darwinism to the fore (figure 2.3).

Huxley and Darwin's friendship was a strong one, and lifelong. They acknowledged births and deaths, the most joyful and difficult family moments alike, sharing high and real emotion on life's turns. Their interactions were sincere, direct, authentic and forthcoming. On the untimely death of the Huxleys' firstborn, Darwin was 'grieved', having gone through it all himself: 'I know well how intolerable is the bitterness of such grief . . . To this day, though so many years have passed away, I cannot think of one child without tears rising in my eyes.' 'God Bless you', he signed off.[64]

The men and the women of the families engaged closely. Charles Darwin wrote to Henrietta about poetry and her beloved Tennyson. She was able to play with him, correcting Darwin for a misquotation, and this from 'a philosopher of your repute . . . If the "facts?!" in the Origin of Species are of this sort – I agree with the Bishop of Oxford.'[65] There was a literary and intellectual exchange across families and across generations. Charles Darwin's learned daughter, also Henrietta, offered a criticism of Huxley's *Lectures to Working Men* (1863): 'You here & there use Atavism=Inheritance.' But she thought this

Figure 2.3: *The Royal Society*, a portrait group of the most distinguished fellows, wood engraving, 1889. Owen (front right) claims the scientific word, but by 1885 Huxley's (front left) X Club of scientific naturalists and Darwinians were in control. John Tyndall, front centre.

incorrect. Atavism normally meant 'resemblance to grandfather or more remote ancestor, in *contradistinction* to resemblance to parents'.[66] Editor of her father's work, well-read Henrietta Darwin was worth listening to. Huxley tried to make light of the correction, but also had to explain: 'I picked up "Atavism" in Pritchard years ago – and as it is a much more convenient word than "Hereditary transmission of variations" it slipped into equivalence in my mind – and I forgot all about the original limitation.'[67] His confusion was telling, and one indication of how familial inheritance over the short term, and Darwin's new biology of generational change over extremely long terms, were coming to be a sticking point.

THE DEATH AND REBIRTH OF DARWINISM

The Huxley–Darwin friendship continued for decades through thick and thin. In 1882, another of Darwin's children, Francis, wrote to Huxley, sadly passing on news. Grief was shared unsparingly.

> My mother asks me to write to you and tell you of my dear father's death. 'He died yesterday afternoon about 4 o'clock; he was not unconscious except for the last ¼ hour. He had an attack in the middle of Tuesday night in which he had some pain which was continuous but not severe. He fainted and soon regained consciousness but remained in a condition of terrible faintness and suffered very much from overpowering nausea interrupted by retching. He more than once said "If I could but die". He got more and more pulseless & the final change came on quickly at the end.' ... She felt his suffering so much. I think there is some consolation, even for her, in the shortness of this illness.[68]

Huxley was one of the pallbearers at Darwin's funeral in April 1882, carrying his friend's body and laying him to rest in Westminster Abbey. Darwin left Huxley £1,000 'as a slight memorial of my life long affection and respect'.[69] And it was Huxley who spoke at the unveiling of the statue of Darwin in South Kensington in June 1885, as president of the Royal Society, alongside the Prince of Wales. After twenty years of controversy, the world now 'discerned the simple, earnest, generous character of the man that shone through every page of his writing', he glowed.[70] The memorialization had begun. Francis Darwin enrolled Huxley into the project of his father's *Life and Letters*,[71] commissioning a chapter on the reception of *The Origin*. Huxley was able to claim in 1887 that ' "The struggle for existence" and "Natural selection", have become household words and every-day conceptions.' It all seemed straightforward to him, a narrative of old battles fought and won, and now all who mattered were persuaded: 'the contrast between the present condition of public opinion upon the Darwinian question ... and the outburst of antagonism on all sides in 1858–9 ... is so startling that, except for documentary evidence, I should be sometimes inclined to think my memories dreams'.[72] But Huxley was out of date. In truth, soon after Darwin died, aspects of 'Darwinism' looked moribund too.

When Julian Huxley was born in 1887 – the year that Francis Darwin and Huxley cooperated to publish a history of the reception of *The Origin* – the moment was marked with the composition of 'The Inheritor'. Leonard Huxley drew down on one of Charles Darwin's core ideas: blending inheritance.

> BABY mine, how strange to see
> Other faces blent in thine,
> Other greatness touching thee,
> Baby mine.

It worked poetically, but it simply didn't biologically. Blending inheritance was proving to be a major problem for Darwin's thesis. If, over evolutionary time, blending inheritance eliminated variation, how then did natural selection – which absolutely required variation – operate? Much later, perhaps with his father's poem at hand, Julian explained the old understanding of biological inheritance 'in Darwin's day': 'biological inheritance meant the reappearance of similar characters in offspring and parent, and implied the physical transmission of some material basis for the characters'.[73] The critique of Darwin was threefold: his assumption that most variations were inheritable; his failure to appreciate that genetic recombination may produce and modify new inheritable variations; and his partial reliance upon Lamarckian principles.

Charles Darwin had early realized his own problem. In the years after the publication of *The Origin*, while Huxley was busy comparing bones in his new role as palaeontologist, Darwin was continuing complex experiments breeding plants and animals and observing generational changes. In 1868 he published *Variation of Animals and Plants under Domestication*, noting the phenomenon of reversion to ancestral types, but not knowing how. He simply did not know the mechanism of inheritance, even though his theory – as he himself stated – depended entirely upon the inheritance of small variations. He was deeply interested in breeding, presumed that 'crossing' led to blending, but then realized the conceptual hole he had dug for himself: how does variation (required for selection to be effective) continue within this blending tendency? His thin solution was a hypothesis of 'pangenesis', that variations can be transmitted as 'gemmules' to the reproductive 'germ cells', bringing a Lamarckian mechanism into the overarching theory.[74] This, for Darwin, might be how variation was sustained. It was an unsatisfactory explanation, and everyone knew it. Darwin only ever claimed it as 'hypothetical', in large part owing to Huxley's particular scrutiny. The incapacity to address this question, and Darwin's own insufficient answer, meant that scientifically speaking 'Darwinism' was in decline towards the end of the nineteenth century. As T. H. Huxley grew old and Julian Huxley grew up, the weaknesses of Darwin's ideas were being brought to the fore.

Next-generation Cambridge-educated geneticist William Bateson killed off blended inheritance in his *Materials for the Study of Variation*, published 1894. Its subtitle immediately signalled the challenge to Darwin's central idea: 'with especial regard to discontinuity in the origin of species'.[75] Discontinuity not gradual blending was the key. Following up with intricate breeding experiments on plants and animals in Cambridge, Bateson was the first to use the word 'genetics' in 1905. From the Greek 'to give birth', it heralded a different kind of creation story altogether. Bateson's work coincided with the rediscovery in 1900 of the results of Gregor Mendel's plant-breeding experiments, originally published 1866: if only Darwin had read them. The problem of inheritance and therefore natural selection began to unravel. Breeding between generations was not about blending characters at all. Rather, characters – the colour of pea plants for instance – recurred in patterned ways over two, three, four generations. Traits might 'hide' in one generation only to reappear in the next; recessive and dominant traits, Gregor Mendel called them. The possibility emerged that new species could be created suddenly by the mutation of genes, 'origin by mutation', the Dutch botanist Hugo de Vries called it, reworking in the light of Mendel's experiments the older idea of sudden jumps in evolution to which T. H. Huxley himself was partial.[76] This came to light in 1900, signalling a whole new era for the science of life, and a whole new century of biological knowledge. Julian Huxley called it 'the Mendelian Epoch', and it was his own.[77]

Julian was catapulted into exciting new Mendelian-inspired laboratory work, witnessing some of the early twentieth century's key biological experiments that on the face of it were undoing natural selection as an idea, but in fact were slowly building it up again, now with new genetic knowledge on a surer footing. He spent July 1912 working at William Bateson's new John Innes Horticultural Institution in Merton Park, Surrey, an experimental breeding institution. His own work was routine – 'mostly picking and checking of peas' – but he appreciated their Mendelian significance.[78] And later in 1912, en route to the opening ceremony of the new Rice Institute in Houston, he spent time in Thomas Hunt Morgan's laboratory, the famous Columbia Fly Room. At the beginning of the century, Morgan had set himself against Darwin's thesis that natural and sexual selection produced new species slowly, and the neo-Lamarckian view of inheritance of acquired characters that some biologists, especially French ones, still favoured. Morgan was initially more convinced that Mendelian inheritance and de Vries' sudden-leap mutation theory were the key. For years he tried to prove de Vries' theory via his fruit flies, experimentally mutating the flies and then breeding them, to track the inheritance of the mutations over generations. But the results forced a rethink of macro-mutation, since his team were finding many small factors ('genes') that were carried on chromosomes in patterned ways. Some, his team showed in 1910, were sex-linked. In the light of these many 'genes' on chromosomes, instead of Mendelism counting against Darwin's theory it was looking possible that Mendelian inheritance could support it. Mutations increased genetic variation in any given population, thus enabling natural selection.[79]

Julian visited Morgan's lab at an exciting time, just when these massively significant experimental results were being noted, digested and organized, in anticipation of *The Mechanism of Mendelian Heredity* (1915), perhaps genetics' foundational text.[80] He befriended a junior member of Morgan's team, and co-author of the book, Hermann Muller, and managed to recruit him to his own new biology department at Rice. It was to be a long intellectual connection, cemented by their firm sense of the politics of biology, their co-authorship of the anti-fascist Geneticists' Manifesto (1939) and their post-war advocacy of a eugenic programme they were to call 'Germinal Choice' (chapter 9).

Julian was intellectually 'finished', then, by the generation who thoroughly rethought Darwinism in the light of Mendelism. His immediate circle comprehended that inheritance was not about blending, nor about reversion, but about new combinations, recombinations, of old genes.[81] By the time Julian returned to Oxford, appointed Fellow of New College and Senior Demonstrator in the Department of Zoology and Comparative Anatomy, he brought his US and German laboratory knowledge with him and taught a new kind of Darwinian selection to the next generation of Oxford students, between 1919 and 1925. Ecological geneticist Edmund Brisco Ford was one of them and recalled Julian to be 'the most powerful voice in developing the selectionist attitude there'.[82] If he had witnessed the initial Mendelian Epoch unfold in his pre-war Oxford studentship, Julian's connections with Bateson's and Morgan's experimental work linked him into a biology circle teeming with fresh ideas, experts and evidence on heredity for a new natural selection, a possible neo-Darwinism. It was not quite the generational change that his grandfather had ushered into British science in the 1860s, but Julian's circle were certainly busy with new societies, new journals and new ideas.[83] But was Julian's own work up to the mark?

Over most of the 1920s Julian was still an active research biologist, yet he was brought up short after a run of journal rejections, including negative reviews from close colleagues. When he left Kings College, London, in 1927, he retained the title Honorary Lecturer in Experimental Zoology, but it seemed his experimental work was not sufficiently sustained or rigorous. Some of his friends were telling him as much, with ruthless honesty. Soon to be appointed to a chair in social biology at the London School of Economics, Lancelot Hogben thought that Julian was sidetracked (and thus perhaps being sidelined) by his popular writing: his mind and his focus were elsewhere.[84] One of Julian's senior colleagues, the marine biologist George Parker Bidder, put it in especially uncomfortable terms: 'for goodness sake do decide which branch of biology you are expert in. A man now cannot be a universal expert ... You must not be led away by the notion of imitating your grandfather.' Quite correctly, Bidder warned that there 'is ten times the amount of biological knowledge now than there was then, perhaps twenty times'.[85] Julian ignored this advice completely, over the late 1920s writing *The Science of Life* with H. G. and G. P. Wells, which was about everything and more. Yet the substantial income derived did help him concentrate on a specialist book for biologists over the early 1930s, one that restored his credibility and (finally) earned him a fellowship of the Royal Society. In 1934 he published *The Elements of Experimental Embryology* with his Oxford student Gavin de Beer.

Still, Julian was being biologically outdone by a phenomenal generation. Of all his brilliant contemporaries – especially J. B. S. Haldane and Sewall Wright – it was probably the statistician Ronald Fisher's population genetics that was singularly transformative. Fisher published *The Genetical Theory of Natural Selection* in 1930, a book dedicated to Leonard Darwin: the mutual affections and scholarly devotions are marked among this inter-war generation. Fisher was entirely clear about the implications of his mathematics: evolution by natural selection was reconcilable with the probability of inheritance of mutations. [86] As Julian later explained with regard to Fisher's work: 'the particulate theory of modern genetics provides firm support for Darwin; only on a particulate basis will Natural Selection be effective'.[87] Not blending inheritance, not old 'atavistic' genes recurring, but mutations, new combinations – recombinations – passed on over generations, and this was now proven to be the case, experimentally and statistically.

As Julian already perceived, and as his colleagues recognized, sustained and intricate experiments were not his strength. Translating biological ideas for public consumption, however, was his clear talent. He might have projected his laboratory credentials (figure 2.4), but

Figure 2.4: Julian Huxley at the laboratory bench, Oxford, 1921.

the fact that the press was interested in catching him in this work itself indicates that he was destined to be a science communicator more than a scientist (figure 2.5). Over the early 1920s he turned Mendelism into something riveting for readers of *The Morning Post*, like an alchemist of ideas.[88] And for three days over August 1925 *Manchester Guardian* readers were vastly improved by a series on the Darwinian theory of evolution.[89] A set of wireless talks the following year, 'The Stream of Life', explained heredity, evolution and the 'betterment of man'.[90] Well may his friends have cautioned him, and well may Bidder have warned him against the professional taint of 'generalist'. Yet it was his capacity to structure and sequence ideas, combined with his more-than-sound knowledge of multiple specialties within biology and zoology, that underwrote his greatest success, *Evolution: The Modern Synthesis*, a book dedicated to T. H. Morgan.[91]

It arose from Julian's 1936 presidential address to the Zoological Section of the British Association for the Advancement of Science on the new work on natural selection and what he considered to be the neglected idea of evolutionary progress. Its positive reception was the immediate prompt to write a large study that synthesized Darwinism and Mendelism. 'The time is ripe', he explained, 'for a rapid advance in our understanding of evolution. Genetics, developmental physiology, ecology, systematics, paleontology, cytology, mathematical analysis, have all provided new facts or new tools of research.'[92] The whole added up to a re-animation of Darwinism in the light of new knowledge of mutation on the one hand and recombination on the other. Darwinism was born again. In a biological field fundamentally composed of the facts of birth, reproduction and death, Julian Huxley put this intellectual story into vital metaphors that he had himself lived through: ailing Darwinism from the mid-1890s; the death of Darwinism in the light of Mendelism after 1900; the rebirth of Darwinism over the 1920s; and, later, the regenerative therapeutics that reconciled Mendelism and Darwinism by the 1940s.[93] The synthesis of Mendelism and Darwinism had a long gestation, but in Julian Huxley's own telling he was its midwife, and, long past its due date, *Evolution: The Modern Synthesis* appeared in 1942. It was a great success, widely recognized as a credit to Julian's capacity to write, organize and collect highly diverse sets of information, and to see a

Figure 2.5: Julian Huxley the writer, the science communicator, 1920.

large picture built from specialist knowledge. Julian brought to the task his family's capacity to craft ideas and words, as well as his own experience in writing complex theses with clarity and liveliness. Along with his brother Aldous, he had an eye and ear for an irresistible title.

Julian Huxley laid down a foundational history – really a narrative – of Darwinism and neo-Darwinism between his grandfather's generation and his own. In doing so, he was one of the original historians of modern biological thought, recording and interpreting the accumulation, rejection and reconciliation of successive theories and evidence. The fact is that the very term 'Darwinism' was under challenge by Julian Huxley's generation. Lancelot Hogben, for example, barely recognized the new biology as 'Darwinism' at all, and said so often, but Julian influentially insisted that it *was* 'Darwinism' that was being reborn from its own ashes, a 'mutated phoenix'.[94] He did a great deal to ensure that the term and the long history of Darwinism was retained in both scientific and lay circles. Far more than a mere chronicler, Julian Huxley set forth Darwin's evolution by natural selection as a revolution again and again, establishing a foundational history.

1958 and 1959 were two big anniversary years for just that. The centennial anniversary of the first proposal of the theory of evolution by natural selection in the joint Darwin–Wallace Linnaean Society papers in 1858 was the opportunity for many published words. Julian prepared a revised edition of *The Living Thoughts of Darwin*, celebrating the year when 'the theory of evolution by means of natural selection was first given to the world'. It was that year, not 1859, Julian insisted, that 'what we may call the biological revolution' commenced.[95] A new edition of *The Origin* was published in 1958, with an introduction by Julian Huxley. That year, Julian also drove forward a co-authored set of reflections by scientists and theologians, *A Book That Shook the World: Anniversary Essays on Charles Darwin's Origin of Species*, collaborating with Ukrainian-American geneticist Theodosius Dobzhansky, religious philosophers Swami Nikhilananda and Reinhold Niebuhr and 'cosmic humanist' Oliver L. Reiser. Even the great book of scientific naturalism, *On the Origin of Species*, was re-enchanted by that generation.

For all Julian Huxley's efforts to secure 1958 as the big year for the

Figure 2.6: Julian Huxley giving his convocation address, 'The Evolutionary Vision', for the Chicago Darwin Centennial celebration, from the pulpit, 26 November 1959.

history of evolutionary thought, 1959 was irresistible: the 150th anniversary of Darwin's birth and 100th anniversary of the publication of *On the Origin of Species*. Like the 1909 anniversary in Cambridge, it was still a familial and intergenerational affair. Many Huxleys weighed in; even Julian's son the anthropologist Francis Huxley was publishing on Charles Darwin that year.[96] Partly because Julian had become perhaps the world's leading popularizer of evolutionary thought, and partly because he was Darwin's bulldog's grandson and proxy, he found himself centre stage. The largest celebration was a massive Darwin Centennial event at the University of Chicago. Julian Huxley delivered a convocation address. It was ceremonial and it was quasi-religious. It was on Thanksgiving Day. The procession down 59th Street was to the chapel, a distinguished group walking together including Muller, Dobzhansky and Darwin's grandson the physicist Charles Galton Darwin. And there, from the pulpit, Julian Huxley delivered his address, 'The Evolutionary Vision', followed by a benediction.[97] The irony escaped few. Thomas Henry Huxley would have enjoyed the headline: 'Clergy still recoil from Huxley speech'[98] (figure 2.6).

3
Malady of Thought: Huxley Minds and Souls

UNQUIET

UNQUIET past, you will not lie asleep,
But leap into my soul and drive a way
Down all the mesh of nerves, secret and deep,
To gain the whole of body to your sway:
Unquiet thoughts, you send the beating blood
To sweep forgotten corners of the brain:
Unquiet Life, against your heaving flood
My little private self protests in vain.

I wish no longer what I had before –
The dusty quiet of a partial death.
All this Unquiet is so much the more
Of Life, to draw with a more vital breath.
If life and quiet will not share my mind,
I'll march with life, and leave my peace behind.

Julian Huxley, 'Unquiet', in *The Captive Shrew and Other
Poems of a Biologist* (Oxford: Blackwell, 1932), 38.

Leonard Huxley had written 'The Inheritor' on the birth of Julian
Huxley, a poem of hope, if one marked by a poignant sense of his own
failed expectations. He looked into his newborn's face and saw the lines
of the fathers and the grandfathers. What Leonard did not know then,
but what Julian soon would, is that the character most strongly inherited was not visage but melancholy. Like his grandfather – more than his

grandfather – Julian succumbed to a paralysing depression. 'Paroxysms of internal pain', as the elder Huxley once described his own malady.[1]

When Julian later put down thoughts on the birth of his own first-born, on 2 December 1920, he had no capacity to translate emotion into iambic pentameter. This was no neat verse, rather scattered and disordered prose. He was hopeless.

> My son, I see you now an infant … It is the eternal wish of fathers to instruct their sons in the art of living. They make so many mistakes themselves, they regret so much of life wasted, such energies dissipated, so many hopes on the rocks. Such exquisite possibilities crumbled into nothing.

Like his father, Julian Huxley considered what his own son would inherit: 'With the father and mother you have, you will, for good or for evil, inevitably be afflicted with the malady of thought. If so, the most entangling & enduring of your troubles will be those that proceed from thinking.' Then Julian started to write about himself and spiralled. 'Your mind … will maybe collapse & leave you shelterless, will be ugly, will be so bewitched as to turn (when least expected) from a palace of an Arcadian bower into a pigsty or a prison.' Julian all but cursed his own newborn. '[Your mind will] burn your feet because it is paved like Hell with unfulfilled desires, will mock you with its puny futility'.[2] He was in despair.

The Huxleys' suffering was grave and intergenerational. There was an unmistakable family history of very marked melancholy. T. H. Huxley's father George ended his days in an asylum. His doctor brother James pursued a career as superintendent of that same Kent asylum, but in middle age was himself troubled.[3] Thomas Henry recounted his own intermittent suffering in diaries, journals and letters, especially on his South Sea journey. His sons, Leonard the classicist and editor and Harry the doctor, both seem to have been spared, Julian later noting that he had never seen his father, Leonard, depressed. As an old man himself, Julian commented that the Huxley genes for melancholia skipped a generation.[4] But this is not true, or it is only so if the Huxley women are set aside. Thomas Henry and Henrietta's daughter, the artist Marian – Mady – 'was going mad … by her look of melancholia' and was treated by the Parisian neurologist Jean-Martin Charcot in 1882, just before her early death.[5]

The next set of Huxleys – Julian, Aldous, Trev and Margaret's generation – suffered with particular intensity. Julian was perhaps more troubled than his grandfather by what was by then called 'depression' and later 'manic depression'. He spent long periods in institutions, sometimes experiencing months of paralysed inactivity and his own paroxysms of internal pain. It is extraordinary that despite and around this inherited state Thomas Henry and Julian Huxley were both so prolific, served in such high office and drove forward their ideas with such success and effective purpose.

Over two centuries of radically changing expertise, experimentation and therapy, the Huxleys did not just experience disordered minds, they also actively thought and wrote about that state of being. Their inquiries led to musings on sense, sensibility and perception, on consciousness and unconsciousness, on evolutionary psychology, psycho-pharmacology and even para-psychology. The working of particularly creative and learned minds turned out to make the Huxleys a living, channelled into bestselling science essays, poetry, fiction and non-fiction. Yet the Huxleys knew better than most that their minds could not necessarily be relied upon. The dark malady of thought was too often unrelieved, neither by rest cures, nor by talking cures, nor by soma in any brave new world.

LOVE-SICKNESS AND DEATH

Julian Huxley wrote of his malady of thought when he tried to compose his scrambled reactions on the birth of his son. His grandfather also used the phrase much earlier and differently. On the island of Mauritius, early in the *Rattlesnake* voyage, Huxley considered that morals and religion 'are one wild whirl to me'. It was the intellect that freed him and allowed him to 'get rid of the "malady of thought"'.[6] Julian was troubled by thinking too much. Grandfather Huxley, by contrast, relied upon structured and disciplined thought to keep his malady in check. If Julian felt burdened by thought, for T. H. Huxley it was his therapy. He wrote from the *Rattlesnake*: 'I had one of my melancholy fits this evening and as usual had recourse to my remedy – a good "think" to get rid of it. It took me an hour and a half walking on the poop however to accomplish the cure.' He was sketching his next paper.[7]

Huxley was wholly given over to self-contemplation, as was his era. Selves – moral and ethical – were core Victorian business. Huxley subjected himself to rugged processes of assessment and critique. As a young man, he set himself goals – learn ancient Greek, read all of Goethe, master classical philosophy – often impossible even for him, and he reproached himself for coming up short. Worthiness and self-worth were utterly bound up with his intellectual programme, his self-education, his learning. This was a punishing and self-imposed regimen.

Huxley was first catapulted into a serious illness as a result of attending a post-mortem as a young medical apprentice, or at least so he recounted. Aged only fourteen, perhaps thirteen, he remembered its effect as a kind of poison that induced 'a strange state of apathy'.[8] This was death at its starkest. Worse than death, this was dismemberment, and Huxley came face to face with what the body and the soul might be in relation to one another. What is left of any human after death? Is the mind – the self – separate from the body as Descartes insisted in the seventeenth century? Is the soul made of the same matter as the body, as radical Joseph Priestley argued in the eighteenth century?[9] Later, Huxley would have definite opinions on all such questions. As a fourteen-year-old he probably already thought too much about it, sinking into a melancholic state.

At the other end of the emotional and bodily spectrum from death – love and desire – Huxley and Henrietta were hopelessly struck, almost from their first meeting. Yet this emotion was also bodily and manifested as a kind of sickness during their long engagement. After only a handful of meetings – not even that – Huxley recalled 'I strongly suspect I was in love without knowing it.' He found himself 'half mad with excitement'. But anxiety over what Henrietta thought, and what was to ensue between them, 'brought on my old nervous palpitation and I became less and less fit for quiet thought'.[10] Julian later diagnosed it as a kind of sickness-from-love leading to an 'acute self-blame for what appear to have been the minor peccadilloes of a normal young man'.[11] Did Julian mean that his grandfather had been unfaithful? It's not entirely clear. But true love could weigh on the soul, Julian mused on his grandfather (and himself). It can manifest as a commitment of one's entire self, resulting in joy and elation or in unhappiness and doubt. Thomas Henry was subject to 'a painful and unbalanced mental state', 'lethargy', 'self-questioning' and 'depression'. At the time, Thomas Henry thought that he kept his mood to himself, feeling simply 'down-hearted' when he sailed northwards from Sydney on the *Rattlesnake*'s first cruise in October 1847: 'as usual chewed my own cud, and said nothing to anybody'. When he felt discontented, he said he reached for Carlyle, and tried to discipline himself by scheduling thoughts-of-Henrietta into his daily routine. One hour before bedtime should do it, allowing the rest of his waking hours to focus on his jellyfish.[12]

Of course, Huxley could not schedule his mind out of its own maladies, no matter how disciplined he sought to be. On the *Rattlesnake*'s third cruise from its Sydney base, this one departing April 1848 north to Cape York by way of the magnificent Great Barrier Reef, he succumbed to a prolonged depression. When melancholy was this deep, Huxley was hammock-bound and immobilized but for reading. Julian later interpreted the slowing journal entries through his own experiences: an ordeal and illness 'scarcely less trying for being intangible, wholly inescapable because it was played out within his own mind and soul'.[13] His grandfather was surrounded by an exuberance of oceanic life that should have had him intellectually energized and excited. Instead, Huxley spent hours reading Dante's *Inferno* in Italian: not literature to raise the spirits. From his hammock, he could barely write any letters to Henrietta, who was herself pining in Sydney. Correspondence slowed and then stopped. Later, Huxley admitted he could hardly remember those months at all, and considered them 'deserts of the most wearisome *ennui*'.[14] In the 1930s, writing his own poetry on 'unquiet', Julian could easily recognize this as 'a black time of struggle and depression'.[15]

Accounts of other cruises over those years seem to have been completely different. Heading up to Cape York again at the end of 1849 and over the first months of 1850, Huxley was filled with excitement and curiosity, recounting in his journal great detail about the landscapes and seascapes through which they passed, and the Aboriginal people with whom they engaged. He was entirely taken out of himself. It was the *Rattlesnake*'s final cruise, leading to what should have been a joyous reunion with Henrietta, and thus a relief for her love-sickness too. But even that homecoming to Sydney was tortured. Huxley looked back over his diary and took account. He recognized that one day he was 'proud and confident' and the next 'sunk in bitter despondency'. He knew that his diary-writing in no way caught his true self, his true wanderings of the soul: 'Could the history of the soul be written for that time, it would be fuller of change and struggle than that of the outward man.' He found it difficult to write partly because of an over-busy mind. As soon as he picked up a pen, he confessed to himself, 'I begin thinking of all things in heaven and earth relevant and irrelevant to the matter under consideration.' Besides, a weight hung over him; 'a sense of evil' he called it. Now that the *Rattlesnake* journey had ended, he and Henrietta were forced to part again, since he must return to England to find his fortune, or at least a modest salary on which they could marry.[16] Henrietta herself endured a long five years after Huxley left Sydney and their reunion in London, 1850 to 1855. She waited for the ships to come in with letters and wrote to him that when they did not arrive, she felt 'heart-sick'.[17] He used the German word *mutig* – courageous – for her. In the long wait we might say that she failed to thrive. She faded away, arriving in London ill and weak. He was shocked.

In their early correspondence Thomas Henry and Henrietta discussed death quite deeply. Even in 1847 she was casting forward to their mortal separation, wishing and willing it to be far distant.[18] And early in their married life they came up against death as a fact, when Noel died. Henrietta lived with an enduring grief, but she did not suffer with melancholia. Rather she looked after Huxley, who did. Friends were brought in, or brought themselves in, on occasion, to assist, recognizing the disability brought by melancholy. Huxley disclosed to Darwin:

> I have for months been without energy & without hope and haunted by the constant presence of hypochondriacal apprehensions, which my reason told me were absurd, but which I could not get rid of – No, I was breaking down; sliding into the meanest of difficulties.[19]

He had a serious breakdown in the early 1870s. Darwin and seventeen other friends pulled together with £2,100 to see him over the difficult period. This amount would 'enable you to get such complete rest as you may require for the re-establishment of your health', Darwin offered.[20] Huxley's periodic depressions were widely acknowledged. His mental state was never actively hidden or fully private.

Julian was even more frank and public. He wrote about his troubles openly as a young man in *Religion without Revelation* and as an old man in his *Memories*. Julian linked his own early breakdowns to the extreme emotions associated with love. His first experience, with two boys at Eton, he found easy to share publicly in his memoirs.[21] It was disproportionately intense, he confessed, not 'common calf-love' but instead 'the Dante and Beatrice type'. He first fell for 'an angel-faced little boy two years my junior (now in the Foreign Office, & going bald). Not content with that, I fell in love (more passionately) with his brother.'[22] In Julian's narrative of the formation of his self, love derailed him, time and again. After his Eton loves, all was going swimmingly at Balliol – his Newdigate prize, his Blue, his presidency of the Scientific Society – and then he fell in love, again, 'my first adventure with the opposite sex', sexual adventure that he had evidently not shared with his fiancée, Kathleen Fordham. All was confusion with regard to this longstanding commitment. In 1908, beginning his final year at Oxford, he wrote in his journal: 'K appeared . . . I had resolved to tell her everything; but it was very difficult . . . What I am to do now, I don't know . . . I pray that I may learn to <u>LOVE</u> her as well as merely <u>loving</u> her.' He was overwhelmed by doubt and about love and, it seems, about sex. 'God bless (whatever that may mean!) her, & me, father & for Mother's sake, & help me be unselfish & unselfconscious.'[23] All as his mother was dying.

Julian had been romantically entwined with Kathleen Fordham for many years by that point. They were most certainly engaged, and to members of the family she had long been 'the bride'.[24] She was *in* the family. In 1911, Julian took her to Eastbourne to visit Henrietta, and to induct her in reverence for the patriarch.[25] Trev was also close to Kathleen Fordham and they enjoyed day trips together.[26] And so it was on the other side of the family, Julia Arnold's side. Aunt Mary Ward congratulated young Julian on his Rice Institute appointment in 1912, content that he would be sailing across the Atlantic in Kathleen's company, at least that was the plan. 'I feel much more cheerful about it than if you were going alone. You and Kathleen will carry your happiness with you.'[27]

And then the engagement dissolved, by Julian's account Kathleen's dramatic decision, taken before the Atlantic crossing. It is not entirely apparent why, but it is clear that his siblings, his parents and what would have been his father-in-law were dismayed. And Julian had shocked himself to the core with failure to deal properly with conflicting 'attraction, loyalty and guilt'. It all turned into a 'depression neurosis', precipitated, he thought, by the discordant pulls of love and duty; not loving enough and not following through on the duty that derived from a promise.[28] His doctor-uncle Harry was called in by Leonard, and Julian entered a sanatorium in Surrey.

In 1913, not only had Julian been committed to Kathleen Fordham, he was also committed to President Lovett of the Rice Institute, who was, with good reason, wondering quite where his star new appointment was. Julian's commencement at Rice was staggered and complicated. He had travelled over for the opening in September 1912 and was back in England by November. The first half of 1913 was spent mainly in Germany, Julian negotiating a year out before a full move to Texas, but as far as Lovett was concerned, he seemed to have gone missing altogether. Lovett was notably invested in Julian, even sending Kathleen – the apparent future Mrs Huxley – trinkets and promises from faraway Houston, luring them both as a youthful couple to the bayou: what was yet to become Rice University's magnificent Mediterranean Revival quadrangle, but was still largely a swamp. All kinds of detail had been negotiated for the soon-to-be newlyweds, including an apartment in Houston, arranged personally by Lovett.[29]

In June 1913, Lovett received a letter of explanation, not from Julian, but from his father Leonard. That summer, Julian was so low from the break with Kathleen, so tortured by the implications – as much for his own character as for her pain – that he literally could not speak for himself. His father spoke and wrote for him. Leonard explained to Julian's distinguished future employer that some three months earlier his son 'had a very bad nervous break down, with insomnia', that he and Kathleen 'have formally broken off their engagement & no definite tie exists between them'. Julian was now in the doctor's hands, under a 'rest cure' and forbidden from writing letters. Leonard explained that he wrote as his son's scribe, yet the communication was really man to man, discussing the young biologist who was reduced and dependent. He was in bed, an enforced diminution. 'The boy', Leonard even called him, in his letters to Lovett.[30]

Decades later – around the time that he was compiling and publishing his grandfather's *Rattlesnake* journal – Julian recounted this episode as his loss of youth.[31] In truth, it was a false start into adulthood, into marriage, into professional science. Julian had failed on a dozen counts to become his own man. Courtship had turned very sour.

It so happened that over those years Julian had also been thinking about courtship of a different kind altogether, an intellectual as well as intimate matter. With his brother Trev, he was watching water birds' idiosyncratic 'courting' behaviour at Tring Reservoir in Hertfordshire. The male and female grebes unusually have virtually identical plumage when finding their mate, and the Huxley brothers were observing their poses, their 'attitudes', in the elaborate process of what he sometimes called their 'love habits'. The courtship ritual of the great crested grebe was to be Julian's signature ornithological study.[32] His paper-turned-book about 'ante-nuptial behaviour' was conceived, researched, presented and written over two tumultuous years. Beginning contentedly enough in 1912 with springtime fieldwork with Trev, Julian's first paper was published in November that year, in *Science*.[33] The two Huxley brothers camped together productively and happily in Aunt Mary's Stock Cottage, near Tring, drawing, watching, following those intriguing birds – 'the pleasantest of holidays', Julian wrote, just him and Trev.[34] When his own courtship ritual went awry in mid-1913 and he was bedridden and reduced, all still seemed rosy for the other brothers. They had a luxurious 1913 summer-in-nature, at least according to Aldous: 'Trev and I rolled very merrily over the Southern Downings ... everything is right in this best of possible worlds,' he wrote.[35] But things started to go wrong. While Julian recovered sufficiently to return to Houston that autumn, and was well enough to publish further from his grebe field work, his depression returned, and the following year he made his way back to England and back to the sanatorium.

The summer of 1914 was tragic on every level. Over the weeks in which Germany invaded Luxembourg and then Belgium, when Britain declared war on Germany and then Austria-Hungary, Julian Huxley was hospitalized again, struggling with depression, though comparatively mild this time by his own account. More seriously afflicted that year, and now in the same institution, was his beloved brother. Poor Julian and poor Trev were both in pain. Physically together, emotionally they were intensely alone.

Trev suffered with the grandfather's malady too. In February 1914, he described being 'lost in a pit an enormous way behind my eyes'.[36] Trev's depression seemed to have descended quickly and circumstantially. He was also brought terribly low by love, another courtship gone wrong, this one apparently a relationship with a housemaid at the Huxleys' London home that broke up after the two parties realized there was no prospect, seemingly cut adrift by class. According to Julian, Trev was reduced by a sense of hopelessness; according to Aldous by the impossibility of his high ideals.[37] Either way, one day he suddenly left the sanatorium and could not be found. On 19 August 1914, Leonard wrote to Julian about the search. Perhaps Trev had enlisted in the army?

> I have nothing definite to tell you, but am coming to believe that T is either planted on some . . . farm – or enlisted – or has lost his memory . . . Where did he go? . . . As for suicidal tendencies, A[ldous] had observed absolutely none in him, tho' often in other people . . . Of course T's silence, if he has enlisted . . . is very painful.[38]

Trev hadn't joined the army or the war. His death was a different kind of 1914 tragedy. He was found hanging from a tree in a nearby wood, and had been hanging there for many days. His death date was recorded as 15 August 1914, four days before Leonard's anxious letter to his eldest son.

The Huxley family was rocked. 'My Trev', Mary Ward called to him in her grief. The only blessing was that her sister was spared. Julia had glowed about her first two boys, the pair. 'You have a goodly heritage', she had written to her Eton boys back in 1904, little knowing what that heritage would bring. 'I don't think there ever can have been children who started with fairer prospects than you boys. Not in money, or influence, as honoured names & good capacities & simple wholesome tastes. So bless you!'[39] And poor, well-tempered Leonard, who first bore Julia's sudden death from cancer in 1908, then managed Julian's struggle in and out of nursing homes, followed by Trev's suicide in August 1914. His new wife, Rosalind Bruce, also bore the brunt of this trauma and tragedy along with the responsibilities of mothering a grieving young Margaret.[40] Julian never spoke of Trev, at least to his wife, and Juliette was surprised that as an old man he wrote about it all so openly in his *Memories*.[41]

LOVE, MARRIAGE, JEALOUSY

Love and its illnesses were to continue to try the Huxleys, as individuals and as a family. For Julian and Juliette difficult personal times unfolded over the inter-war years and coincided, for better or worse, with a science of 'sexology', then at its high point. Julian was drawn into this medical, biological and psychoanalytic conversation, intellectually and politically. In 1938, for example, he contributed to a strange book, *Love, Marriage, Jealousy*. Chapters on love as an instinct and as a pathology sat alongside chapters on sexual anatomy and physiology, on 'the happy marriage' as well as 'the unhappy marriage', and even on 'the question of the mother-in-law'. Is a reform of marriage necessary? That was the book's foundational question.[42] It was one of thousands of compilations that had come to form 'sexology' as an exciting cross-disciplinary field, and it was hardly a surprise that Julian Huxley was part of this conversation, or that his intellectualizing of sex, love, marriage and jealousy transferred between his private and public worlds.

Sexology had been gaining ground from the late nineteenth century. One strand simply sought a more straightforward discussion of heterosexuality and reproduction, possibly including medical instruction on contraception. Where is sexual physiology, one bemused reader asked T. H. Huxley of his *Elements of Physiology*, back in 1879.

> Is there any plausible reason unknown to us, for not describing the lovely and practical manner of creating mankind? Or exists there another portion of your work, containing Sex only? For Science's sake, you could not omit it, but where is it to be found?

The disappointed Dutch reader signed off 'CJH Van der Brock and wife'.[43] Another strand of sexology was more thoroughly political, psychological and radical. The German physician Magnus Hirschfeld published his global study of homosexuality, *Die Homosexualität des Mannes und des Weibes*, having established the Scientific Humanitarian Committee in 1897 to decriminalize homosexuality. At the same time Havelock Ellis and John Addington Symonds wrote their own book on homosexuality, first published in German and then in English, and censored.[44] The 'great work on sex', Julian later acclaimed the study, wondering why it was still 'under prurient guard' in the British Library in the mid-1930s.[45]

As an evolutionary biologist, Julian linked human and animal sexual behaviour from the beginning. 'Courtship' led to 'marriage'. In 1916 he noted all the varieties of human marriage – promiscuity, polyandry, polygamy, monogamy – evident in different bird species, analysing (what came to be called) gender roles. Hens and toucans enact 'gender' dimorphism, in which 'a woman's place is in the home', he wrote disapprovingly. Grebes and herons and swans, by contrast, all minimize sex differences and 'sex consciousness'. He liked this symmetry. These birds 'come more near to the ideal of the women's movement of today; in them both sexes share the duties of the pair more equally, and in all activities realize themselves equally and to the full'. For both birds and humans this seemed preferable to Julian, noting the 'chivalric, medieval idea of women's place' vis-à-vis 'the ideal behind the better part of the woman's movement of today'. Julian Huxley fully expected the non-human animal world to offer models for human social organization that might at least be considered. He was interested, for example, in the models of 'temporary marriage' and 'seasonal marriage' that he observed among grebes and among other passerine birds.[46]

While grebes paired loyally, at least for the season, Julian and Juliette's pairing was often in peril, an openly discussed arrangement in which both had affairs, but with different entitlements and effects. In the year that any number of high-profile sexologists met in London for the World Congress for Sexual Reform, 1929, Julian decided to sexually reform his own marriage. For at least four years, he had an intense relationship with the German-born New Yorker Viola Ilma (figure 3.1). Of many, it was probably the affair that most broke Julian and Juliette. Separated and ultimately poised to divorce, the long battle was agonizing for all three of them, and for the wider family. Julian was in his forties, ageing at a time when his medical and scientific colleagues were announcing strange and awful methods of biological and surgical 'rejuvenation'. Testicular extract injections, bilateral vasectomy, even transplantation of human testes were the scientific news *du jour*, promising renewed heterosexual vigour for middle-aged men.[47] Julian chose the more conventional route, an affair with a much younger woman. But twenty-year-old Viola Ilma was the thinking man's, the literary man's, middle-aged affair. She was a political spokeswoman for youth.

Viola established a magazine, *Modern Youth*, with a rule that no contributor was to be aged over thirty, well and truly discounting her lover. But she was as besotted with Julian as he was with her, and so in the very first issue, despite proclaiming 'Under Thirty!' with an exclamation point, she got him in on the thinnest of invented technicalities. Each month, she decided, the new magazine would publish a story written by some noted author *rejected* before he was thirty. Buried in the first issue is 'Hansen's Tomb' by Julian Huxley, a third-rate short story he had written in 1915.[48]

Youth was political in the 1930s. Energized by the vitality of young fascists, Viola Ilma toured Europe, by one account meeting Baldur von Schirach, the Nazi youth Führer, Bertrand de Jouvenel, the youthful French fascist, and young José Antonio de Rivera, the former Spanish dictator's son. 'Their visions of power captivated her,' one critic commented.[49] In Germany, under Hitler, she saw 'youth with something to live for' and wrote a book about this vitality.[50] What was Julian thinking? He was about to write his anti-Nazi, anti-fascist *We Europeans*.[51] With impossible irony, Viola harangued her now forty-five-year-old lover in notes written on the letterhead of *Modern Youth: The Mouthpiece of the Younger Generation*. 'Julian why have you ruined so much of my life? Why weren't you brave enough to play the game of life with me? God I want to die.'[52]

Aldous intervened. Leonard did too. After Julian's history of love-sickness and depression and haunted by Trev succumbing, they tried to set Julian straight. But it was Juliette, this time, who suffered, the next Huxley to succumb to love-sickness. 'This dark, hungry and bleeding

Figure 3.1: Viola Ilma, 1934.

thing called love is inhabiting my heart and hurts like an unending childbirth.' What an extraordinary writer she was. She saw George Riddoch, the neurologist, the 'nerve specialist', who confirmed that Juliette 'was in a neurasthenic state'. Her own account is anguished.

> Is there no escape, my God . . . I am sick of this battle. I am sick of this pain I am sick of this heart which cannot die entirely, nor live a little. I am sick of this burden on my shoulders, I am sick of the past, I am sick, sick, sick.

Like poor Trev, love, life and death were too painfully one. 'Sick unto my soul'. But unlike him, she kept her own life. 'NO. I won't die. I will not die. For I will live.'[53] Her account of this time, which she herself kept and deposited among her personal papers, is insightful, tortured and painful even to read, let alone to live through. In a calmer moment she wrote to her neurologist, 'Nowadays, in the dearth of religious help, doctors have to receive this human stuff called souls . . . And patients expect even more from doctors than they did from priests and God.' Juliette's confession was love: 'I loved a man. Simple.' Using the title of one of Julian's recent books on evolution and heredity, *The Stream of Life*, she explained his importance and so his betrayal. 'The stream of life, the tissue of life, the pattern of life, the giant and unutterable birth of life, Julian'. Their young son Francis gave Juliette two love-birds in a cage, telling her that the birds die if they are alone. Do they, wondered Juliette? 'Is love, then, to love-birds, so essential a link with life?'[54] Wanting to rid herself of love for Julian, perhaps like the love-birds she found that she could not (figure 3.2).

The whole sorry affair dragged on and on. In August 1931, Juliette told Julian that he needed to make a decision – 'you will know within you what is right' – but that she would 'choose to remain alone should the case arise'.[55] A year later, August 1932, Julian wrote to her from the Athenaeum Club remorseful in theory, but not in deed. 'Juliette . . . it was as if I had killed something in you, & I felt like a murderer'.[56] Julian often asked Aldous to mediate and to carry messages, but his little brother was increasingly compromised and pressed responsibilities back onto Julian, especially, he said, when Juliette had been making so many sacrifices. When Julian and Viola were coming to

London together in January 1932, Aldous refused to carry the news to Juliette. 'She is really in a horrible state of body & mind . . . bad to the point of being dangerous (for her state is such that a new shock might precipitate almost any consequence).'[57] The brothers had been there before.[58]

Aldous was trying to rescue his brother from a kind of insanity. He cabled Julian legal advice on both divorce and on libel in the United States. Leonard also strongly advised Julian 'to take a clean break from V'.[59] In fact everybody weighed in. Aunt Ethel – Leonard's youngest sister, who had married John Collier – told Julian she worried about his boys, and for Juliette, who was 'having to do a rest cure'.[60] Aldous was most direct: Juliette and the boys were the greater good, so sacrifice the lesser good – Viola. 'All this seems to me to be plain Jeremy Bentham.'[61] Aldous did not think much of Viola, damned by the wordsmith as 'nice' – too many times in one letter.[62]

When Julian opted to return, if fleetingly, to Juliette, they discussed the terms. Writing from Grindewald, where he was walking and climbing and reassessing himself, again, he continued their discussion on just what kind of sexual commitment would work:

> I still hold to what I said about marriage – that is it a mistake to try to make it bear the whole weight of sex in all its multifarious aspects; & that if we were brought up to a broader view, instead of to an ideal of monogamy that is part romantic part moralistic, we should be able to lead fuller lives & with certainly no more heart burn than at present.

He rejected her accusation of 'adulterer': 'I cannot pretend to be ashamed.'[63] A little shamelessly, however, he theorized and intellectualized their tortured circumstance publicly:

> The concept that men and women should belong exclusively to each other during a lifetime marital partnership seems to be shaken by various kinds of experimental relationships. It is debatable whether jealousy is an innate element of human nature or only an acquired phase of conduct which can be shed by men and women if the moral code should change in a different type of civilized society.[64]

Julian Huxley was not good at boundaries. When offered by the publisher Methuen the opportunity to fantasize about a new world

Figure 3.2: Juliette Huxley, 1934.

built from his own principles, *If I Were a Dictator* (1934), Julian wrote that he would install trial marriages. All first marriages would be provisional for the space of one year, 'and, provided they are childless at that date, may be dissolved on request of either party' (the temporary 'marriages' of passerine birds). He would create Courts of Domestic Relations, and add sex education clinics to health centres 'where advice on all matters concerning sex, both as regards the frustrations and the realizations of life and personality which it makes possible, shall be freely available'.[65]

Three decades later, for all his other affairs, Julian was still processing Viola Ilma. In 1965 he was still – or again – asking Juliette's forgiveness: 'I <u>know</u> you are half my life, and that without you it would be not only miserable but meaningless . . . & I go and make you miserable. Juliette, I'm <u>not</u> inhuman or a bully – but clearly I am a fool & carelessly insensitive.' In tortured letters he committed himself to becoming a far better person, indeed trying to follow Aldous's example, to become 'more of an integrated & peaceful being'.[66] He knew he had acted shamefully with regard to Viola Ilma all those years before, indeed 'stupidly'. It must have been some comfort to Juliette, yet the anger and devastation stayed raw. But ultimately, the pair remained joined perhaps like the two love-birds, caged (figure 3.3).[67]

In any event, by September 1934, all the heat had gone. While the year before Viola had dedicated her love for Julian 'until death', now she was measured and calm: 'I think of you . . . I have a warm kindly love which will always be yours . . . How is Juliette?'[68] Well might she ask, since Juliette was about to become embroiled in another wild piece of Huxley love-sickness. By 1936 Viola was out, and May Sarton, the American poet, was in. Julian and May Sarton met at the historian of medicine Charles Singer's home in Cornwall. She was twenty-five; he was turning fifty.[69] But after a short affair, Sarton fell far more deeply in love with his wife. It seems to have been only intermittently even passingly sexual, but it was certainly charged. Juliette has been described as May Sarton's principal muse for what is a very significant opus of poetry. In 1947, on 22 June, Julian's birthday, May Sarton wrote 'The Leaves' about the three of them:

Figure 3.3: Julian and Juliette Huxley, portrait by Wolfgang Suschitzky, 1939.

From then on it was clear
We would be three, not two,
However separate each shadow
One grieves, one lives, one gives,
But I do not care or know
Which is he or I or you.[70]

Long after, Sarton recollected the Julian–Juliette–May sequence as 'a triangle in which all three behaved in exemplary ways'.[71]

It was not a one-sided 'freedom' between Julian and Juliette. Towards the end of November 1933 Juliette had still wanted Julian to make a choice between herself and Viola, but in the meantime she had been seeing a man, happily enough, when he suddenly proposed to her, and it was this that made her press Julian for a more singular commitment.[72] Some of her affairs seem to have been relatively open. Juliette had 'a true and tender love' with Alan Best, the Canadian natural historian and sculptor, assistant curator at the London Zoo.[73] Yet the honesty and freedoms between Julian and Juliette were not equivalent. The affair with May Sarton would evidently not be accepted by Julian under the agreed 'principle of freedom', or at least so Juliette thought. This was a secret relationship, and one suspended by Juliette after the war, when May Sarton threatened to tell Julian all.

Years later, after Julian died in 1975, Juliette Huxley and May Sarton were cautiously reunited in London, Juliette insisting on the intermediary presence of her son the anthropologist Francis Huxley. There was something delightful to him (and presumably to his mother) that May Sarton fell far more deeply for Juliette: 'It was Juliette she loved more than Julian,' Francis wrote in a publication of Sarton's love letters, 'much more, in the end, than anyone else.'[74] The two women thereafter had a long correspondence, frank and close, not least about Julian, what he meant to Sarton and what she meant to him. Sarton's letters to Juliette were published, and with Francis Huxley's blessing formalized in an affectionate foreword that described a final, moving encounter between the women in 1994, when Juliette was ninety-seven and May Sarton eighty-two. Francis agreed to the disclosure of family intimacies since Juliette had fully expected the traumas of the Huxley marriage to come out anyway. She wrote to May Sarton:

I have tried to keep the Image – to protect him as best I could. Francis thought it a pity, and a diminution of the truth. So if in 25 years' time, a curious new image surfaces, he [Francis], in any case, will not resent it. Nor will Anthony [their other son] for that matter.[75]

For her part, May Sarton expressed deep admiration for Julian and Juliette's marriage: 'What did come through to me as a young person beholding you and Julian together was an enchanting foam of laughter and teasing, of constant exchange between you. You did, together, create a magical atmosphere. How I admired it!'[76]

At one point, soon after his death, Juliette asked May Sarton what she really thought of Julian. There was an honest, perhaps too honest, response. For a start, as a writer Sarton was dismissive of Julian's shallowest of autobiographies. That was no news to Juliette, who had also derided the volumes, calling them a 'guide bleu' – a travel guide – to his life.[77] Juliette's own autobiography is a far superior piece of writing, the comparison caught even in the titles: *Memories* for Julian, *Leaves of the Tulip Tree*, for Juliette. What happened to the Newdigate poet, the Mary Ward-trained architect of sentences? May Sarton did not hold back to ageing Juliette about her husband's personality either: 'His depressed accusations against himself were as far off the mark as his euphoria,' May analysed.

> The ego was not robust enough, so to give it temporary satisfactions he permitted himself too much popular stuff … His need for romantic love affairs (but this is so very common among human beings) was the need for a flattering mirror which a wife cannot be if she is honest.

But even in the middle of all that drama, she thought him too easily bored, and wondered why. 'Perhaps the flawed ego was so rapacious because it had not learned what the psychiatrists call "to relate" to other people. Julian did not really listen, or very rarely.' May Sarton was upfront about her own dead lover and Juliette's dead husband: he had a 'sick ego'.[78]

TREATING ACCIDIE

As a family, the Huxleys had much love and death to withstand. Sometimes, and for some of them, this combination of love and sickness was unbearable. At other times, it was the 'expression of emotion', as Charles Darwin had put it long before, to think with and to think through, in evolutionary, psychological and genetic terms. The Huxleys were cursed with maladies of thought, but these men and women of letters also transformed the malady into words and ideas and theses. Aldous perhaps found the best word, deep inside his beloved English canon. Chaucer's 'accidie' in 'The Parson's Tale' was 'a very precise description of this disastrous vice of the spirit,' Aldous explained in 1923. It makes a man 'hevy, thoghtful and wrawe'. Accidie was the Huxley condition. 'From accidie comes dread to begin to work on any good deeds, and finally wanhope, or despair.' How did the Huxleys comprehend, manage and treat accidie's power to 'forsloweth and forsluggeth'?[79]

Thomas Henry Huxley's generation lived through the best and worst of nineteenth-century therapy. Chaucer's accidie had become melancholy, lunacy and insanity, and it had become part of the great modern phenomenon of carceral care. This was the asylum age, and some of the Huxleys were trained in the management of lunacy, as well as suffering from versions of it. At the beginning of the nineteenth century, a therapeutic revolution that called itself 'moral treatment' aspired to rest, quietness, managed exercise and, depending on the patient's class, disciplined therapeutic employment. It also demonized physical restraints. This coincided, more or less, with the County Asylums Act, 1808, which heralded the building of public asylums, initially aiming to relieve prisons and workhouses of their chronically ill inmates. As more institutions were built and larger numbers of people were committed, any ambition for a new moral therapy in the public asylums was superseded by the logistics of simple confinement. This was a problem addressed by the Lunacy Act, 1845, and the resulting Lunacy Commission, which inspected asylums, holding them to a minimum standard of care. Institutions for the insane came under humanitarian watch. Scores of asylums were built, some beginning to take in a small number of paying and private patients; not the rich, but those able and eager to distinguish themselves from pauper status, while unable to afford care at home or in a fully private institution. The latter were sanatoria for the monied, just as troubled, but who for their wealth were subject to sometimes faddish and expensive modes of therapy, anything from fruitarian diets to hypnosis to massage therapy.

Aligning with their rising family fortunes the Huxleys' connection to institutions for the mentally ill began at the public county asylum end of the spectrum and improved into the world of private sanatoria and 'nursing homes' by the time of Julian and Trev's troubles. Barming Asylum in Maidstone, Kent, was opened to patients in 1833, originally the Kent County Asylum. Two decades later, Thomas Henry Huxley's unfortunate father, George, entered the asylum, and the teacher was brought to a low end, dying there in 1855. Still, old George Huxley's passing was perhaps better than most in the asylum because his son was in charge. James Huxley was the medical superintendent of Barming Asylum between 1846 and 1863. He took their now-mad father into his institution, but as neither 'pauper' nor 'private' patient (the latter first accepted in his asylum from 1850). James kept him off the formal books, and George Huxley probably lived and was cared for between James's superintendent's house and the wards.[80]

James Huxley was busy not just running his growing asylum in Maidstone, but also managing the new system of asylum inspection. He accepted his post just after the establishment of the Lunacy Commission in 1845, which scrutinized county asylums, diminishing the authority of the medical superintendent and subjecting his decisions to those of Official Visitors. They came to inspect the buildings, the books and the patients monthly.[81] On one such visit, 15 April 1852, Dr Huxley was prevented from attending 'in consequence of the sudden death yesterday of his mother'.[82] The family buried Rachel Huxley nearby, her mad husband attending. James Huxley would only have been relieved to avoid the Official Visitors that day, since he resented their inspections deeply, and was making a name for himself as one of the Lunacy Commission's fiercest critics, constantly seeking more independence for asylums, including his own. Not infrequently, he simply refused to comply with the Commission's requests and published venting articles about their meddling.[83] But this was not the only problem for T. H. Huxley's brother James. Behind the Lunacy Commission was the long tradition – humanitarian, Quaker and Evangelical – of opposition to the use of restraints in asylums. The 'non-restrainers' were a powerful lobby by this time, and Dr James Huxley was a thorn in their side. Evidence of restraint was what the official visitors were looking for, in the main. Sometimes they found the Barming Asylum inmates 'remarkably tranquil, and quite free from mechanical restraint'. But when they scrutinized James Huxley's journal, they would find otherwise. One patient in 1856, for example, was 'restrained almost constantly during the day between the 19th of February and the 19th of June . . . on account of great violence. And that two females also reported to have been restrained.' Towards the end of James Huxley's long term as medical superintendent, in 1861, 'Another female in no 12 ward was secluded and also was restrained by hand cuffs . . . constantly subjected to this restraint for 6½ hours every day . . . Her hands are fastened behind her back.'[84] The Barming Asylum was one of the few places that continued to countenance the practice. Yet James Huxley cared rather more than this would imply. It turns out that he was as politicized as his brother: 'It is poverty that crams our public asylums with many inoffensive persons who need not be . . . therein if they had other homes.'[85] He sent a clear if un-diplomatic message (and this is in his printed and official annual report) that there was too much lunacy law, and that the Lunacy Board was intent on simply filling its asylums. The Commissioners, he wrote, would 'presume upon their functions and intermeddle'. He used the forum to warn other superintendents to 'guard well the respect due to themselves and to suffer no encroachments'.[86] Huxley's brother was becoming more and more aggrieved, setting himself deeper and deeper against those whose responsibility it was to scrutinize and authorize his institution. This Huxley, like the other one, did not want to be told what to do, especially in an institution he considered his own. Alas, unlike his famous brother, James Huxley's heel-digging against authority was ineffective.

In any case, James Huxley was descending into insanity himself. How he might have self-diagnosed, we don't know. For years, he had been asked to record the 'supposed cause of insanity' in his institu-tion's books. Some had 'melancholia' due to 'adversity' or 'anxiety in business' or 'loss of money'. A twenty-year-old wheelwright from Maidstone had acute mania due to 'religion', while a fifty-two-year-old lodging house keeper was melancholic due to 'disappointment in love', both resonating with the larger and longer Huxley family story.[87] They were hardly alone in their tragedies. After all this diagnosing and managing and fighting, James Huxley's own mind failed, and he retired in 1863, aged only forty-three. It was an eventful year. In August, the other Huxley brother of that generation, George Knight Huxley, died, also afflicted with mental troubles,[88] all while the youngest and most famous Huxley was cementing his place with the instantly renowned *Evidence as to Man's Place in Nature* (1863). James Huxley, former alienist, published medical man, now sufferer, went on to live for many decades in a modest Maidstone house adjacent to his old asylum, receiving a pension of £300 a year from the Kent County Authorities until his death in 1907, aged eighty-seven. According to his obituarist he inhabited 'the quietest and most secluded of lives', perhaps succumbing to accidie's power to forsluggeth.[89]

In the meantime, Thomas Henry Huxley was struggling not only with his own intermittent depressions, but with those of his daughter Marian. Rather than recapitulating the unfortunate family association with lunatic asylums, this now-monied Huxley was able to secure alternative private therapies for his favoured and talented daughter. Her first episode seems to have struck in 1882, the year she sat for portraitist-husband John Collier. He captured the face and soul of sadness (figure 3.4).

Mady seems to have been chronically melancholic over her years of pregnancy, birth and the infancy of daughter Joyce in 1884, although she did paint through those years, exhibiting at the Royal Academy.[90] At some point help was sought from the celebrated Parisian neurologist Jean-Martin Charcot, who had recently established a neurology clinic at the hospital to which he was attached for decades, the Salpêtrière. Charcot was famed for his understanding and treatment of 'hysteria', including the claim that susceptibility to hypnosis was a characteristic of hysteria. John Collier took Marian to Paris in November 1887, but she died there suddenly, apparently of pneumonia, and aged only twenty-seven. At least as far as Huxley was concerned, Charcot was far from the first point of therapeutic intervention. Rather

Figure 3.4: *Marian Collier* (née Huxley) by John Collier, 1882–3.

he was the latest and the last to attempt to relieve Mady's suffering. 'Charcot thought he might be able to do something for her,' he wrote to a medical friend. 'For myself, I have been without hope for months past & I have little doubt that the final ending of the poor child's case is the best that could be hoped for.' He and Henrietta were heart-broken all over again, 'this shipwreck of a life that started as fairly as ever we could wish'.[91] The latest untimely death in the family was too much for Huxley. It followed his early retirement from the Royal School of Mines and from his prestigious presidency of the Royal Society, prompted by his own depressions.

Marian Huxley escaped both the kind of public asylum that her Uncle Jim ran, and the private sanatoria that her nephews Julian and Trev were to enter in the early twentieth century. Their circumstances as well as their illnesses were different. But there were also therapeutic continuities over the Huxley generations. One was the 'rest cure'. For Thomas Henry Huxley, a combination of rest and activity saw him through his troubles at home, under Henrietta's care. For his young grandsons, the 'rest cure' became formalized and institutional. Widely practised by 1900, rest cures usually involved many weeks' regimen of compulsory bed rest, total inactivity and a carefully prescribed diet, followed by a highly managed incremental increase in exercise and stimulation. In June 1913, the summer that Julian broke from Kathleen and failed to show at Rice University, he entered a private hospital for a formal rest cure. By August, reported his father, he was doing well, 'promoted from bed & couch to walking once across the lawn'. By September this had increased to vigorous exercise, he was 'walking 8 miles in 2 hours and had put on weight'. His recovery plan was still cautious, however. On discharge he was to spend two weeks in the Chiltern Hills, followed by a short walking tour through the Lakes, and if all went well he might return to Houston in November.[92]

In and out of private hospitals and nursing homes over the years, Julian was at other times managed with diversional therapies intended to pass the time, but in truth these never did. 'The time is leaden here,' he wrote to Juliette from one hospital. This intellectual giant was reduced to making a woollen rug, and, even more diminished, to updating Juliette on its progress. 'I did a few stitches of the rug – but it is an unsatisfactory arrangement – very short bits of wool that don't pull through.'[93] Julian the poet could barely think or write. 'I can't concentrate and see what words to use.'[94] Focus was very difficult: 'Devil says <u>no, can't</u>: Freudian negativism Main thing, can't concentrate,' he tried to explain to Juliette. He was simply unable, at those times, to be Julian Huxley the writer, the thinker: 'Thoughts slip out of my fingers all the time – don't obey me.'[95] At another point he was more lucid, his mental state in part signalled by capacity to put words together to describe the experience. This was neither a rest cure nor a talking cure, but rather a writing cure. On rereading another attempt to explain to Juliette what was going on, he found it 'pitifully inadequate'. 'One would have to be a bit of a novelist <u>and</u> full of energy to give an adequate description of what is going on in one's mind on an occasion like this, & to show all the relations between different aspects of one's thoughts & feelings.'[96]

If the medically trained Victorian Huxley brothers lived through the age of asylums, Julian was subject to the full gamut of twentieth-century therapeutic innovations, over many hospitalizations from 1913 to the 1960s, of which rest cures were perhaps the least invasive, but he thought also the least effective. His rest cure at a sanatorium near Godstone, Surrey, he later described as having no special regimen 'apart from lying in bed . . . and given tonics; the value of electro-shock treatment had not then been discovered'. He had hypnosis sessions with a psychiatrist in Primrose Hill; he was administered intravenous vitamin injections; he was treated with electric shocks repeatedly, although he found it impossible to concentrate afterwards: 'I still can't follow a line of reasoning', he told Juliette after one session.[97] In handwriting that itself revealed pain and mighty effort, he explained that even if he read the lightest novel, 'I forget who the characters are between one dose & another.'[98]

Notwithstanding intermittent hospitalization, much of Julian's care and therapy was at home in the hands of Juliette, much as Huxley's had been in Henrietta's. It took a heavy toll. During one episode in their late years, Juliette committed the weight of this care to her journal. Over February and March 1966, she watched Julian decline again. He was 'tired and depressed', and they went for a very slow walk. She gave him a massage, watched television with him, and then read to him. She was anxious and knew worse was to come, having seen it all before; a 'refrain', she called it, as Julian literally got slower and slower until he all but stopped. 'What am I to do,' she asked herself by March. 'I lay & wept with misery. J came in & we lay together in this sadness and helplessness.' As Juliette described his state of mind, body and soul, he lay down all day, silent and inert.[99] In such times it fell to Juliette to cancel Julian's appointments, speaking obligations and lecture trips.

Like so many others, generations of Huxleys undertook journeys and separations – changes of air and circumstance – to deal with their troubled minds and souls. In the nineteenth century this had a bodily and physical significance. For young T. H. Huxley, melancholy was not just experienced in, but in some measure was caused by, the 'doldrums' through which the *Rattlesnake* sailed. At the other end of his life, his retirement to the sea was in part a health- and climate-related decision and destination. Airs, waters and places themselves had the capacity to induce melancholia or, hopefully, contentment. By the twentieth century states of mind were comprehended rather more as an interior matter. But while we might imagine great changes in treatment, much stayed the same. Julian had spent a lifetime removing himself to health-inducing places, to relatives or friends, for extended stays. Juliette too. In the early 1930s, when she was so tortured by his affair with Viola, Julian manufactured her complete removal to the care of friends in Baghdad, confiding to them that she had succumbed to 'lots of acute depression', and that she had a terror of disclosing this fact either to herself or to friends. Julian – to his credit, whatever his immediate conduct – was never private or disingenuous about his illness. He was always forthcoming, and at least in this context was able to offer, with considerable honesty, that Juliette's 'neurasthenia' was something he had also experienced. 'I've been thru that myself & know how hellish it is.'[100]

STATES OF CONSCIOUSNESS

Love, courtship, birth, illness, death. Every family does it, one way or another. Not every family, however, intellectualizes such matters through physiology and psychology, evolutionary biology and theories of emotion. The signature Huxley capacity to synthesize made the world of the mind, and even the soul, an intriguing intellectual domain, if sometimes a tortured experiential one. In this regard, the enduring Huxley approach was to refuse Cartesian dualisms. Julian declared at one point that there should be 'no dualistic split between soul and body, between matter and mind, between life and not-life, no cleavage between natural and super-natural'.[101] At the very least, mind and body were the stuff of a continuum of matter, not the stuff of irreducible difference.

Thomas Henry Huxley thought so too, even as a seventeen-year-old. In 1842 he tried to systematize external and internal ways of knowing and feeling. 'My scheme', he noted, 'is to divide all knowledge in the first place into two grand divisions': objective and subjective know-ledge. One of his diagrams-of-all-matter extended from an external world to a mental world, from 'not self' to 'self' or 'Ego (subject)', from 'matter' to 'sensation'.[102] The continuum of external and internal knowledge was experienced as sensation, literally through olfactory, gustatory, visual, auditory and tactile senses, the varieties of which formed different 'states of consciousness'. He was trying to make sense of a whole made up of physiological parts and bodily systems, all in relation to external environments.[103] Young Huxley noted in 1840 'a long argument with Mr May on the nature of the soul and difference between it and matter. I maintained that it could not be proved that matter is *essentially* – as to its base – different from soul.' He was beginning his philosophy of mind, his thinking through metaphysics, as well as his 'agnostics', what could and could not be proven. 'We can-not find the absolute basis of matter: we only know it by its properties; neither know we the soul in any other way. *Cogito ergo sum* is the only thing that we *certainly* know.'[104] How are mind and body to be com-prehended? How are inner selves and outer bodies connected? *How* exactly do we feel?

The mind had long been comprehended as a non-physical entity. But in Huxley's era, the mind itself was being renaturalized through a physi-ology of the senses. His mid-1850s lectures set out and explained 'perception' as a 'Physiology of Sensation and Motion – smell, taste, hearing, touch, sight, locomotion; on voice and other means of inter-communication'.[105] And like many in the middle of the nineteenth century he was trying to think through the physiology of hypnosis and mesmerism. These practices were less about accessing a suggestible interior, and more about affecting levels of consciousness, not least in various attempts to minimize pain. Hypnosis and mesmerism were early forms of analgesia as well as anaesthesia, *Pain of Body and Mind Relieved by Mesmerism*, as the title of one 1861 book had it.[106] One of Thomas Henry Huxley's apprenticeships had been with prominent mesmerist Thomas Chandler, and so he knew that the related phenomena of hypnosis, mesmerism, anaesthesia and analgesia were also adjacent to loss of consciousness, unconsciousness, even loss of life.[107] Huxley had observed with great interest an early experimental application of ether as an anaesthetic in Van Diemen's Land in the 1840s. The more successful compound chloroform he later discussed in detail at the Royal Commission on Vivisection, indicating that it had diverse effects upon people: from those completely stupefied and 'lying like logs' through to inciting the most violent movements, 'as if of pain, occasionally with groans and screams – and yet insensible to pain'.[108] In Sydney, young Henrietta endured a terrifying early chloroform anaesthetic. She herself called it near death. The family was also sometimes paralysed with grief-from-death that manifested as a kind of unconsciousness. On the sudden loss of their son Noel, Henrietta went 'out of her mind'; 'unconsciousness often came over me,' she remembered.[109] In the end Huxley saw mental events as brain events: 'the human body, like all living bodies, is a machine, all the operations of which will, sooner or later, be explained on physical principles'.[110] Even thoughts 'are the expression of molecular changes in the matter of life which is the source of our other vital phenomena'.[111] T. H. Huxley is most closely identified with the mind-body idea of 'epiphenomenalism', the claim that physical and chemical phenomena precede and cause mental events. It was a Huxley line of inquiry that was passed on with stellar results. Andrew Huxley (1917–2012), Julian and Aldous's half-brother, received a Nobel prize in 1963 for discoveries 'concerning the ionic mechanisms involved in excitation and inhibition in the peripheral and central portions of the nerve cell membrane'.[112] Indeed, in the Huxley tradition of delivering the prestigious annual Romanes Lectures at Oxford, first Thomas Henry (1893), then Julian fifty years later (1943), Andrew Huxley's lecture (1982) was on 'The Body, the Physical Sciences and the Mind'.

For early Darwinians, if emotions and mental worlds were molecular and physiological sensations, they were also therefore subject to evolution by natural selection. The implications of understanding minds, sensibilities and consciousness within scientific naturalism, and thereafter and therefore as part of evolution, were profound. Darwin knew it, working with great interest on the evolution of emotion, feeling and sensibility across animals and humans.[113] Already in the 1850s, he had perceived the significance of evolution for psychology: 'In the distant future I see open fields for far more important researches. Psychology will be based on a new foundation, that of the necessary acquirement of each mental power and capacity by gradation.'[114] It was a singular passage on humans in *The Origin*, and Darwin was excited but cautious. Huxley, by contrast, barrelled into evolutionary psychology, along with his friend Herbert Spencer, both thriving on the idea. Like Huxley, Spencer was insistent that the human mind was natural matter, subject to natural laws.[115] It was a proposition talked over late into many a night.

Emotions were intellectually exciting, but they were also a Huxley downfall. This family experienced too much of what Darwin analysed, recorded and pictured in *The Expression of the Emotions in Man and Animals*, a book that included photographs of the insane from just the kind of asylum that James Huxley managed. The full range of emotions which Darwin tabled were if anything over-felt by generations of Huxleys. This family had its fair share of 'joy, high spirits, love, tender feelings, devotion', but also too much 'low spirits, anxiety, grief, dejection, despair'. It would be fair to say that Thomas Henry and his brother James succumbed to 'hatred and anger ... sneering and defiance ... disdain and contempt', while Julian and Trev succumbed to 'helplessness or impotence' and perhaps another of Darwin's ranges, 'reflection, abstracted meditation, moroseness'.[116] As far as their personal and family lives were concerned, it was hard for any of the Huxleys to imagine how the capacity not just to feel, but also to analyse emotion could possibly be an evolutionary advantage. Emotions overcame them, on occasion towards life, on occasion towards death. Intellectually, though, the Victorian Huxley helped establish the parameters of evolutionary psychology that his grandson pursued with far greater vigour. It was Julian who much later argued that the human mind was decisive in evolutionary and ecological terms; its capacity to think forward and to pass knowledge on intergenerationally – including knowledge about minds and bodies – was the advantage that humans held over all other creatures.

PERSONALIA AND PSYCHOANALYSIS

The year that Thomas Henry Huxley died, 1895, a book was published by two Austrian neurophysiologists, Sigmund Freud and Josef Breuer, the first of whom was soon to be hailed by the English as 'the Darwin of the mind'.[117] *Studien über Hysterie* partly set out ideas on hypnosis as treatment for hysteria, after Charcot, as well as analysing susceptibility to hypnosis as a *symptom* of hysteria. Freud might have been credited as a new-generation Darwin, but really he was more akin to Thomas Henry Huxley. Like Huxley, Freud was medically trained. Like Huxley, Freud spent many of his younger years dissecting human and non-human animals, especially brains and spinal cords, comparing many kinds of vertebrates, and even one of Huxley's favourites, the crayfish. They both were wholly enchanted by meta-questions on how we think about physiology and emotion, sensation and sensibility, minds and bodies. And in Freud's mind and via his hand, it all eventually became a famous twentieth-century phenomenon, 'the unconscious'.

The bulk of Freud's work on the unconscious was pursued in the generation between the grandfather's death and the grandson's entry into adulthood. *Three Contributions to Sexual Theory*, for example, was published in 1905, when Julian was between Eton and Oxford, trying to manage his own individuation, perhaps unsuccessfully so. Unsurprisingly, it was the twentieth-century Huxley grandsons – Julian and Aldous, and, as it turned out, the great-grandson, Francis – who were most shaped by the great turn towards a new kind of interior, towards analytic psychology. Their lives spanned the heights of Freudian and post-Freudian psychoanalysis, then anti-psychiatry, as well as a period in which psycho-pharmacology in all manner of regulated and unregulated modes became popular. They experimented in both the experience and the thinking-through of new kinds of interior worlds.

A confessional chapter of Julian's *Religion without Revelation* (1927) was richly titled 'Personalia'. This presented a Huxley genealogy, mixed with Freud-inspired family analysis and self-analysis. Julian pointed to his grandfather Arnold's troubles as much as his paternal grandfather's. Indeed, in his *Memories*, Julian again signalled his own 'instability' as inherited less from his Huxley than from his Arnold line. To comprehend it all he reached for the Austrian: 'Freud tells us that a Father-complex acquired in infancy is the chief or sole reason for our personification of the forces of Nature as a personal God.' While Julian declared himself to be free from a father complex, he did introduce Grandfather Huxley as more culpable, unsurprisingly so. The family intellectual tradition and, more accurately, expectation may have produced what psychotherapists call 'an inferiority complex', he explained. This led to painful oscillations 'between self-distrust and self-assurance, despair and elation, so familiar to many growing minds'.[118]

Julian's finely honed self-absorption plus his intelligence and conceptual sensibilities were made for Freud. He was certainly obsessed with his ego as well as his 'previous ego', as he put it in 1917. That year, he was already musing over his very self, his 'I'. 'Before there <u>was</u> a central I'. That ego presented to the world fully formed as 'aspiring poet – cloistered scientist – athlete – bird-watcher – passionate lover of mountains and hills, woods & flowers & streams'. But then his list of identities suddenly turned:

> Sisyphus-Climber towards Spirit – doer of penance – weeper over Sin – worshipper of strong and great men – a being afraid of simple realities like Sex & Anger & Religion . . . one day an ascetic, the next thinking only of funny stories or a silly game . . . the figure of hopeless despair; & of would-be suicide.

Julian was confessing all this to Kathleen Fordham, whom he had devastated in 1913, crashing into his own depression. He told her he had few friends, that those from Oxford to whom he was close were dead, killed at the front, and while he loved Aldous dearly, he was 'remote & Olympian'.[119] Clearly Julian needed someone more appropriate than spurned fiancée Kathleen with whom he could work through his ego. Juliette appeared that summer.

Often enough, Julian's self-assessment was less Freudian than a Victorian-style study of character, in the manner of his grandfather. On a train journey through Italy in 1936 he wrote personal assessments and self-directed instructions, much as T. H. Huxley had done, though for Julian this was more about being than doing. He sought contentment, as he put it, a desire for 'being instead of becoming'. It was after the affairs with Viola Ilma and May Sarton, and he suddenly appreciated 'what was the matter with me'. The insight was perhaps not a stunning one, though honest enough: youth that was now 'past and gone'. There was a better way now: not to seek to recover youth, but a 'capacity for deep but undemonstrative enjoyment, for being able to live & experience fully when opportunity presents itself, unhampered by weakness'. He sought all that without his previous hectic intellectual activity, without awful conflicts and without 'memories of my emotional upheaval'.[120]

Psychoanalysis helped Julian Huxley's science as well as his psyche. He wrote about 'mental embryology' as an emerging field of knowledge, as young a science as his grandfather's physical embryology had been a century before. His 1943 Romanes Lecture, 'Evolutionary Ethics', translated the study of embryological development into a kind of developmental psychology, tracking from the zero of a human ovum: 'The Freudians call it the primitive super-ego', while he called it 'the proto-ethical mechanism'. The lecture is utterly representative of its moment, one half a meditation on derailed Nazi ethics, the other half on developmental psychology and psychoanalysis. He was reading Melanie Klein at that point, and in the light of her work Julian identified the second year of human life as the time of critical individuation of the human self from mother, 'or mother substitute'. Mother is chief object of love and is chief source of power, the primal conflict that generated infantile 'anger, hate, and destructive wishes'. He saw this as an adaptation, part of the peculiar condition of human infancy: 'Out of this primal conflict there grows the beginnings of ethics.'[121] Thereafter comes an individual ethics, social ethics and evolutionary ethics. Momentarily, for Julian, the simplistic (grand)father-induced inferiority complex became something far more sophisticated, focused on the mother, for better or worse. As arbiter and creator of productive guilt, the mother fully replaced the cultural guilt that used to be produced by God the Father out of original sin. His grandfather's light rendition of his own mother's power – a fifty-seventh birthday ditty – had turned into something far more terrible, for both mothers and infants (figure 3.5).

152

Figure 3.5: Thomas Henry Huxley's birthday ditty, '"O Alma Mater",
Mother dear, now in thy 57th year'. By the psychoanalytic 1920s,
the power of the mother had turned into something far more terrible.

Julian kept up with British-based psychoanalysis, especially with the pioneer Marjorie Brierley's work. She was interested in child analysis and considered it a British specialty. Julian appreciated the Freudian idea of adult sexuality constituted by the long development of drives, active from infancy, not an adolescence-onwards formation. It was object choice in infancy that was Brierley's focus too – father or mother. Julian put an exclamation mark against her account of Freud's mid-1920s understanding of infantile sexuality and object choice between the ages of three and five. 'The notion that fixation in the Oedipus phase was the precondition of neuroses became firmly established during these years.' Mental diseases, he read, could be classified 'by the relative depth of regression behind the Oedipus phase'. He also noted – underlined – Freud's idea that 'the self-reproach of melancholics' was recognized to be a reproach 'against a former external object now identified with the self'.[122] Julian was entirely clear that Freud offered insights into the self, the unconscious and ways of being that offered a great advance over what was available to his grandfather. He found a good deal in the concept of repression: 'I also dealt more thoroughly than my grandfather with the difficult problem of conscience, whose origin too usually involves some degree of repression and unconscious guilt.'[123] Julian even introduced basic Freud into his manifesto for Unesco. He wanted to set the organization on a sure and secular footing, through 'scientifically proven facts'. These were fivefold: evolution; chemical combination; the microbial causation of disease; Mendelian heredity; and finally, the 'facts of psychological repression and dissociation'.[124]

HUXLEY PSYCHES IN HIGH MODERNITY

Towards the end of his life Julian Huxley insisted more and more on the genetics of mind and behaviour. Over the 1960s he worked to (re)turn conceptualizations of psychiatric disorder to a biological domain, and to a genetically oriented psychiatry that might even have an evolutionary and adaptive significance, he thought. The prospect also returned him momentarily to something like research. He joined forces with the world-leading evolutionary biologist Ernst Mayr and two psychiatrists, Humphry Osmond and Abram Hoffer, to co-author a paper on schizophrenia in an evolutionary perspective, published in *Nature* (1964).[125] Both Mayr and Huxley pushed against psychoanalysis at this point, at least as far as psychoses were concerned. This group hypothesized a gene for schizophrenia, but their key research question derived from evolutionary biology: why had schizophrenia not only persisted, but retained a relatively high incidence (1 per cent of a population), when people with schizophrenia had a documented low fertility? Huxley argued that there was some selective advantage that must be present alongside its disadvantages.

This work with Mayr, Osmond and Hoffer signalled, if anything, a late-in-life turn away from psychoanalysis, but certainly not a turn away from the greater Huxley quest for a philosophy of mind. The older the Huxley brothers got – Julian and Aldous – the more they co-operated on the dissemination of social and psychological insights (figure 3.6) It was a family affair that was coming to include the next generation as well, especially Julian and Juliette's second son Francis, who was to be the most fully immersed in psychoanalysis of them all. If his brother, Anthony, pursued the emotionally safe and sustaining world of botany, gardens and plants, Francis was as mired in the human mind as it was possible to be. Studying first zoology and then anthropology at Balliol College from 1946, the 'Marginal Lands of the Mind' was Francis's emerging specialty, this chapter contributed to one of his father's books on humanism. This was a new trio of Huxleys – Julian, Aldous and Francis – each offering their thoughts on minds and behaviour, on ways of becoming; 'potentialities', Aldous called it.[126] Aldous was already well known for his spiritual practice with Swami Prabhavananda through the Vedanta Society of southern California, and for his experiments and reflections on the mind and on perception.[127] But Francis Huxley's anthropologies of magic and possession, of psychic healing in the Amazon and in Haiti, on shamans and shamanism through time and place, exceeded even Uncle Aldous's ventures into other ways of being and knowing.[128] And yet, while this was a line of work inspired by Aldous, we might also see Francis Huxley's anthropologies of the mind as a mid-twentieth-century rendition of Thomas Henry Huxley's abiding interest in states of consciousness and in

Figure 3.6: Julian and Aldous, the speaker and the listener, 1958.
Taken in Julian and Juliette's home, 31 Pond Street, Hampstead,
by Wolfgang Suschitzky.

sensibility. Indeed twentieth-century Huxleys occupied the entire con-tinuum of approaches to 'senses' and 'sensation', from Andrew Huxley's physiology of nerves, to Julian's evolutionary psychology, to Aldous and Francis's cultures and substances that altered perception.

Francis Huxley's anthropologies of possession and magic aligned with aspects of Julian's thought too. In 1965, Julian organized a con-ference on ritualization, sponsored by the Royal Society. It concerned the human extension of the animal ritual ethology that had begun with the grebes so many decades earlier. Francis was roped into the exciting rethink and spoke on 'The Ritual of Voodoo and the Symbol-ism of the Body'. Some aspects of ritual carry 'what Piaget called sensory-motor symbols, springing from sources hidden to the rational mind ... ritual can be seen as a way of resolving the mind–body dichotomy, by acting out the force of those images which the body proposes to the mind'.[129] Also invited to Julian's conference was the giant of mid-century psychoanalysis, R. D. Laing, who found ritual-ization a rich and profitable idea, not least in understanding analyst–patient relations. It was at his father's Royal Society gather-ing that Francis Huxley first met Laing, and they became lifelong co-workers in the psychoanalytic space. Francis was the latest Huxley to consider 'The Body and the Mind', a talk given at Laing's invit-ation. 'The Liberating Shaman of Kingsley Hall', Francis titled his obituary of the famed analyst.[130]

Julian was interested intellectually in the high modern extension of perception in a Darwinian context. In 1958 he wrote: 'Administration of drugs like mescalin and lysergic acid are revealing wholly unex-pected possibilities of behaviour and subjective experience in ourselves and other mammals; electrical stimulation is mapping the human cor-tex and showing us the material basis for memory.'[131] Julian took offence when journalists asked about his own habits with LSD. Yet it was a fair question, since the Huxley family was widely connected, the personal and the professional folding together. While it was Julian who intermittently suffered in psychiatric institutional care, it was Aldous who introduced him to this other world of experimental psych-iatry. Julian had met his co-author on the schizophrenia paper, the psychiatrist Humphry Osmond, through his brother. Mescalin brought them together.[132] Yet it was an even thicker Huxley family connection, since over 1957 and 1958, while undertaking an anthropological study of Weyburn Hospital, Saskatchewan, Francis Huxley participated in Osmond's research on the therapeutic value of LSD for alcohol addiction.[133] Francis, like (and with) his uncle Aldous, was involved in LSD therapeutics and clinical experimentation as well as exploring the potential for new perception.

The whole field of LSD usage and research was less illicit than intriguing over the 1950s. Indeed, it could be positively reputable. Osmond, for instance, administered mescalin to English politician Christopher Mayhew, whose experience was filmed by the BBC. In 1953 Osmond visited Aldous Huxley's Los Angeles home, administered a dose of mescalin and monitored Aldous's reaction and experiences. For Aldous, this led to the book *The Doors of Perception*, and more. Osmond proposed the term 'psychedelic' at a meeting of the New York Academy of Sciences in 1957: the word meant 'mind manifesting'. Aldous Huxley's remarkable mind was manifest until his very last breath, and into an afterlife, when, in 1963, on his written request, his psychoanalyst second wife Laura injected him with intramuscular LSD as he died. She called it a sacrament.[134]

These are the two organs that recur in Aldous Huxley's opus: the eye and the brain; sight and mind. These were deeply connected in the author's favoured metaphors of sight and insight, eyes and eyelessness. For close-to-blind Aldous, seeing was an art as well as a function, linked to consciousness, to perception and to memory. Eyes and the mind were close. And so Aldous's book on therapies to improve sight – *The Art of Seeing* – morphed into a study of psychotherapy and the pursuit of insight. *The Doors of Perception* and *The Art of Seeing* were from Aldous's mind and hand, but for the Huxleys intergenerationally, the challenge of mental health, the links between evolution and sensation and the mysteries of mind and body were a shared intellectual compulsion. This stretched back in time to T. H. Huxley's map of the senses, how the brain interiorizes the external through olfactory, gustatory, visual, auditory and tactile perception. It stretched forwards in Huxley-time even to 1990, when Francis Huxley offered his farewell talk at R. D. Laing's memorial, offering 'The Eye – Seer and the Seen'.[135] These Huxleys explored modes of perception and ways of thinking about character, the self, the ego across the great changes in modern psychology and psychiatry. They produced inquiries into multiple states of consciousness between life and death in traditions as variant, but nonetheless related, as evolutionary psychology, Freudian psychoanalysis and psycho-pharmacology. Experiencing, mitigating, managing their own unreliable minds and emotions was one enduring task. But for generations of Huxleys, *thinking* about emotions, minds and bodies was itself a wonderful compulsion. All along, the malady of thought was dominant and recessive, accidie their brilliance and their burden.

PART II

Animals

4

Creatures of the Sea and Sky

THE BIRDS

TO most of us, a bird's a feathered song
 Which for our pleasure gives a voice to spring.
 We make a symbol of its airy wing
Bright with the liberty for which we long.

Or we discover them with love more strong
 As each a separate, individual thing
 Which only learns to act, or move, or sing
In ways that wholly to itself belong.

But some with deeper and more inward sight
 See them a part of that one Life which streams
Slow on, towards more mind – a part more light
 Then we; unburdened with regrets, or dreams,
Or thought. A winged emotion of the sky,
The birds through an eternal Present fly.

Julian S. Huxley, 'The Birds' (1922), in *The Captive
Shrew and Other Poems of a Biologist* (Oxford: Blackwell, 1932), 56.

Dead and alive, animals were everything to the Huxleys, from marine invertebrates to the great crested grebe, from orangutans and chimpanzees in London's Zoological Gardens in the 1860s to the okapi gifted by Belgian King Leopold, the Congolese novelty of Julian's zoo in the 1930s. The Huxleys had a long and intimate involvement with creatures domestic, zoological, stuffed, fossilized and wild. Thomas and Julian were both zoologists, but of very different orders, spanning the great development in the field over the nineteenth and twentieth centuries. Their ways of knowing animals were as new for their times as they were different from one another. As young men in their early careers – Huxley in the 1850s and Julian around the First World War – their expertise lay respectively with creatures of the sea and creatures of the sky. Through and with these animals they pushed forwards the new zoological philosophies and methodologies of their ages. Young Thomas Henry took his sea creatures – strange jellyfish, tiny aquatic protozoa and plant-like hydrozoas – observed them whole and then looked at their dissected parts through his microscope in order to perceive the smallest unit of life, the still-mysterious cell. Huxley's microscopic work on sea creatures harboured expansive meaning, however. He was a master at linking the micro to the macro, particular morphologies and physiologies to general, evidence-based principles.

Young Julian, however, looked to the trees and to the sky. From the beginning to the end, his first zoological love affair was with birds. And rather than observing their structure and cells, Julian observed their behaviour and their 'rituals'. It was a new kind of field research altogether, a blend of the long tradition of amateur ornithology with a Darwinian interest in animal expression: not dead taxonomies, but life in the wild was what drew him in. Between these Huxleys, we can see zoology transforming from morphology to ethology, from taxonomy to ecology, and from the primacy of dead animals to the enchantment of live ones.

T. H. HUXLEY'S MARINE WORLD

It was the turn of events that put Huxley on the deck of the *Rattlesnake* that suggested marine invertebrates for his initial work as a natural scientist. He owed a lot to creatures of the sea, and to the *Rattlesnake* voyage, even if he was miserable for much of it. Four years immersed within nautical, maritime and marine worlds was a foundational experience that deeply shaped his research, his future and his reputation. It was a nautical world of sails and cordage, shanties and hammocks, mates, officers and crossing the Line. Huxley should have been more integrated than he was. He lived among his fellow naval men reluctantly – the officers of the wardroom, the midshipmen, the petty officers, the ordinary and able seamen – often with irritation and sometimes with plain condescension. He viewed them all, over long periods, through his own dejection and depression. Theirs was a maritime world as well as a strictly nautical world, a mass of men not just living on the sea, but becoming part of the sea. The *Rattlesnake*'s men would bathe from the ship, when the opportunity arose; becalmed near the Equator, for example, as Captain Owen Stanley depicted in his watercolour *Becalmed Near the Line – Hands to Bathe* (figure 4.1).

There, they were joined occasionally by other mammals: whales who sometimes played and 'amused themselves', and at other times were slain.[1] Beneath was a less-human and less-mammalian marine world, and that was Huxley's domain. Sea creatures of all kinds surrounded the ship, on the surface and at great depths, and the ocean yielded its bounty. Fish of all species were hauled up as food, and from makeshift tow-nets all manner of invertebrates were aquired. This was a bountiful natural history to be inspected, dissected and collected, for both MacGillivray – the official naturalist – and for Huxley – the emerging comparative anatomist. Turning 'opportunity into knowledge', he observed many species closely and carefully.[2] Although he voraciously read the work of earlier anatomists and naturalists, it was already axiomatic for young Huxley that he see it all for himself, if he possibly could.

And yet Huxley's zoological concentration waxed and waned over the various *Rattlesnake* cruises. In his depressions a lot passed him by. Sometimes he watched pelagic life taken in the tow-net but had no 'mental energy' to investigate. One animal that eventually seemed to cut through Huxley's mental fog, however, was the unlikely jellyfish. There was an abundance of them near Port Essington in the Torres Strait, but how should they be classified? It was a good question, and a difficult one.[3] While mammals were easy enough to distinguish from other phyla, other vertebrates were classified with more difficulty. It was all those worm-like, plant-like, pulp-like *invertebrates* that really challenged inquiring minds. By Linnaeus's eighteenth-century benchmark, a sixfold system divided animals into birds, mammals, amphibians, fish, insects and 'vermes' or worms, the latter:

Figure 4.1: *Becalmed Near the Line – Hands to Bathe. Voyage of the H.M.S.* Rattlesnake, *1846–1849*, by Owen Stanley.

animals of slow motion, soft substance, able to increase their bulk and restore parts which have been destroyed, extremely tenacious of life, and the inhabitants of moist places. Many of them are without a distinct head, and most of them without feet. They are principally distinguished by their tentacles (or feelers).[4]

It was an inadequate catch-all, the 'leftover' class, for which Linnaeus was often criticized. By the 1830s, the term 'zoophyte' (animal-plant) was in fashion, an attempt to refine the description of these in-between creatures. The English natural theologian Charles Bell raised the status of zoophytes to one of four divisions of the animal kingdom, quite unlike Linnaeus's. These simplest invertebrate animals were homogeneous pulps resembling a plant, but which were 'moveable and sensible', like the sea urchin or the hydatid, the hydra or the 'medusa', the jellyfish.[5] Although the term zoophyte was to disappear, it was still used over the first half of the nineteenth century. Charles Darwin did so in his account of the *Beagle* journey, and Huxley relied on it too, throughout his 1850s publications based on *Rattlesnake* specimens. Through these strange zoophytes, Huxley made his early zoological name.

They came in all shapes and sizes and were often easily accessible in the upper layers of tropical seas. Huxley looked closely at the salpa, a tunicate with a long gelatinous body. He dissected the pyrosoma, also a tunicate, sometimes called a 'sea pickle', an intriguing conglomerate of hundreds of tiny individuals. They can be luminescent, and Huxley described their ethereal moonlit delicacy: 'I have just watched the moon set in all her glory, and looked at those lesser moons, the beautiful Pyrosoma, shining like white-hot cylinders in the water.'[6] He looked at a radiolarian which could also be bioluminescent, naming it *Thalassicolla punctata*, 'a new zoophyte', as he subtitled one of his early articles.[7] Huxley was early interested in more complex marine invertebrates too. He dissected echinoderms, including those known as starfish, sea cucumbers and sea urchins. He looked at molluscs, a large phylum with thousands of species known, even in Huxley's time. The cephalous molluscs that he was interested in included cuttlefish, squids and octopi. He dissected and described the Portuguese man o' war. Julian Huxley was to be badly stung by one in West Africa in the 1940s.

It was the jellyfish that secured his earliest publication success. The animal had already been the object of extensive study by the 1840s, when Huxley started to look at the medusae scooped up in the *Rattlesnake's* tow-nets. Jean-Baptiste Lamarck had described perhaps fifty species, dividing them into six genera. François Péron, who sailed in the Baudin expedition to the South Sea (1800–1803), by Napoleon's scientific order, also did a great deal of anatomical work on jellyfish, catching them in some of the same waters through which Huxley later sailed. Passing the 'fine and bold' Van Diemen's Land coastline in February 1848, Huxley began to compose his own thoughts on the jellyfish. He already planned for himself that a medusae paper would be a 'turning point'.[8] It was also a testing point. If it went well in London circles, he would consider himself fit for future duties as a student of zoology. If it went poorly, he would abandon those plans and pursue some other means to support his own and Henrietta's future, to which he was already committed. Fortuitously, he was able to send his paper to the Bishop of Norwich, FRS, Captain Stanley's father, who communicated it to the Royal Society on his behalf. It was published in the *Philosophical Transactions* in 1849, but without Huxley's knowledge.[9] As Julian later commented, his grandfather sailed back to England unaware 'that his zoological name was already made'.[10]

The medusae paper anticipated his return, but most of Huxley's studies on marine invertebrates were written up and published over the early 1850s, during the lean years between his return to London from the *Rattlesnake* voyage and his post with the School of Mines. It was a struggle to get them into print, and especially a struggle to fund a distinctive and illustrated – that is to say expensive – book. Nonetheless, Huxley was becoming known among top-tier naturalists for his marine invertebrate work. George Allman, botany professor in Dublin, wrote a testimonial for Huxley's unsuccessful application for the Chair of Natural History in Toronto. 'Beautiful discoveries', he glowed, recommending Huxley's work on the fauna of the tropical oceans.[11] Richard Owen recommended him as a zealous observer 'on the form, structure, and habits of the rare animals of those seas'.[12] And while none of those posts eventuated, Huxley accumulated significant accolades for the zoological work he had undertaken on the *Rattlesnake* journey.

Why was it all so important? These animals held a larger significance than their simple form might imply. Marine invertebrates had been a classificatory problem for Linnaeus, Lamarck, Péron and Cuvier. That was interesting enough, and classification was certainly still confused and confusing in Huxley's 1850s. But taxonomy in and of itself was the zoological drive of an older generation. Huxley was engaging with debates that were more complex and fresher: on the individuality of organisms, on embryology and on the reproduction and development of organisms over their lifetime. Huxley's dissections and descriptions of zoophytes were undertaken with a new methodology based on microscopic observations of *cells*. This was German-rooted scientific inquiry, proposed in the 1830s by the botanist Matthias Schleiden and the physiologist Theodor Schwann. They considered cells to be the basic unit of all living things; and that existing cells produce new cells. Remarkable. The microscope and the experience and capacity to use it well were essential for this kind of observation of life, and it was in minute matter that living truths were to be found, Huxley declared in the 1850s, not in the 'osteological extravaganzas' of fossil museums. He thought that 'the smallest fact is a window through which the Infinite may be seen',[13] and although he got those facts wrong in an early article on 'the Cell Theory' (1853), later correcting himself, he did get it right in his microscopic studies of jellyfish and polyps.[14] They were composed of two cell-layers, or membranes, that could be equated with the two layers of the early embryo, he showed. Huxley did not yet know it, but this insight was to become the foundation on which Ernst Haeckel later developed his 'Gastraea Theory'; that all multicellular animals can be traced to a hypothetical two-layered ancestor. In Julian Huxley's time this was applauded as 'the very basis of philosophic zoology'.[15]

This 1840s and 1850s debate over life's fundamentals made Huxley's jellyfish, squids, protozoa, molluscs and tunicates gathered in the warm seas off the Great Barrier Reef, the Torres Strait and the Louisiade Archipelago, the unlikely matter for the adjudication and illumination of the basis of life in every literal sense (figure 4.2). Each time Huxley or Haeckel dissected a jellyfish and observed cells through their microscopes, they looked to creation, to origins, to genesis. How, precisely, did life come from prior life? How was each

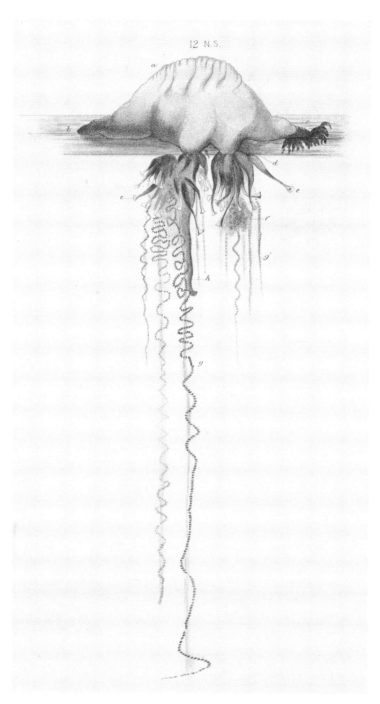

Figure 4.2: *Physalia*, from Thomas Henry Huxley,
The Oceanic Hydrozoa (1859).

organism connected to, or individuated from, its parent? Did single cells hold the complexity of their own later developed organism already within them? This examination of the beginning of (each) life was about the order of life as a whole.[16] And it was a discussion of creation closer to genetics than it was to Genesis. They weren't yet seeing those chromosomes, but they were looking for the cellular key to the reproduction of individual organisms. Huxley noted, for example, that the substance of the 'spermarium' in Hydrozoa differentiated 'into minute, clear, spherical vesicles of about 1/4000th to 1/5000th part of an inch in diameter'.[17] Such tiny observations were presented in oversized plates. It was a developmental and physiological account that spoke to the reproductive significance of 'germs' or the cells of every organism.[18]

Huxley's marine work culminated with the *The Oceanic Hydrozoa* in 1859, a quarto-sized extravaganza, produced with the early encouragement and assistance of his friend and colleague Edward Forbes, who preceded Huxley as palaeontology lecturer at the School of Mines. The large book was published a full ten years after returning from the *Rattlesnake's* South Sea voyage, a delay that Huxley put down to Captain Owen Stanley's sudden death in Sydney: the *Rattlesnake* officers lacked a leader on return to London, someone who would promote the expedition and its findings. In the meantime, 'eminent German observers' had continued to refine the field of cellular theory, of embryology and of recapitulation.[19] He felt he was missing out on the opportunity to make his mark on the debate of the decade. To make matters worse, by the time *Oceanic Hydrozoa* was finally published – in 1859 – it was completely overshadowed by Darwin's book also published that year.

Huxley dedicated *The Oceanic Hydrozoa* to Royal Navy Arctic explorer and fish-specializing naturalist Sir John Richardson; with uncharacteristic and definitely false modesty from 'a beginner in zoology'.[20] By 1859, he was hardly a beginner. People across the world had already started to send Huxley all manner of curious sea creatures, signalling a growing reputation as a specialist in marine invertebrates. A New Zealand friend sent him a preserved Pearly Nautilus, for example, originally from New Caledonia. This gave him an opportunity not just to dissect and observe a rare creature, but also to work over and pointedly improve Richard Owen's intellectual terrain. Owen had published a *Memoir on the Pearly Nautilus* decades earlier, in 1832, and Huxley set to work revising the master's observations with improved microscopic precision, measuring apertures, sacs and heart chambers to one-eighth of an inch.[21] Now with a secure position in London, Huxley far preferred that specimens came to him, than he go to the specimens, and in fact the *Rattlesnake* voyage was to be his only sustained period of what we would now call field research. Indeed, after years on the *Rattlesnake*, the deep sea was the last place Huxley wanted to be.

The coast, however, was another matter. He was a contented beachcomber, and rockpools suited him far more than marine life on the open seas. In 1851, he holidayed with close friends Dr George and Ellen Busk on the coast at Ipswich; he wrote to Henrietta that Ellen Busk was 'almost as enthusiastic a Naturalist as her husband'.[22] In the summer of 1854 they holidayed again at Tenby on the Welsh coast, Huxley and Henrietta returning there for part of their honeymoon. All kinds of surface-dwelling zoophytes washed up on coastlines, and yet even there Huxley was not interested in documenting the environments in which the creatures lived; he was perfectly happy to remove them from their terraqueous world, dissect them and compare them. This was not biogeography, Darwin-style, nor was it ecology, Julian-style.

In 1855 Huxley was appointed to a commission whose role was to survey the fauna and flora of coastlands of the British Isles. This was the first of many public commissions on which he sat, most of them about the sea in one way or another. He contributed to a sequence of government inquiries dealing with the regulation of fisheries and fishing, especially in Scotland. They sometimes took him to islands and villages and coasts that he loved: Skye, Ardrishaig, Tarbert, Greenock. In 1862, he was appointed to a Royal Commission on trawling for herring off the coast of Scotland, and again in 1884 another commission on net and beam trawl fishing sought his expertise.[23] He was on call for advice on the deep sea as well and offered it willingly, but only from London's terra firma. This began in 1857, when HMS *Cyclops* was commissioned to find the shortest route for an electric telegraph cable between Ireland and Newfoundland. It was the opportunity to undertake deep-sea hauls, and Huxley was invited to issue instructions to his maritime counterparts. He offered detailed recommendations to the ship's surgeon on dredging, on the preservation of specimens, on methods of observation.[24] And the 'mud' that came up from the bottom of the ocean was sent to Huxley for analysis. Was it living or not-living? In fact, he confirmed, it was the discarded shell of the plankton *Globigerina*, which covers so much of the sea-bed.

The deep sea was very much the business of scientists on HMS *Challenger*, which sailed in December 1872. This was an expedition dedicated to research on the world's oceanic depths; a circumnavigation of the world ocean and a step change in oceanography. Again a land-based adviser, Huxley took to the expedition with excitement. He was very happy to explore the deep sea from a distance, including the technicalities of dredging and trawling, how the *Challenger* was vastly to extend the 100 fathoms that had, more or less, been the limit up to that point. First at 600 fathoms and later from an extraordinary 2,600 fathoms, the *Challenger* scientists trawled up more 'Atlantic ooze' from an ever deeper sea-bed. 'The Problems of the Deep Sea' was followed by three Huxley articles on the work of the *Challenger* expedition, in which he gave his further account of bivalve and univalve molluscs, of starfish, sea urchins, sponges and more.[25] Huxley was intrigued by the prospect of determining a zone of zero life.[26] Where might the limits of life lie? The deep sea yielded as much for Huxley over the 1860s and 1870s as the surface layers had in the 1840s and 1850s.

There were, on occasion, birds as well as fish in Huxley's world. At the end of a day's sail in the Louisiade Archipelago, he had looked to the sky and remarked on the great parrots, white cockatoos and kingfishers; but he was more interested in what he saw in the sea, in his next glance: 'queer little leaping fish (Chironectes)'.[27] Birds probably interested him most in their ancient form, so ancient that they had been found fossilized in rocks. Something very special had been uncovered in Germany in 1860 in this regard. What came to be called the *Archaeopteryx* was viewed by many as a kind of original bird, part avian and part reptilian. Was this the earliest known bird perhaps a transitional species between dinosaurs and modern birds? Taking yet another opportunity to correct Owen's work, he considered this fossil 'more remote from the boundary-line between birds and reptiles than some living *Ratite* are', that is, older in species terms than large, flightless birds like the ostrich.[28] Huxley was becoming clear that birds, reptiles and dinosaurs were connected, eventually fully persuaded by the palaeontologist Othniel Charles Marsh's US discovery of the Cretaceous toothed birds *Hesperornis* and *Ichthyornis*.[29] Huxley's interest, then, was not in birds for their interesting or beautiful selves, but in birds because they were like reptiles, and even like dinosaurs. This is how he taught 'birds' to his students, 'so essentially similar to Reptiles in all the most essential features of their organization, that Birds may be said to be merely an extremely modified and aberrant Reptilian type.'[30] In the late 1860s, he wrote to Haeckel that 'the road from Reptiles to Birds' was the question most thoroughly engaging him at present, how 'wings grew out of rudimentary forelimbs'.[31]

JULIAN'S AVIAN WORLD

If the grandfather had an off-centre way into the avian world, dealing with not-quite birds, conversely the grandson dealt with not-quite fish. Julian spent time at the Marine Biological Station in Naples immediately after his Oxford graduation. Grandmother Henrietta very much enjoyed this Huxley recapitulation: 'Pater & I were there in 84 or 85 ... I must write to Dr Dohrns & tell him you may be coming I know he will welcome Pater's grandson – a descendant of "the Happy Family" as he named us.' Unaware how much Huxley's name weighed on Julian's mental health, she was unknowingly rubbing salt into the psyche's wound: 'you have his example as a great heritage'.[32] In fact, Julian had thus far followed his grandfather's zoological footsteps rather literally. Just as Huxley had begun his study of life with the tiniest marine organisms, so Julian's biology began with the very smallest and the simplest: at Oxford, he took as his special subject protozoa.[33] He pursued this work in Naples, along with a study of sponges, but such seemingly passive and simple animals never held his attention for long. One water animal that did intrigue him, however, was the axolotl, precisely because it metamorphosed and it was strange. Not a walking 'fish', the axolotl is properly an 'amphibian', the term that had been invented to describe animals of both aquatic and air-breathing life. The axolotl was betwixt and between, a little like the plant-animal zoophytes, perhaps, and certainly like his grandfather's other favourite, the water-land crayfish.

In 1920, at New College, Oxford, Julian was still a zoologist at the bench, a habitat he was shortly to exchange for the writer's desk. For the moment, though, he was experimenting with two axolotls in a tank of water, following up previous German and British studies on the amphibians: experimentally forced to breathe air, the axolotl's metamorphosis to salamander could be artificially induced.[34] But could an axolotl be induced to metamorphose from water to land by other means and other materials – by ingestion of a mammalian hormone, for example? Julian decided to test this by feeding his axolotls thyroid of ox. It was a bizarre twist on an already established experiment showing that feeding tadpoles with mammalian thyroid made them metamorphose into frogs 'precociously'. He watched his precocious Oxford axolotl's gills and fins reduce, eventually to just 'vestiges'. And behold! In only three weeks, the animals climbed out of the water onto a platform and walked as air-breathing creatures. It was a modest Frankenstein moment: an experiment with the transformation of life. Undertaken over December 1920, it was written up on Christmas Eve for his grandfather's journal *Nature*. Julian sought a strange zoologists Christmas gift: 'I should be grateful if anyone possessing Axolotls, whether young or old, would give me the opportunity of purchasing some, as they are at present very difficult to obtain in the market.'[35]

Julian spent vacations roaming many coastlines, including with his grandfather around Eastbourne, but instead of peering into the tidal mud of estuaries, or the rockpools of shorelines, he looked up. He spent the spring of 1911 on the Welsh coast nominally studying protozoa, but his attention was drawn elsewhere towards those busy shore birds – redshanks, sheldrakes and curlews.[36] Like a salamander himself, Julian the zoologist was moving from water to land. And yet, in truth, he had been looking skywards all his life.

In his boyhood, life at Prior's Field and in the Surrey countryside had been filled with birds and the watching of them. He took long walks with Trev. 'We found 2 deserted thrush's eggs in the copse & Trev found another old nest.' They removed these nests to the garden of Prior's

Field School, found some hen's eggs to put in it and tricked little Aldous into thinking it was a cuckoo's nest.[37] Julia and Leonard taught them all to look into the trees. What are those tunnel-shaped nests, nine-year-old Julian asked Leonard. Longtailed Tits, the father thought. 'The eggs ... are white freckled with pink ... & very small.'[38] And while he was away at Eton, Julia kept her eldest son informed on the progress of the various nests around Prior's Field. She and Leonard monitored a lapwing's nest one spring: 'The wren's nest looks finished, but I never see the bird.'[39] Aldous too, kept Julian up to date, writing in the winter of 1902: 'It is very frosty weather; & I can stand on the pond. We have been giving crumbs to the birds.'[40] The birds of Surrey and Hertfordshire were the objects of Julian's juvenile ornithology. And while at Eton, he wrote home with a long list of seventy species he had observed, trying to be thorough. One in particular took his early fancy.

> [O]ut there is the most interesting (I think) of all the birds about here, the Great Crested Grebe. There is a family – mother & father (who has a larger & brighter crest than his spouse) & three little chicks, who are striped on the neck with black & white ... Further on is a beautiful group: four Grebes.[41]

As a schoolboy, Julian made a note of the difference between the male and the female. Little did he then know that the fascinating point of his later famous work on the great crested grebe was their monomorphism; the male and the female were hard to distinguish.

The study of animal behaviour had long been any naturalist's second nature – the kind of observation that Julia, Leonard, Trev and young Julian enjoyed around Prior's Field, at Stocks or around Eton. But there was a new system and complexity to field studies of birds that Julian early refined, and that his grebes enabled. His observations were as precise as any serious naturalist's, but it was his *interpretation* of bird behaviour and the questions he asked of that behaviour which heralded the new method, 'Ethology'.[42] Although it was often designated a post-Second World War enterprise consolidated by the Austrian Konrad Lorenz and Oxford's Niko Tinbergen, those ethologists themselves signalled Julian Huxley as a pioneer.[43] But Julian himself nominated Darwin's *The Expression of the Emotions* (1872) as the origin point for connecting observational study of behaviour with evolution by natural selection, the latter remaining key to ethology through his own generation and beyond. How are functions adaptive, and related to natural selection? What is the benefit of a certain behaviour? And how are behaviours learned over the lifetime of each animal? Ethology is conceptual as well as observational: Tinbergen developed ideas about 'instinct', Lorenz about 'imprinting', and before them both, Julian Huxley offered the concept of 'ritual' and 'ritualisation', as he watched those grebes dance.

Julian's earliest presentation on bird behaviour was to the Decalogue Club at Balliol in 1907, in which 'The Habits of Birds' were explained as natural *and* sexual selection.[44] Darwin had flagged sexual selection in the *Origin of Species,* but developed it fully, in relation to both animals and humans, in *The Descent of Man* (1871). Against considerable disagreement, he always maintained sex difference and sexual selection to be central to selection and evolution.[45] While Thomas Henry Huxley managed to ignore sexual selection almost completely, Julian spearheaded its revival in the early twentieth century and influenced its credibility in ethology. In an early field trip to north Wales, perhaps his first systematic observation of 'courtship', he watched the redshank males' displays and the females' rejection of most of them: 'it shows how Darwin's idea of Sexual Selection was undoubtedly right in certain cases, such as this – the hen has the power of choice . . . Not only has she got it, but she exercises it a great deal'.[46] The excited field observations were relayed to Kathleen Fordham, in 1911 (figure 4.3).[47]

The grebes forced Julian to refine, revise and then strongly qualify the sexual selection thesis.[48] If grebes' display was the same in both sexes, how did this fit with the difference between male and female that sexual selection was meant to produce? He developed the idea of mutual selection, suggesting that characters were selected continually by both male and female, to the extent that they transferred between sexes, and the sexes became similar to each other.[49] And yet what he observed was a 'courtship' display that seemed to function neither for direct mate choice, nor as a stimulus for mating. Rather, he decided, these displays were a kind of expression of emotion that was self-limiting – that did not lead directly to mating – but served some other purpose. It produced a kind of species-beneficial seasonal

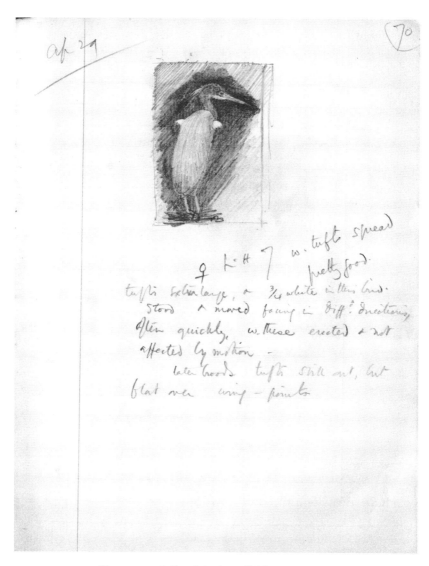

Figure 4.3: Julian Huxley's fieldnotes, 1911.

'marriage': 'constancy – at least for the season – between members of a pair'.[50] In the end, the displays were not to be interpreted as sexual selection, but as natural selection. A decade later, in Spitzbergen, he watched red-throated divers and again could not quite reconcile what he saw with sexual selection. The birds had already selected one another *before* display rituals took place. And again he saw the displays as 'self-exhausting', and not directly related to mating. Their function, perhaps, was a longer-term 'emotional' one. He dipped back into Darwin and rescued another of his early ideas, 'the development of an epigamic character [one attractive to the opposite sex] is dependent upon the emotional effect which it produces upon the mind of a bird of the opposite sex'.[51] Julian was far more open to the study of the animal mind, after Darwin, than his grandfather, but to satisfy zoologists of his grandfather's physiological ilk, he explained this also as the actions of muscles and nerve endings. Julian's half-brother Andrew Huxley was much later to receive his Nobel prize for the study of nerve impulses, but physiology was simply insufficient for Julian. He went on to assess these rituals as 'psychology': from what 'courtship' rituals did the grebes derive 'pleasure'?[52]

Watching his grebes, Julian sought to analyse not just what they did, but how they felt. He sought to interpret their behaviour as a kind of cultural ethnographer, even psychologist, the former not his term, though the latter was. Julian thought the various strange attitudes might best be thought of as 'symbols' that progressed from being useful or functional, to becoming what he called 'mere ritual': 'I mean that they are not so much associated with readiness to pair as with the vague idea of pairing in general. Thus associated with pleasurable and exciting emotions, they may become the channels through which these emotions can express themselves.'[53] These 'special courtship acts' between the near-identical looking mates, Julian declared, were ultimately less functional than 'ritualised'.[54]

Although the term 'sociobiology' was more than a generation away, young Julian Huxley already moved seamlessly between human and animal, especially when it came to understanding emotion, after the work of Darwin. On the one hand, when he asked his readers – 'what do we know about love?' – a good deal of his answer was ethological, about the behavioural and even emotional life of animals.[55] On the other, he deployed a reverse sociobiology, speculating on the meaning of grebe courtship rituals by thinking about human behaviour. People 'express their emotion' by a range of ritual actions, he explained: holding hands, putting their arms round each other's waists, or by kissing. He also explained the idea of 'self-exhausting' ritual through the example of human kissing, or possibly this is how he comprehended the grebe behaviour in the first place: 'The kiss exhausts itself, but leaves a memory, a trace and desire to repeat itself.'[56]

We might know this from our own experience, Julian wrote in his famous 1914 grebe paper, suggesting that he knew it from *his* own experience, presumably with Kathleen Fordham, or possibly with his Eton loves, the brothers. Julian was upfront about a human–bird behavioural continuum. Grebes choose their mates like humans. If only the reverse were the case, and Julian had chosen his mates as well as the average great crested grebe: much suffering would have been spared.

In his postscript to the *Zoological Proceedings* paper he wrote: 'in the intensely-felt affinities of man and woman – in that condition known as "falling in love," where the whole of the subconscious mental activities become grafted on to the inherited sexual passions, the whole past of the mental organism is summed up in the present'. When one chooses a mate from among thousands, what are the strange emotions, rituals and human attitudes that say: 'that one being, and no other, is the being that I desire for my mate'? Further, we know the feelings and behaviours of mate-finding, the emotion of the human love-ritual, through its representation in literature, he explained. There was no keeping this Arnold-Huxley from the literary canon. We might look to Dante, he suggested in his article on the grebes, or to the lovers in the George Meredith novel *Richard Feverel*, or to Romeo and Juliet. Even more imaginative – even bizarre – for an ornithological study published in the Zoological Society's *Proceedings*, Julian reached back to the classics, suggesting the usefulness of Plato's epigram to the handsome Athenian poet Agathon, 'Lovers' Lips': 'Kissing Agathon, I had my soul upon my lips'. And so, the science of love for Julian Huxley was traceable between species, with equally weighted evidence magnificently drawn from the man-poets of classical Athens and the sexually monomorphous water birds of Tring Reservoir.[57]

In 1912 the Huxley brothers visually recorded the grebes as pencil sketches in rough field notebooks. Just a few years later, in 1915, Julian travelled with a Rice University student to watch birds in Louisiana with a quarter-plate camera. In his short time at Rice, he was enchanted by birds that were entirely new to him: the hovering humming bird capable of flying in reverse, the little blue western gnat-catcher and the grey and pink crested fly-catcher, which displayed a new kind of courtship altogether. He made mountains of lantern slides, especially of his favoured water birds, egrets and herons. Julian had the opportunity to observe them closely, especially on Avery Island, Louisiana, owned by the millionaire Edward McIlhenny, who grew peppers and made a sauce called Tabasco. Thousands of herons bred on a lake he had especially built, with a punt and birdwatchers' hideout in the middle. Heaven for a Huxley, and Julian's thesis was confirmed: for birds in which secondary sexual characters are equally developed, 'mutual courtship' takes place, with the birds displaying to each other.[58]

A few years later Julian led an Oxford expedition of biologists to the Arctic island of Spitzbergen. He was watching water birds again, photographing them, and his lantern slides enabled teaching, research, lecturing and publishing for many years to come.[59] This expedition included zoologists of many specialties, yet they were all, as a group, interested in selection as well as in behaviour. They were also developing and applying an idea of 'ecology', building on studies of population dynamics, food, reproduction and death rates, across many species. A few were interested in how these principles might translate to human populations, but one of Julian's best students stuck with non-human animals. Charles Elton had been an apprehensive freshman at Oxford. Intimidated by his tutor with the famous name, he was reassured when Julian gave him 'not a dry manual, but a book by a philosopher & psychologist about aesthetics and science'. In Spitzbergen, they shared a tent in the snow. Elton said he was 'honoured' when Julian asked him to make some observations on the red-throated divers and was later grateful for Julian's support for his new approach.[60] In 1927 Elton's work emerged as the now-classic *Animal Ecology*, outlining the principles of ecological study of animal behaviour, feeding, reproduction and death rates, in specific environmental contexts, in ecological niches.

It was the kind of interrelated holistic philosophy of nature that enchanted Julian Huxley. If evolution had explained to his grandfather's generation that everything was connected in the deep past, ecology explained how everything was connected in the wonderful present, and that included humans. Ecology begged a completely different methodological approach to anything grandfather Huxley had practised, and when Julian edited and commented on the *Rattlesnake* journals in 1935 he couldn't resist layering it with his own new ecological vision. Explaining the significance of surface-layer jellyfish in the mid-Victorian context of cell theory, individuation and recapitulation, Julian indulged in his own interpretation as well, noting their coloration, their adaptation towards transparency, which was all about their surface-dwelling 'environment', not a term his grandfather would have recognized. The organism in relational context is what fascinated Julian. Since they must not sink far below the surface, he explained, their tissues are diluted with water, 'so that the specific gravity approximates to that of the surrounding medium, and sinking becomes less rapid'. Most jellyfish are translucent or transparent, 'crystal clear' in a way that effected concealment and protection; a kind of invisibility. And, noted Julian, this very transparency had permitted his grandfather easy direct observation of organs and structure, a sort of evolutionary bonus.[61]

Birds – and especially grebes – stuck with and to Julian for his whole life, even if sustained and scholarly fieldwork diminished as he grew older, rather as Huxley's had done. Over time, Julian kept his birds to the fore as a science communicator rather than researcher. In June 1950 he delivered a talk on 'Bird Courtship and Display' at the Congress of the Swedish Academy of Sciences, receiving an honorary degree from the University of Uppsala. He recycled it as 'Bird Display and Bird Behaviour' in 1956, his lecture for the Royal Society on the occasion of the Queen Mother being awarded an honorary fellowship. He brought along slides and stuffed birds 'with fantastic plumage, evolved by the operation of selection to stimulate the opposite sex, to proclaim ownership of territory, or to cement pair-bond between mates'.[62] Charles Darwin would be pleased that natural selection and sexual selection had gone somewhere. And in his own minor ritual of courtship display, he presented to the Queen Mother a tail feather of the male Argus pheasant, apparently taken from the Natural History Museum.

NATURALISTS FOR THE WORLD

With their deep, but very different kinds of knowledge about creatures of the sea and sky, both Huxleys sought to educate the world. They communicated not just information on those creatures, but through them foundational information and education on the study of life itself.

One of Huxley's most successful Victorian 'lectures to working men' was on an unlikely marine crustacean, the lobster. In what was to become a signature technique of explaining the largest scientific questions through a specific animal instance, he lectured, in 1861 'On what there is in a Lobster to be explained by any theory of the Universe or Cosmogony'.[63] Analysis of the lobster could explain the scope and divisions of zoology to laymen and women. First, morphology (form and structure, including anatomy, development and classification); second, physiology (including functions and actions, animal bodies-as-machines impelled by 'the laws of the molecular forces of matter'); and third, distribution or the absence, presence and density of species across the globe. He proceeded to textually dissect the lobster, explain its parts, what can be observed if we remove the legs, what we see in the tail, the jaw and the rings of the lobster's body. This all showed 'unity of plan, diversity in execution'.[64] But the lobster had not always been thus. 'It was once an egg, a semifluid mass of yolk', subdividing into sections that were forerunners of the rings of the body. What could be observed in the lobster was that the smallest and earliest parts are initially alike, but as they develop they become 'distinguished'. The complex is everywhere evolved out of the simple.[65] This was a lesson in evolution, but not, as we shall see, in natural selection.

Years later, the idea turned into a book, *The Crayfish* (1880), through which everyman was introduced to zoology again, and in more detail. It aimed to teach people to observe for themselves. A crayfish is easily obtained, and its parts are easily separated. Huxley imagined his reader with *The Crayfish* on one side of the table and the actual crustacean on the other, a step-by-step guide. He walks his reader through kinds of knowledge about the crayfish, from the simplest form of common knowledge to the kinds of observations an old natural historian would have produced about the crayfish (Buffon is his example), towards contemporary cellular morphology, physiology, distribution and species varieties. Finally he introduces evolution as the overarching pattern of explanation.

A 'common knowledge' observation, for example, is that the crayfish walks forwards on four pairs of jointed legs, but can also swim backwards, using a kind of hard fin at the end of its body. Any natural historian would observe this, as would any fisherman. They would also note that its skeleton is on the outside of its flesh, an 'exoskeleton' that makes it unlike the endoskeletons of mammals. Dissected, within that skeleton might be observed three large reddish teeth in its stomach; blood that is almost colourless; a heart that still pulsates when the shell is removed from a recently killed animal. *How* it does that, however, moves beyond natural history and anatomy, into physiology, 'the elucidation of complex vital phenomena by deduction from the established truths of Physics and Chemistry, or from the elemental properties of living matter'. Thus he explains the morphology of the cray's tissues, cells and nerves, and its development from an egg: 'an animal not only is, but becomes.' This development he called 'a process of evolution', by which he meant each egg's evolution into an individual cray.[66]

Then came comparative morphology: how the crayfish is like and unlike other *Arthropoda*, and like or unlike *Mollusca*, and how similar but different animals are distributed across the world. All living things are different, but they are also allied, was his axiom. Enter evolution. Zoology's largest question is: why do animals of particular structure, and in particular locations, exist? There are two possible hypotheses, he suggested. The first is a force outside nature, 'what is commonly termed Creation'. The second is as part of the 'usual course of nature' – by some 'doctrine of Evolution'.[67] Even – perhaps especially – this, he instructed his everyman-zoologist to assess calmly by available evidence. Direct evidence of Creation, he said, is unobtainable. That's the problem. And yet available evidence for evolution was not yet conclusive either. It is conceivable, he wrote, but not yet proven, that 'all plants and all animals may have been evolved from a common phys-ical basis of life, by processes similar to those which we every day see at work in the evolution of individual animals and plants from that foundation'.[68] The fossil material collected up to that time relating to the crayfish was 'too scanty to permit of the tracing out of all the details of the genealogy of the crayfish'. Nonetheless, it was equally important to recognize that every piece of evidence that was currently available *did* thus far accord with 'the requirements of the doctrine of evolution'.[69] Nothing, so far, countered it, and therein lay its weight. Still, there was by no means consensus on Darwin's *particular* theory of evolution by means of natural selection, neither popularly nor among zoological experts. In 1880 even Huxley was not entirely convinced by natural selection as the process of evolutionary change. His book, instructing people on every kind of zoological knowledge of the crayfish, neither mentions nor even hints at natural selection, still less sexual selection.

Huxley may have been a reluctant popularizer of science, but he was an effective one.[70] Charles Darwin spotted his talent early, in 1864 suggesting he write a general book, noting the interest with which his wife Emma had consumed Huxley's 'lectures for working men'. 'Now with your ease in writing & with knowledge at your fin-gers' ends', suggested Darwin, 'do you not think you could write a "Popular Treatise on Zoology"?'[71] *The Crayfish* might have been an introduction to zoology, but it was not quite the popular overview that Darwin had in mind. For all his millions of published words, Huxley never quite wrote that all-encompassing zoology book, but his grandson did: *The Science of Life*.

'So you are the grandson of the Crayfish,' Julian was once greeted in a German laboratory.[72] He was, and the inheritance was direct and true: both the desire and the capacity to communicate natural sciences. Julian did so initially thanks to the driving force of one of Huxley's own students, Herbert George Wells (figure 4.4). Having just completed his *Outline of History*, Wells found all human history too limiting and now wanted to write about all life, in all time. The problem was that his 1880s biological education was deeply pre-Mendelian. Wells knew he needed a co-author whose knowledge was up to date, who was learned in the new ecological approaches and who was partial to thinking

Figure 4.4: H. G. Wells posing as his teacher T. H. Huxley, Royal School of Mines, 1880s.

about the inorganic, the organic and the human world as an interconnected whole. Julian had only recently (September 1925) secured the coveted professorship of zoology at Kings College, London. After the comings and goings – the false start at Rice, the interruptions of the war and then his lectureship at Oxford – the London chair offered high status and a secure future. But almost immediately H. G. Wells made overtures about a too-tempting 'Outline of Life'.[73] And by April 1927 Wells had somehow persuaded Julian that he should simply abandon his academic work and write with him full time.[74] At one point, Wells asked Huxley about involving J. B. S. Haldane, another biologist-turned-writer, but the final line-up was H. G. Wells and his son, G. P., or Gip, Wells and Julian Huxley.[75] The three of them wrote in Wells's house in Essex, often with Juliette present. Wells 'worked, talked and played with a sort of fury', Julian remembered. At other points of its long production, Julian and Juliette took a writing-retreat chalet at Les Diablerets, along with Aldous and Maria and their son Matthew. Soon D. H. Lawrence and his wife Frieda turned up, and it was Garsington Manor all over again, though less interesting for the women, who spent mornings in the chalet typing up what their husbands had written the day before. The worst was in store for Maria Huxley, who had to type up the draft of what would become *Lady Chatterley's Lover*, based, not subtly, on Ottoline Morrell.[76]

The published outcome of the Wells/Huxley collaboration was stunning and successful, the first introductory text on 'life' from an ecological point of view, published the same year as Charles Elton's classic, *Animal Ecology* (1927). But the writing process was tortured. Julian learned that thoughtful Arnoldian instruction on similes and stanzas only went so far. Not that his Arnold aunt, Mary Augusta Ward, was any slouch (twenty-six novels and ten non-fiction books by the time she died in 1920), but Wells was a writing machine. Or better, he was a writing manufacturer who pressed his factory hands – Julian Huxley and Gip Wells – to the limits of their endurance and capacity. Wells constantly sought the most efficient and direct line from the idea to the printers. 'Book IV I await with impatience,' he pestered Julian. 'It ought to be in the printer's hands now.'[77] Wells was appalled at Julian's ad hoc process and he expected far more – 'You don't work like a professional author at all' – and was dissatisfied with his young co-author's lack of a secretary and typist.[78] Wells relentlessly pestered Huxley for copy and testily warned him against distractions. He struck aggressive editorial lines through draft after draft of Julian's over-long, over-drawn chapters.[79] 'Nonsense man, nonsense,' he wrote of one of Julian's attempts.[80] But eventually it came together, and Wells became quite excited. How about sections on the conquest of the air and the conquest of the deep, Wells suggested, quite possibly thinking of Huxley's deep ocean worlds. That would make good reading, the master knew. He wanted it to read like a novel, as much the 'History and Adventures of Life' as a Natural History of Life. And it does: it is a poetic study of natural wonders that still delights and enchants.

For Julian, it was all an informal apprenticeship of a different order to his grandfather's medical apprenticeship. T. H. Huxley would have scorned writing instruction from an older master, certainly one to whom he had to defer, and who occasionally referred to himself as 'God the Father' of the authorial trinity. But Julian absorbed it all and learned a great deal about process, context, syntax, sequence and audience, even as he offered to Wells essential content from the latest in ecology, reproductive sciences and the new genetics. Wells taught him to write popular books for a reader 'who is just as intelligent as you are (but does not possess your store of knowledge)'. And then he turned up the pressure: 'Shakespeare, Milton, Plato, Dickens, Meredith, T. H. Huxley, Darwin' all wrote for such a reader.[81]

The magnificently illustrated *Science of Life* began, in a way, with Thomas Henry Huxley. The frontispiece of volume 1 was his world, a rich watercolour of a microscope: 'the key to the knowledge of life' (figure 4.5). Behind the microscope a human skull on a stack of books – the props of many a Huxley portrait – and in the background a line-up of primate skeletons, recalling the famous frontispiece of *Man's Place in Nature*. It looks like a small homage to Thomas Henry Huxley, who would certainly have appreciated that image. But what would he have made of the rest of the book? Turning the page to the Table of Contents, he would have been bemused at the very least. What on earth was the logic of this book's structure? For a start, it was bookended by humans. Volume 1 began with the human body as machine. Volume 3 ended with human minds, including a foray into

It is not too much to say that all modern biological knowledge is founded on the microscope, of which we figure here one of the most perfect and beautiful of modern examples. The lenses in the tube can render objects as small as one hundred-thousandth of an inch visible, and the fine adjustment up and down can make and measure movements of one twenty-five-thousandth of an inch. Below the stage, which carries the object under examination on a glass slip, a complex system of lenses can vary, intensify, and focus the light on the object.

FRONTISPIECE—Facing Page 1

Figure 4.5: Frontispiece to *The Science of Life*. The microscope as *The Key to the Knowledge of Life*.

spiritualism that Huxley would not recognize in a science of life, and a discussion of human futures with biological possibilities like grafting animal bodies onto human bodies. The crayfish and the axolotl might grow a new limb – that was perfectly comprehensible to T. H. Huxley – but seeking to introduce this capacity for regeneration into humans by grafting animal tissue? It is a measure of fifty years of biological sciences to imagine Huxley's response to what Julian thought 'imminent': 'it may even prove possible to operate directly on the germ-plasm, for the geneticist can already produce mutations by means of X-rays,' Julian wrote excitedly.[82] In between this wild discussion of human possibilities were all the creatures that T. H. Huxley recognized, yet even in this matter his student and his grandson didn't build their work systematically, neither working from small organisms to large ones; nor backwards from complex phyla to simple. It didn't work outwards conceptually from a single organism, like *The Crayfish*. Instead it meandered across connections, from human bodies to jellyfish, courtship in birds then lizards, multiplying insects then dinosaurs, and back to humans and their capacity to artificially propagate. And what were these concepts of partnership, habitat, parasitism, coloration, niche and pattern in life? *The Science of Life* was structured by ecology, by a 'fresh way of regarding life, by considering the balances and mutual pressures of species living in the same habitat'.[83] This was a kind of swaying balance of diverse species that Haeckel had *called* ecology in T. H. Huxley's lifetime, in 1878, but which had since developed into refined understandings of complex systems, a 'web of interrelations'. '*Life is one thing,*' *The Science of Life* declared emphatically, wearing its holism proudly.[84]

When it came to the descriptions of animals, however, the Victorian Huxley would have been zoologically comfortable. The vertebrates, the arthropods and the 'leftover' animals on which he had made his name were all there: molluscs, polyps, jellyfish, sea anemones. A microscope-enabled photograph of Huxley's precious minute hydra replaces his delicate mid-Victorian pencil drawings. Their classification had changed too. They used to be called 'zoophyta', because they are plant-like animals, and they used to be classified with Echinoderms (starfish, sea urchins) in the phylum *Radiata*, the twentieth-century authors corrected, but it was now recognized that they were distinct.[85]

Yet more importantly, and just like in Huxley's work, these animals provided the opportunity to explore key questions about individuality and reproduction, recalling mid-nineteenth-century debate on cells and life. The hydra is a 'loosely organized cell-community', each fragment able to reconstitute itself into new hydras. Its reproduction is the strangest thing. It can develop buds on its side, each bud then able to develop mouths and tentacles, and break away from its parent. Yet it can also reproduce through the eggs and spermatozoa that appear on the same 'individual', which join and produce hydra embryos.[86] And there is the lobster again: 'it can part company with a limb in order to save itself, and it can slowly grow a new one'. A sea-squirt cut in half can grow back into two sea squirts, with all missing organs reconstituted. The very idea of individuality, even of the 'self', was brought into question biologically and philosophically: 'the elaborate self-centred individual is not by any means an indispensable part of the scheme [of life]'.[87] Thomas Henry Huxley had helped set the foundations for just that conceptual and observational insight.

On the face of it, T. H. Huxley might also have contributed to the sections on the deep sea and on various fishes. There were his herrings, the species commercially critical to northern fisheries. At one point, T. H. Huxley wrote thousands upon thousands of words on herring morphology and physiology. He was fascinated by the herring's mouth, its silvery air bladder just under its spine and its throat cut by four deep and wide clefts.[88] And on and on. Wells would have cut thousands of words to perhaps fifty if he had had the chance. Far more interesting was the ecological work on the herring that Julian Huxley brought to *The Science of Life*. The point of the herring for Julian was neither anatomical nor physiological, but illustrative of the new idea of food chains, 'linked up together in an interlacing meshwork'. If T. H. Huxley had explained what herrings ate in order to understand a peculiar jaw and throat, Julian displayed a system-web representation of the multiple species involved in sustaining just that one species.[89] This would soon be called the herring's marine 'ecosystem', a powerful new word invented by Julian's friend the ecologist Arthur Tansley. This was the generation inspired to conceptualize webs, connections, balances and systems.

Julian's birds are in *The Science of Life* too. Deep inside volume 3, the courtship and display of animals is explained, the most thorough treatment of birds in the entire work. Julian re-used his photographs of Louisiana herons to show 'the ceremony of nest-relief' and 'the ceremony of twig presentation'. And of course the elaborate courtship of the great crested grebe was richly illustrated. Instead of a photograph or a drawing from Julian's old notes, a watercolour was commissioned from Gilbert Spencer, an artist patronized by Ottoline Morrell at Garsington (figure 4.6).

After he wrote *The Science of Life* Julian never held another academic position. He was now a science communicator, armed with systems, sentence structures and similes bestowed by H. G. Wells as much as his Huxley father and grandfather. Juliette sometimes urged him away from popular writing. 'You <u>must not</u> become a journalist. You are far too good for that,' she worried. 'Your time has come for a big work – your best work, & you must prepare for it, & not fritter your mind on popularizing.'[90] Still, their household was dependent on his writing, every bit as much as Huxley and Henrietta's had been. It was easy enough for Julian to turn his bird watching from serious ethology into consumable nature-writing. The readers of *Country Life* could enjoy 'Egrets in Louisiana', and readers of the *New Statesman* could learn about 'Bird's Play and Pleasure'.[91] The return on every word was decent. Even in the 1960s, he insisted on a fee for republication of extracts of the *Rattlesnake* diary. 'The copyright is mine,' he curtly reminded conservationist Nigel Sitwell in 1964, then editor of the World Wildlife Fund journal, *Animals*.[92]

If scientists appreciated Julian's *Modern Synthesis*, it was media visionaries who were struck by the richly visual romp that was *The Science of Life*. So many were inspired. Walt Disney diversified into the natural history genre during and after the Second World War, producing a 'true-life adventure' series, possibly triggered by the Wells and Huxley 'history and adventures of life', even their 'spectacle of life', the organizing titles in their book. Disney flattered Julian, telling him he regarded his autographed volumes of *The Science of Life* – signed by both Huxley and Wells – among his 'proudest possessions'.[93] David Attenborough, too, was clear that the volumes had 'a huge effect on me as boy'.[94]

Julian's decision to leave behind the laboratory bench to write biology

THE COURTSHIP OF THE GREAT CRESTED GREBE

The crowning ceremony in the very elaborate courtship of the Great Crested Grebe. The two birds dive below the water, and when they emerge again each carries a bunch of pond-weed in its beak. They then swim together, breast to breast, and rise as shown out of the water, rocking from side to side, and violently churning with their feet in what is almost a dance. Finally they sink slowly down again into more normal attitudes, and throw the weed away. This is not generally followed by pairing, but is simply an expression of mutual esteem. (Painted for this work by Gilbert Spencer.)

FRONTISPIECE—Facing Page 223.

Figure 4.6: *The Courtship of the Great Crested Grebe*, watercolour by Gilbert Spencer.

with H. G. Wells launched a career not just as an author, but also as a science broadcaster and filmmaker. Nature films were a new kind of documentary in the 1930s and 1940s, and Julian did a good deal to establish the genre. He set up Zoological Film Productions in agreement with Strand Films, producing a twelve-part series including *Zoo Babies* (1938) and *Monkey into Man* (1938).[95] Another, *Free to Roam* (1938), was filmed in and about Whipsnade Zoo, not an ordinary zoo with cages, but one in which animals could be studied 'in their natural surroundings'.[96] Many of these were short educational films to be shown in schools, but Julian had already engaged a wider audience producing *The Private Life of the Gannets* in 1934. He won an Oscar for it in 1938: the Academy Award for Best Short Subject, One-reel.[97] Located on the small rocky island of Grassholm off the coast of Wales, the kind of coastal landscape that his grandfather sought out, the film was noted for innovations in technical photography – its close-ups and aerial shots. It finished with a dramatic sequence of gannets diving into the ocean waters. Often credited as 'supervisor', and sometimes also narrator, Julian Huxley contributed to the inter-war creation of the genre, the industry and the appetite for British natural history films. These included *Animals of the Rocky Shore* (1937), which recalled Thomas Henry Huxley's molluscs, and *The Earthworm* (1936), which recalled Charles Darwin's late-in-life obsession.[98] This was all far more than an indirect anticipation of David Attenborough and the BBC's later phenomenal success and impact with Natural History Films. They worked together closely at one point. Attenborough's first television production was presented and narrated by Julian Huxley, about the rediscovery of a rare order of fish, the *Coelacanth* (1951). Another of Attenborough's earliest productions with the BBC was *The Pattern of Animals* (1953) series, which Julian narrated, poorly enough according to Attenborough. Three episodes were signature Julian Huxley: communication between animals, camouflage and coloration, and courtship displays, and they discussed the series in Julian's library, superintended by John Collier's portrait of Thomas Henry Huxley. It was after that that Attenborough stepped in front of the camera, with the *Zoo Quest* series (1954–9), a more sophisticated television rendition of Julian's short educational *Animal Kingdom* films of the 1930s, many also filmed at the London Zoo.[99]

When Julian died in 1975, Attenborough sent warm sympathies to

Juliette, indicating his 'immense debt to Sir Julian'. 'That I was lucky enough to know him as well was a great privilege for which I am deeply grateful.' Well into the 1980s, David Attenborough and Juliette Huxley continued their conversations, often centred on discussion of Julian and his creatures. She congratulated Attenborough on *Life on Earth*, and he modestly claimed it very much as a team effort. What a 'thrilling' time they had, making the series. All compliments were a bonus, and 'too good to be true'. Flattering, he added that compliments from Juliette Huxley were to be especially treasured, 'from someone who played such an important part in bringing *The Science of Life* into existence'.[100]

5

Animal Politics:
From Cruelty to Conservation

The very word 'biology' was still under construction over T. H. Huxley's lifetime. And it was under challenge. He kept notes on theologian-philologist Frederick Field's problems with the term. Field was one of hundreds – thousands – to take issue with Huxley, who was ever more provocatively high-profile after his invention of 'agnostics' in 1869. Field did not appreciate that term at all. But he also took pains to complain about 'biology'. 'Bios' from Greek suggested life, certainly, but was ill-used with respect to animals, complained Field. In a more correct and subtle interpretation of the Greek original, it meant *contemplation* of life 'in the moral aspect'. For this reason, zoology and botany should not be subsets of 'biology'. The latter should be 'reserved for Man, and with him, not as a *living*, but as a Rational, social and accountable being'. He thought that 'the term *Biology* recently imported into the scientific vocabulary is a BLUNDER' and suggested instead, for the study of non-human animal life, 'zoticology'.[1]

Field's complaint was disingenuous, however, because the nature of other animals' sensibilities, their capacity also to contemplate, even their possible moral scope, was by no means a settled question in the late nineteenth century. Indeed, such matters have never been settled. Over Huxley's lifetime animal sensation and sensibility was a particularly acute physiological, theological, moral and philosophical problem, and a well-known one. 'Biology' stuck as the agreed descriptor for the study of life across plants, animals and humans, but early biologists, including Huxley, routinely pursued the study of animal consciousness within their ambit. That very consciousness made human–animal engagement of any kind an ethical and political matter.

Across scientific and non-scientific domains alike, a common enough Victorian inclination was to feel for, and even with, the suffering of other animals.[2] For both Huxley and Darwin, if in different ways, the very study of 'feeling' and 'sensation' across multiple species was compelling. Charles Darwin empathized across species and was unsettled by the pain sometimes inflicted on innocent animals in the interest of medical and scientific experimentation. For Huxley, however, animals' comprehension of pain was rather more interesting than troubling. Analysing the functions of their nerves, brains and spinal cords, he found animal consciousness to be a fascinating physiological matter. The animals in Huxley's life were far more likely to be specimens than pets; more likely to be dead than alive; and more likely to be physiologically than emotionally intriguing. His engagement with philosophies of mind with regard to humans drew from his longstanding interrogation of animal sensation. Over the 1870s especially, Huxley inclined his physiology towards philosophy, joining a new debating club, the Metaphysical Society, in which theologians and scientists together discussed concepts of sensation and death, machines and bodies, animals and souls. This combination of credentials placed him centrally in one of the great politico-moral contests of the 1870s between the proponents and opponents of the vivisection of animals for the production of medico-scientific knowledge.

Animals were key to both Huxleys' professional worlds, and so it is unsurprising that the politics of human treatment of, and relations with, animals also shaped their work and their influence. Their vital dates, 1825–1975, spanned the beginning of the Royal Society for the Protection of Cruelty to Animals and Jeremy Bentham's pronouncements on animal cruelty in the mid-1820s all the way to philosopher Peter Singer's Bentham-inspired *Animal Liberation: A New Ethics for Our Treatment of Animals* published the year of Julian's death.[3] Between them, Thomas Henry and Julian Huxley embodied and enacted the modern shift in animal research and animal ethics from imperial natural history, in which animals were collected, and then pinned, stuffed or pressed, to the modern phenomenon of the 'zoo' in which live animals are protected and enjoyed, and onwards to early environmentalism, the 'conservation' of wild species and their habitats.

The Zoological Gardens was a constant in these changing Huxley worlds, opening its gates to successive generations of animal–human interaction. The diverse purposes, fortunes and cultures of the Zoo (as the Regent's Park site came to be called, starting an international trend) are clear indicators of evolving human perceptions of what captive animals are for. It was a place that the Huxleys shared intergenerationally, the London location for live animals a counterbalance to the dead and fossilized animals of London's natural history collections at the Royal College of Surgeons, in the British Museum and later (from 1881) the Natural History Museum in South Kensington. These were T. H. Huxley's native habitats. Julian lived inside the noisy Zoo in the late 1930s. He was far more interested than his grandfather in how animals interacted with *each other*, irrespective of humans, and how they did so in *their* natural habitats. He folded nineteenth-century interest in animal emotion and consciousness into his own early twentieth-century specialty, animal behaviour. In his later life, it was not Regent's Park but Africa that provided Julian the context in which to serve as ecologically responsible 'trustee' to animals: the conservation of African wildlife *in situ* was his major project. However, it was less individual animals that drove his activism in the manner of 'animal rights', and more concern over populations of animals caught in dangerous and endangered economies and ecologies. Preserving wild animals and their habitats became Julian's mission, in cultural and natural contexts as diverse as Tring Reservoir and Serengeti National Park.

ANIMAL SENSE AND SENSIBILITY

T. H. Huxley loved physiology most, the only part of his training as a doctor that really interested him, he later recounted.[4] He enjoyed dissection of the crayfish so as to reveal its anatomy, but that was intellectually basic compared to the physiology of life and the conception of animal *feeling*. Everybody knows, he wrote in his *Crayfish* book, that if we put our hands into its tank the cray will swim backwards and away. If we place meat in the tank it will walk forwards and devour it. Why? Because it is aware of danger in the first instance, and in the second, the cray knows the meat is good to eat. His point, of course, was to ask: what do we mean by the crayfish being 'aware' and 'knowing'?[5] Huxley's keen interest in the physiology of nervous systems opened up inquiry into movement, sensation, reflex, instinct and ultimately emotion, memory and consciousness in animals. These were complex philosophies as well as physiologies, and he used it all to underscore his uncompromising scientific naturalism.

Huxley followed Descartes into the world of animal consciousness, and unsurprisingly so, since the seventeenth-century French philosopher's work on animals-as-automata was physiologically as well as philosophically argued. Through Descartes, Huxley investigated the idea of 'reflex' in non-human animals, something like, but not reducible to, Aristotelian 'instinct', both distinguished from human reason. Animal reflex might produce both sensation and perception, but Descartes thought animal consciousness was limited: capacity for thought or reasoned understanding was solely human. Huxley appreciated the physiological argument, and the work on nerves, muscles, brain and reflexes, but by the 1870s, as an evolutionist, he disagreed with Descartes's conclusions. In his Presidential Address to the British Association for the Advancement of Science in Belfast in 1874, 'On the Hypothesis that Animals Are Automata, and Its History', he argued that animals are, at the very least, 'conscious machines'. A strong trait like consciousness could not appear *de novo* in one species, it must have traces in earlier and related species too.[6]

It was a lively debate and one in which Huxley engaged for decades, partly in ongoing conversation with Herbert Spencer. Walking, visiting, talking and dining, the two of them thrashed out ethics and evolution, including the difficult question of human relations to other animals. Spencer's *Principles of Ethics* dealt with animal ethics, what he called 'sub-human justice' and its relation to 'human justice'.[7] Spencer derived the second from the first on the same evolutionary principle through which Huxley had earlier argued for animal consciousness: 'Human life must be regarded as a further development of sub human life . . . human justice a further development of sub human justice.' Part of a continuous whole-of-life, justice nonetheless 'becomes more pronounced as organisation becomes higher'.[8] Huxley agreed, certainly with regard to animal sensation. This was a continuum, and sensation varied between creatures of the sea and of the air, between mammals and indeed, he thought, between varieties of humans.

In the early 1870s, when Huxley began lecturing on animal consciousness, and Spencer was writing his ethics books, Darwin was also setting down his thoughts on shared sense and sensibility.[9] Over that decade, these now-triumphant scientific naturalists were putting biological twists not just on the Cartesian philosophy of animals-as-machines, but also on far older scriptural debates about whether or not animals have souls, and whether those souls were immortal. This was one of many questions that the new Metaphysical Society broached. Established in 1869, this dining and debating society aimed calmly to consider theological questions through science, and scientific questions through theology.[10] One of its number called it a 'Religious Museum', since it boasted one of everything:[11] a learned Anglican bishop; an inquiring Catholic bishop; a non-trinitarian 'arian'; one deist; and so forth over the doctrinal spectrum. In Huxley, they gained the world's first 'agnostic'. Animals for their own sake were probably the last thing the Metaphysical Society was concerned with. But their learned discussion on the nature of souls, life and death did serve to foreground fascinating questions regarding the boundaries between animal and human, physical and metaphysical. It was at an early meeting of the Metaphysical Society that Huxley described his own philosophy, rejecting unverifiable spiritual knowledge. Of course he held that there *was* no 'metaphysical' state that we can know, yet the Society threw back at Huxley great scholarly challenges. They discussed the theory of a soul, the relativity of knowledge, the nature of death, the elements of sensation and 'rank in animated life'. The eminent physiologist William Carpenter followed up Huxley's Descartes address with a paper 'On the Doctrine of Human Automatism'. It was all a heady mix of theology, philosophy and science, moulding Huxley's 1870s intellectual responses, shifting his scholarly ambitions towards ethics in a rigorous way and helping to hone his ideas and arguments with regard to animal sensation specifically. He provoked them all with 'Has a Frog a Soul?' Reaching once more to those intriguing amphibians, classically subject to the vivisector's scalpel, Huxley tried out his secular and scientific postulata. He reported experiments that showed a reflexive response in frogs that stayed the same when their spinal cord was cut, and when portions of their brain were removed. He argued that, without a brain, the frog could not be conscious and yet could still do the same sort of things that it could do before. Where, then, is the seat of conscious-ness? Are sensation and consciousness equivalent? Incidentally for many in the Metaphysical Society, but of real import for others in the 1870s, the means of demonstration were drawn from a great deal of cutting-up of frogs. For anti-vivisectionists, the question of the soul of the frog was secondary to the question of the pain felt by the frog, especially in experiments seeking to establish just that question of animals and sensation.[12]

When Huxley praised Descartes as 'a great and original physiolo-gist', he was also praising a well-known vivisector.[13] Cutting into live animal bodies for the purpose of physiological or medical knowledge was one of the most acute political controversies of Huxley's lifetime. It reached a peak during the 1870s but had long crossed philosophical and medical domains. Two generations earlier, Jeremy Bentham had indicated that 'the question is not, Can they *reason?* nor, can they *talk?*, but, Can they *suffer?*' Yet this did not mean that the utilitarian objected to medical experiments, arguing that vivisection was in the end justified by the human benefit that it might yield. The year that Huxley was born, 1825, Jeremy Bentham wrote to the editor of *The Morning Chronicle*:

> I never have seen, nor ever can see, any objection to the putting of dogs
> and other inferior animals to pain, in the way of medical experiment,
> when that experiment has a determinate object, beneficial to mankind,
> accompanied with a fair prospect of the accomplishment of it. But I
> have a decided and insuperable objection to the putting of them to pain
> without any such view.

It was all about intention and outcome. 'To my apprehension, every act by which, without prospect of preponderant good, pain is knowingly and willingly produced in any being whatsoever, is an act of cruelty.'[14] Fifty years later, when anti-vivisectionist politics was at its height, such a view was more difficult to proclaim publicly: discomfort over animal pain, no matter what the intention, was beginning to trump any pos-sible benefit for humans. Yet Huxley sustained Bentham's position, more or less, even as he sat on the Royal Commission on Vivisection in 1875. It was a debate that brought various philosophies of conscious-ness, mind, feeling and pain to the realities of physiological learning, and to the intention and benefit to humans of experimenting on live animals. Bentham and Huxley's compromise eventually won the day, becoming the law of the land in the Cruelty to Animals Act, 1876. This outlawed inflicting pain on animals through vivisection, but, like Bentham's proposition, did not outlaw animal experimentation itself. It was deemed acceptable when such animal pain might prolong or save human life.

The Cruelty to Animals Act was prefaced by a complex sequence of far more starkly anti-vivisectionist bills, originally driven by a feminist–Quaker alliance led by Frances Power Cobbe, founder of the Society for the Protection of Animals Liable to Vivisection. Cobbe had recently published 'Darwinism in Morals' (1872), including a critique of *The Descent of Man*, yet she and Darwin were friendly enough, linked not least by their love of animals.[15] What did he think of her recent article in the *Quarterly Review*, 'The Consciousness of Dogs', she asked him at one point.[16] They may have both loved their dogs, but regarding her major anti-vivisection efforts, including her mobilization of petitions, Darwin was agitated. He wrote to Huxley: 'I received some time ago a foolish paper from Miss Cobbe about vivisection which I did not sign, & this morning I have received a duplicate asking for my signature, which I shall refuse.'[17] He was exercised precisely because Cobbe was proving so successful in gathering support against vivisection, from fellows of the Royal Society, from the Archbishop of York, from a suite of medical men, from members of parliament. She was busy drafting uncompromising anti-vivisection bills, and Huxley and Darwin were particularly concerned that these would stymie advances in physiological knowledge.[18] The bill that was eventually successful was largely a result of Darwin and Huxley's joint work to offset Cobbe's influence, but it sought also – mainly on Darwin's part – to balance a deep concern about cruelty and pain with the needs for physiological experimentation.[19]

A Royal Commission was established to gather perspectives and evidence, on which Huxley served between 5 July and 20 December 1875. The humans investigated in the Royal Commission included physiologists, popular vivisectors, medical doctors, veterinarians and public entertainers. The animals at issue ranged from frogs, cats and dogs to lobsters and goats. There, Huxley rejoined some of his Metaphysical

Society acquaintances, including the editor of the *Spectator*, theological writer Richard Hutton. The two of them had been at philosophical odds, especially over Huxley's increasingly robust agnosticism. And if Hutton was the anti-vivisectionist on the commission, it was soon clear that Huxley intended to defend the interests of proper physiological advancement.

Huxley brought good expertise to the Royal Commission on Vivisection, combining his medical training and his experimental physiology. He knew the benefits and pitfalls of experimental physiology, and he also knew when experiments were unnecessary, as he saw it: the difference between 'rational' treatment (learned by experiment) and empirical treatment (learned by experience).[20] He often served as a voice that countered the anti-vivisectionist position, not so much by being pro-vivisectionist as qualifying some of the anxiety. For example, Dr John Anthony was questioned in October 1875, an ageing doctor who had been dissector for Charles Bell in the early part of the century, preparing specimens for the eminent natural theologian's lectures. Anthony was asked to nominate the most severe experiment he had been engaged in, or heard of, and explained the cutting open of a guinea pig, tying one fallopian tube and leaving the other, not a procedure either he or Bell had in fact performed, he claimed, but reportedly performed elsewhere. Huxley's response was to ask whether this procedure was any more or less painful than the common agricultural practice of spaying a sow? Far less, Dr Anthony responded, and so the anti-vivisection position was forced a step back.[21] Huxley was keen to enter a complex understanding of pain into the commission's proceedings. Pain varied between species according to their higher and lower complexity, and even within species, even indeed among humans. Every physiologist knows that the pain of childbirth in 'civilised women' is agony, but in 'the savage woman' offers little inconvenience. Well might we ask the apparent epistemological question how every physiologist 'knows' this. He called it 'the hyper-aesthesia of civilisation', the idea of heightened sensations, not necessarily a good thing, that modern society effected.[22]

Several of the anti-vivisectionist witnesses saw Huxley himself as thoroughly implicated in animal experimentation, not least because of his description of vivisection in *Lessons in Elementary Physiology*

(1874). The reforming politician Lord Shaftesbury, whose lunacy and county asylum laws had been the curse of alienist brother James, had very publicly raised the seeming endorsement of, even instruction in, vivisection in Huxley's book. Several witnesses read directly from the incriminating pages to the Royal Commission. Huxley had written, for example:

> If the *trunk* of a spinal nerve be irritated in any way, as by pinching, cutting, galvanizing, or applying a hot body, two things happen: in the first place, all the muscles to which filaments of this nerve are distributed, contract; in the second, acute pain is felt, and the pain is referred to that part of the skin to which fibres of the nerve are distributed.[23]

Strongly, if disingenuously, he rejected the implication that this was endorsement of painful vivisection, insisting that it was simply a description of others' experiments.[24] But the fact is he habitually rehearsed previous experiments of all kinds, so as to see for himself. It was, after all, imperative that experiments be repeatable.

'Seeing for himself' was Huxley's personal requirement for checking and claiming facts, and Shaftsbury's accusations were well grounded. In his essay on Descartes and animal automatism, for example, Huxley detailed German experiments on frogs that had been undertaken in 1869: spinal cords dissected; acetic acid applied to its skin; optic lobes of brains removed; the 'foremost two-thirds of the brain entirely taken away'. It was all the work of physiologist Friedrich Göltz, who was in the habit of removing cerebral hemispheres from frogs and dogs, and was a published proponent of vivisection.[25] As for Huxley, just the year before the Royal Commission he verified the claims, presenting this work to the Belfast meeting of the British Association: 'I have repeated Göltz's experiments, and obtained the same results.'[26] He was hardly agnostic on vivisection.

Darwin was questioned on 3 November, indicating that he had never experimented on living animals, but that he also believed that such experiments had been, and would continue to be, essential for useful physiological knowledge. Most experiments could be achieved while the animal is insensible to pain. But I am no physiologist, he demurred. Experimenting without anaesthetic 'deserves detestation and abhorrence'.[27] And then the witness withdrew. It was short and to the point, and Huxley let the chair, Viscount Cardwell, question his friend, who had gone to the trouble of travelling up from Kent.

At the very least, Huxley sat on the fence with regard to animal experimentation. He was certainly not outspoken on cruelty, as was Darwin. In the Royal Commission itself, Huxley suggested that humans and animals had distinct pain experiences, because humans anticipated it, a line of argument that followed classic distinctions between animal 'instinct' and human 'reason'.[28] He broadly, if reluctantly, endorsed a set of principles that the British Association had approved in 1871: that no experiment which can be performed with an anaesthetic should be performed without it; that painful experimentation is not justified simply for the purpose of demonstrating a fact already known; and that experimentation (for facts unknown) should be performed only by experienced and well-equipped researchers, maximizing the chances of successful result.[29] Perhaps the latter was Huxley's justification for repeating Göltz's experiments.

From this safe position, Huxley questioned suspect vivisectionists strongly, some of them quite distinguished. The Scottish neurologist David Ferrier, for example, experimented on monkeys' brains, inducing contortions of the face and limbs, possibly just for amusement. Ferrier had a long track record of experimental physiology in high-profile institutions, but in this case he dissected in the forum of an open lecture to lay men and women, and this rendered his intention questionable. The problem was that the 'grotesque' contortions elicited laughter, apparently from the lecturer as well as his audience, and, it was charged, were intended to do so.[30] Ferrier was asked whether he aimed simply to give an exhibition, a 'sensational' lecture. No, the neurologist replied: this was a valid instruction to the public on the effects of electricity upon animals when in a state of anaesthesia.[31] Clearly vivisection for entertainment was beyond the pale, but for Huxley in cases such as this, the objection lay as much with the disrespect for experimental science as for the animal itself. As it turned out, Ferrier was the first to be put on trial after the 1876 Cruelty to Animals Act was passed, and subsequently acquitted.[32]

Huxley had a good deal of time for other professional experimentalists. He engaged closely with entrepreneurial physiologist Charles Brown-Séquard, who was alarmed by the sudden rise of anti-vivisectionism in Britain: so alarmed that he intended to visit in 1874 to deliver lectures – and live experiments – specifically 'to show by a number of examples that science owes much to vivisections'. He had two lectures and experiments in store, like a physiological travelling showman. The first would demonstrate how one single experiment 'can give an immense number of new facts, lead to many new theories and confirm or disprove already promulgated doctrines'. The other would show how much physiological knowledge about the nervous system had been progressed by vivisection.[33] What did Huxley think of that, he pressed London's most famous physiologically trained educationist? Huxley knew how controversial such unmediated propaganda for vivisection would be in mid-1870s London. The political context would not permit it. But there is no question that personally he was thoroughly intrigued with the possibilities of animal experimentation – and with anaesthesia – and entirely captivated by the work on nerves and sensations with which Brown-Séquard was then engaged.

In this way, Huxley operated far more effectively as the defender of physiology on the Royal Commission than as the defender of animals. He was affronted by the anti-physiology stance of many anti-vivisectionist witnesses: those who refused to acknowledge that useful physiological knowledge had been derived from animal experimentation;[34] and those who suggested that knowledge of anatomy and normal function – gained through the vivisection and dissection of animals – did not assist in diagnosis or therapeutics of disease and abnormal function.[35] Huxley also pressed hard those witnesses who suggested that physiological facts need not be learned by students and practitioners through their own vivisection of animals under anaesthesia. Like chemical experiments, good education in physiological knowledge required hands-on learning, not textbook description at one remove, Huxley insisted. This was the difference between 'seeing' physiology and 'merely talking about it'.[36] He prompted one witness, for example: 'It is somewhat difficult to convey to persons who have not actually been engaged in the study of science, the intensity and the reality of the difference between knowing anything of your own knowledge, and knowing it from being told.'[37] At the same time, such physiology might and should be conducted 'with a much greater economy of pain'.[38]

Huxley's content-loaded 'questions' throughout the Royal Commission on Vivisection were so directed and opinionated that he functioned as much as a witness as a neutral commissioner. Anti-vivisectionists certainly thought so, targeting him and turning his own evolutionary 'Darwinism' around. They charged animal experimentalists to stand prepared themselves 'to be dissected alive'.[39] The future possibility that *humans* might be subject to vivisection was an argument enabled by evolution: 'it may be, that in the mysterious evolutions of matter ... those who have practised, and those who have permitted Vivisection, may find that they have inherited the appalling destiny that they have inflicted.'[40] Huxley stood warned, they thought, but really he was as disdainful of the thin and unsophisticated response as he was of the politics of anti-vivisection itself.

Over time, at least in Britain, anti-vivisectionists were unquestionably more successful than the experimentalists. Cobbe won out over Huxley. There was another Royal Commission that reported in 1912, just after another *Protection of Animals Act* (1911), and just as Julian Huxley was embarking on his own zoological career.[41] His title at Kings College London was 'experimental physiologist'. He certainly undertook experimental work, and held a vivisector's licence to do so: vivisection did continue, though now heavily regulated, including at the London Zoo. Julian was later held to account for carrying the licence itself. The fact was noted by some early objectors to his 1930s reforms as Secretary of the Zoological Society of London, and Julian very publicly surrendered the licence and made a show of banning vivisection at the London Zoo.[42]

ANIMALS IN CAPTIVITY: THE LONDON ZOO

The London Zoological Society was born just after Thomas Henry Huxley, in 1826. It opened its Gardens and living specimens initially to naturalists and gentlemen-menagerists who subscribed, and two decades later to the public, who paid per visit. In 1850, just back in London from the South Sea, Huxley was as yet so little known that the Zoological Society of London incorrectly addressed its ticket of admission to its Regent's Park Gardens to Mr T. R. Huxley (figure 5.1). It would be an unthinkable mistake even five years later, so meteoric was Huxley's rise. He was a regular visitor over the years, but Regent's Park was more than a destination, it was his neighbourhood. His lodgings from April 1851 were at 1 Hanover Place, Clarence Gate. When Henrietta arrived from Australia, he set her up, with her parents, at 8 Titchfield Terrace, Regent's Park, 'close by me'. Then they moved to 14 Waverley Place in 1855.[43] His brother George Huxley and many friends also lived nearby. The Zoo became part of his walks and talks, both neighbourly and scholarly, especially with Herbert Spencer, who lived at 64 Avenue Road. Alfred Russel Wallace moved into St Mark's Crescent, Regent's Park; George Romanes lived at 18 Cornwall Terrace; and George Henry Lewes and 'George Eliot' lived together at The Priory, 21 North Bank, Regent's Park, from 1864. Towards the end of an era of global natural history collecting, this Regent's Park concentration of zoologists and philosophers was as local and metropolitan as it was possible to be.

Also established around the same time as the Zoological Society was the Society for the Prevention of Cruelty to Animals, first meeting in Old Slaughters Coffee House, St Martin's Lane in 1824. This was a different set of animal lovers to those whose society supported the Zoological Gardens, though many of them enjoyed large private menageries. Driven by such anti-slavery Evangelical giants as William Wilberforce and Thomas Fowell Buxton, the initial victims for this Society's protection were domestic English animals: the bull who was baited; the cock who was made to fight; the fox who was hunted; and the cart-horse who was whipped. These were animal lovers more of Charles Darwin's ilk, growing up in a landowning family with much-loved 'pets'. Huxley's struggling childhood and youth in London and Coventry allowed for no such luxuries. Animals existed to work, to be eaten, to be studied, dissected and exhibited, not to be loved and petted. Later, three cats and a dog lived in Huxley households.

Human design and ambition for animal welfare has an unpredictable history. We can rarely presume what different generations imagined as best 'protection' for animals. Those who early advocated 'the rights of animals' were not necessarily inclined to the London Zoo.[44] The Reverend John Styles, for example, in his *Animal Creation: Its Claims on*

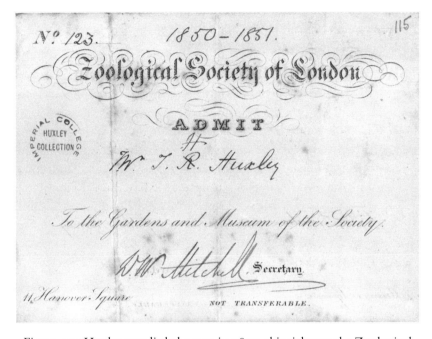

Figure 5.1: Huxley was little known in 1850, this ticket to the Zoological Gardens and Museum admitting Mr T. R. Huxley.

Our Humanity (1838), considered the Zoological Gardens simply an 'exhibition for the mere indulgence of the most vulgar curiosity'. Yet counter-intuitively he endorsed museum-style taxidermy, the world that Huxley would occupy: 'A collection of animals in cases is a better instrument for the study of the animals, than a living menagerie under the circumstances of the Zoological Gardens.' Perhaps the gentler races of animals might be kept in zoos, Styles continued, but

> leave the rhinoceroses, the elephants, the lions and tigers, eagles and condors, to their own forests and mountains, to the use of the noble powers which nature gave them, and the enjoyment of those free, wild, and magnificent scenes of nature in which alone they can be said to live.[45]

Huxley did see elephants, tigers and condors in cages when he walked through the Regent's Park Zoo, as an unknown in 1850 and then a rising zoological celebrity. He could appraise their anatomy closely, although ironically he would agree with animal protectionist John Styles that more could be learned once the creatures were dead, and later dissected, mounted or skeletonized.

From the beginning, the Zoo was both a place of animal collection and natural history learning and a place of entertainment and indulgence. John Styles was correct. The same animals, doing the same thing, could serve all these human purposes. 'Jenny' the orangutan, for example, was a particular favourite of Richard Owen and his wife, amusing them by taking tea 'in the most ladylike way'.[46] The original 'Jenny' (there were several with the same name) was purchased in November 1837 and lived until May 1839. Charles Darwin observed her closely in 1838 (figure 5.2). The great naturalist was struck by her closeness to humanity, jotting down notes which much later formed *The Expression of the Emotions in Man and Animals* (1872).

> Let man visit Ouranoutang in domestication, hear expressive whine, see its intelligence when spoken [to]; as if it understands every word said – see its affection. – to those it knew. – see its passion & rage, sulkiness, & very actions of despair . . . and then let him boast of his proud preeminence.

It was an early and still deeply private entrée into the dangerous conclusion: 'Man in his arrogance thinks himself a great work, worthy of the interposition of a deity. More humble and I believe true to consider

THE FEMALE ORANG - UTAN.

Figure 5.2: One of a series of orangutans kept in London Zoo
from 1837, and named 'Jenny'. This is likely the 'Jenny'
with whom Charles Darwin interacted in 1838.

him created from animals.'[47] Jenny, indeed, was instrumental in early formations of the idea of shared human and ape descent from extinct primates, which Darwin theorized and published only after his friend Huxley had presented his version of the same idea in *Evidence as to Man's Place in Nature* (1863). It is not incidental that Huxley's book was illustrated with skulls and skeletons, and Darwin's with drawings and photographs of live creatures, both human and animal.

It is a great mid-Victorian irony that the apes observed by Huxley and Darwin, who inspired new theories on 'man-like' apes and 'ape-like man' in nature, spent their captive lives actively made to be human-like. Darwin observed Jenny's child-like emotions through the child-like clothing in which she was dressed. Later, Jenny grew into 'human' adulthood by taking tea, the performance that Richard Owen and his wife so enjoyed, along with countless others. Domesticating apes and monkeys through the English ritual of tea was already a longstanding amusement and became a standard, and by the Victorian public even expected, part of a zoo experience. It borrowed from an eighteenth-century tradition of displaying and domesticating chimpanzees. As early as 1738, one chimpanzee brought to London from Angola was presented in *A Description of Some Curious and Uncommon Creatures*: 'The Chimpanzee was very pretty Company at the Tea-table, behav'd with Modesty and good Manners, and gave great Satisfaction to the Ladies . . . it would fetch its little Chair, and sit in it naturally, like a Humane Creature, whilst it drank Tea.'[48]

The Chimpanzee Tea Party became one of the London Zoo's biggest drawcards, right up until 1972, that is, throughout Julian Huxley's term as secretary. Chimpanzees were dressed and domesticated with a table, tablecloth and crockery, performing a tea party daily.[49] There was a double or even triple sense in which the tea parties anthropomorphized the chimps, not just into human 'adulthood', but into another form of human childhood: the tea party was a game that human children played when they pretended to be adults. Thus chimpanzee tea parties were enjoyed as 'chimpanzees imitating children imitating adults in a tea party'.[50] Both at Regent's Park and later at Whipsnade Zoo, they were wildly successful as far as human adults and children were concerned (figure 5.3). The London Zoo held a dual

Figure 5.3: Julian and Jacqueline's tea party. Jacqueline was born at the London Zoo on 28 November 1937, but her mother refused to feed her.

identity as a place for science and a place for human entertainment as strongly in Julian's time as in his grandfather's.

The younger Huxley was an interesting and almost obvious appointment for secretary of the London Zoological Society, an experienced and committed educator for public science, with not just his own zoological credentials but also those of his grandfather to boast. He was a good appointment for an institution that felt itself august.[51] The Zoological Society's offer placed Julian and Juliette in the heart of London: the family could reside in an apartment within the Zoo's grounds. Julian thus occupied 'Regent's Park' even more centrally and unequivocally than had his grandfather. On weekends, and over the summer, Julian and Juliette, and their sons Anthony and Francis, also enjoyed living at Whipsnade Zoo in Bedfordshire, where they kept two Icelandic ponies and Julian's camera caught long summer days for family albums. Proximate to Mary Ward's grand house, Stocks, and indeed to Tring Reservoir, the family rode the ponies over the same hills that Julian had enjoyed and explored with Trev, Aldous and Margaret at the beginning of the century.

The difference between T. H. Huxley's natural history museums and Julian's London Zoo, a 'living museum', is plain. While the grandfather tended towards the description of dead animals, the grandson sought in the London Zoo not just a menagerie of live ones, but the materialization of a philosophy of their relation to generic biological principles: the new ecology. Still, there was always a continuity between zoos and menageries, hunting and collecting, live animals and stuffed ones, behavioural studies and anatomical studies, a crossover between the study of nature and a culture of animal aesthetics. This extended even to interior decoration. Julian Huxley took liberties in his earliest weeks as secretary, newly resident in the Zoo's grounds. In July 1935 he commissioned a dressed and mounted polar bear skin, from E. Gerrard & Sons in nearby Camden Town. They advertised as 'Naturalists, Taxidermists, Furriers', a telling sequence, the continuum of uses of dead animals for London's humans of the 1930s. They undertook 'artistic and decorative natural history', 'educational and nature study' preparations, as well as 'skin dressings, rugs and fur preparations' to adorn living rooms or glamorous ladies' shoulders. Receiving a cheque for £6.0.0

from Julian Huxley, the taxidermists dressed and mounted the polar bear as a rug.[52]

As secretary, Julian acquired a much-needed salary of £1,200 per annum, the first since his Kings College professorship. It was earned by looking after people in an organization as well as animals in cages, and it was the humans who proved to be more difficult. The board was a problem, initially presided over by the Duke of Bedford, 'aged and eccentric' according to Julian. He had mixed feelings about the Duke's credentials: on the one hand an important figure in the retrieval of Père David's deer from extinction (the breeding herd still at Woburn Abbey), on the other responsible for introducing the American grey squirrel into England.[53] Julian tried to persuade the Duke to allow the Zoo also to breed a herd of the endangered Chinese deer, offsetting the risk of disease or death at Woburn, the sole herd on the planet, and was scathing about his refusal. Julian the zoologist took exception to the board as 'wealthy amateurs, self-perpetuating and autocratic', sounding very much like a young T. H. Huxley, self-superior to an entrenched establishment.[54] But he should have known that keeping the board onside was part of the job, part of his £1,200 per year. Instead, the tension was more or less immediate, as he (too) swiftly set about a process of reform that sought to shift the Zoo's organizational philosophy broadly from his grandfather's generation of morphology and taxonomy towards his own generation's ecology and ethology. He wanted to offer a modern vision of biology to visitors rather than sustain a living version of the Natural History Museum's stuffed animals. Instead of focusing on species, he tried to introduce living and arranged versions of concepts like coloration, niche and natural selection. Julian sought to authorize and implement a scientific philosophy that recognized processes rather than objects, function rather than structure, dynamics rather than stasis.[55] He tried to rebuild the Zoo's purpose, display and infrastructure out of the kind of synthesis, synopsis and connection-oriented studies of animals that his generation were pioneering.[56] This all proved far easier to put into print in *The Science of Life* than to bring to life in the London Zoo.

This was a landscape which animals inhabited often from their birth until their death, in which a handful of humans lived and into which throngs of other humans passed daily. Julian was intent on updating its design, for all these animals. He was troubled by the facilities at the Zoo, its architecture and enclosures. The small scale of the gorilla cage was a problem, as were the lion and tiger cages, the former 'moping' and the latter 'pacing' continually. He was intent on comparing the latest trends in zoo design and swiftly set about gathering comparative information. In 1936 he took a trip through Europe, comparing zoos in Paris, Liège, Bremen, Frankfurt and Munich, an interesting time to be in Germany, and he learned a good deal about new philosophies of enclosure. His notes from that tour display a great deal of attention to cages without bars, to the use of ditches not fences, versions of the eighteenth-century ha-ha wall, and to the simulation of a wild ecosystem. At the French-American ornithologist Jean Delacour's estate near Dieppe he was inspired by free-roving animals and birds, especially the gibbons, and subsequently a gibbon island and a flamingo island were built at Whipsnade, emulating this private-menagerie freedom as much as he could in a public zoo.[57]

For Julian, the London Zoo was the public communication of *The Science of Life* in three dimensions and with all senses activated. He introduced a touch-experience with the petting zoo, 'Pet's Corner'. It included both domestic animals and pettable others, including a lion cub, a python, a chimpanzee and a giant tortoise. Seemingly a shallow display, expedient for the Zoo's popularity and visitor numbers, Pet's Corner in fact aligned with his intention and ambition that the Zoo display animals by their range of function, including by animal 'purpose' in relation to humans, guide dogs, for instance. Julian also worked on zoo sounds, recording animals and playing the tapes back to observe their reactions experimentally. Was animal language a case of true speech? It was an old question considered through new media. Out of these recordings came *Animal Language: How Animals Communicate* with text by Julian Huxley, sound recordings by the broadcaster and recent refugee from Germany Ludwig Koch, and photo illustrations by the Hungarian animal photographer Ylla. And characteristic of his excitement about new media, the gramophone record *Zoo Voices* was released as a 'sound book' in which one can hear the calls of sea lions, exotic birds, the big cats and 'the howling pack' – wolves, dingoes and hyenas.[58] Through and over them all we hear Julian himself, the great trustee of evolution, narrating. In his time at the Zoo, the press, film and broadcast media were heartily welcomed, the public communication of zoology as high on his agenda as working men's lectures had been on his grandfather's. He had a grand but unrealized ambition to build a cinema at the Zoo that would screen the new genre of nature documentaries and even anything Walt Disney might put out with cartoon animals in it. Through all this daytime activity, at night Julian was writing *Evolution: The Modern Synthesis*. It was a ghostly reinhabiting of his grandfather's frenetic day-and-night work in and around Regent's Park.

The undoing of Julian Huxley at the Zoo took place over the war years. In September 1939 he was in Dundee, at a meeting of the British Association for the Advancement of Science, when war was declared. He returned quickly to London, since many animals had to be relocated, and the Zoo was shut down. Poisonous spiders and snakes he had killed, some large and dangerous animals – the elephants for example – he moved to Whipsnade, a little like the evacuation scheme for London children. In fact Julian and Juliette considered offers of North American evacuation of their own boys, Anthony and Francis, but Anthony was on the cusp of adulthood, about to enter Trinity College, Cambridge, and Francis was already out of the way, at Scottish Gordonstoun, the school itself evacuated to Wales.[59] In coming years both of them were called up and served, Anthony at home in an air force plotting room, Francis in the navy, in the belly of the beast at Normandy. In the meantime, their father entertained and diverted an anxious, stressed, tired British wartime population as a key player in the latest BBC invention, *The Brains Trust*. It was a London-based war for Julian Huxley, regularly enough in the shelter built at the Zoo, sharing it occasionally with old friends like Jack Haldane. The bombs came close, too close – the Zoo's buildings on fire at one point and directly hit at another.[60]

In November 1941 the Rockefeller Foundation invited Julian to lecture across the United States, not yet in the war, and Huxley was granted three months' leave. From the perspective of the Zoological Society's board, however, he disappeared. A little like President Lovett of Rice University many years before, they wondered quite where their secretary was. His opponents took the opportunity to undermine his position, in his absence abolishing the very office of secretary, and thus the serving officer, Julian Huxley, was dismissed in

March 1942. A bitter wrangle ensued: Julian fought back, rallying considerable support within and without the Zoological Society, but unsuccessfully, and he formally resigned in August that year. Huxley was a reformer but characteristically was too swift. He did implement a great deal at the Zoo, and in a short space of time, never short of energized ideas. But he was fickle and impatient about their implementation. He wanted things changed quickly and thought he could drive it all through on his vision, at his pace. Achieving the directorship of Unesco shortly afterwards went some way to offsetting the humiliation. Yet the whole breathless process was to be recapitulated when his directorship there, too, was short-lived.

It was a diminishing end to a fine office for a Huxley, yet over time the family retained close and affectionate connection with the London Zoo. Julian had been discarded, but once that indignity dissipated, he was a regular enough visitor, as his grandfather had been. Julian and Juliette held their Golden Wedding anniversary party at the Zoo,[61] and he returned often with various projects in hand. One was to bring the talented photographer Wolfgang Suschitzky to create stunning animal portraits and to work together on *Kingdom of the Beasts*, a Thames & Hudson coffee-table book.[62] He was surprised by a vicious review in *The Times Literary Supplements* which took exception to Huxley's self-appointed trusteeship of and for animals. His popularization of natural science was assessed as crude, as 'the Hollywood of Science'. The reviewer also objected to Julian's sense of superiority over other animals, not just the sin of his anthropomorphism, but his own anthropo-centricity. 'While he refuses to allow human traits to be *detected* in other mammals . . . he is not ashamed to *impose* human laws and values . . . The combination of resentment with condescension produces a sort of meanness in some biologists, which comes out in Dr Huxley's remarks.' And then it got personal: 'It would be instructive to discover what personal sense of guilt the biologists are seeking to purge, by inflating humanity, denigrating its relatives and claiming priority for themselves.'[63] Julian Huxley did feel that he was responsible for nature, that he had to lead and redirect human trusteeship for the future of animals. Not everyone appreciated his stance, however, or what could be, and was, read as a self-righteous pomposity.

Julian also retained a connection with the Zoo through one of its later and long-serving secretaries, primatologist Solly Zuckerman, council member, then secretary, then president, between 1955 and 1984. It was a far more sustained association than Julian's, but this did not stop the ageing zoologist from pestering Zuckerman with endless lists of deficiencies and improvements still to be made both at Regent's Park and at Whipsnade, as if he still had a say, and as if he had not been unceremoniously removed from influencing Zoo futures. Julian persisted, irritatingly, about design and publicity, and constantly asked why his brilliant idea for a cinema had still not materialized. Why are the sheds and shelters at Whipsnade so hideous and why are there so many domestic animals? Why in fact are there no longer any chimps at Whipsnade? And on and on: why have the flamingos been removed from the island; why are no wild horses visible?[64] When the Zoo was modernizing again in 1960, he pressed Zuckerman further, this time on his plans for a new elephant house and a new rodent house. Had the Society looked at these?[65] By 1967, he had backed off a little, but poignantly so. Zuckerman respectfully invited him to attend the Queen's visit to the Zoo that year, but Julian – honest about his mental illness as always – had to decline. 'I had a so-called nervous breakdown last year, & am still having to refuse all formal occasions ... I am getting slowly better, but it takes a long time.' He affectionately recalled showing the young princesses around the Zoo decades earlier.[66]

WILD ANIMALS

While the evolutionary framework for understanding animals and birds in the wild, in zoos, in the fields and ponds of Hertfordshire was handed to Julian by his grandfather's generation, the political and intellectual context that gave it meaning in the twentieth century could not be more different. For Thomas Henry Huxley in the late 1840s, collecting animals in the Empire's lands and seas, dissecting them *in situ* and displaying them later in London's museums and lecture halls were complementary modes of biology and zoology. We know retrospectively that T. H. Huxley's own South Sea voyage was a late chapter in the long modern story of ecological imperialism. Yet for Huxley himself this was all a promising and optimistic moment of global gathering: a little later, more persuaded by evolution, he would consider it essential evidence to turn that theory into a fact. The more thoroughly species from all quarters of the earth, from the depths and shallows of the seven seas, from the deepest past and immediate present could be identified, the more the evidentiary puzzle of evolution could be pieced together and verified. Species that had existed at one point in the planet's long history, but were no longer living, were a matter of evolutionary nature for Huxley. A century on, his grandson took evolutionary evidence for granted and saw species extinction as a *political* matter, part of human-induced global destruction that threatened habitats and species. After the Second World War, the effects of ecological imperialism were only too evident, as were the effects of political imperialism. Julian Huxley's conservation efforts were linked to, and enacted within, the great mid-twentieth-century shift from a colonial to a post-colonial world, especially on the African continent. For Julian Huxley, the global wildlife commons needed protecting, not raiding and looting.

'Conservation' started at home and with birds, in his grandfather's time, though without his grandfather's input. Indeed it started with the great crested grebe. The Plumage League, the Selborne Society for the Preservation of Birds, Plants and Pleasant Places and the Royal Society for the Protection of Birds were linked late nineteenth-century organizations that aimed to curtail the trade in birds' feathers for millinery accessories and as fur-substitutes. The plumage trade by that point was already an international one, and highly profitable.[67] The grebe was of special concern; the crest was in demand, and so was grebe 'fur', the soft under-pelt of its breast feathers. These could only be taken by killing the bird, and numbers fell dangerously, almost to extinction. Thomas Henry Huxley would never have associated with these early 'preservation' societies: they were amateur and they were formed in the main by women. Julian Huxley, however, drove forward 'preservation' and 'conservation'.[68] In the early 1930s, he worked closely with Max Nicholson, who had helped found the British Trust for Ornithology, and when Julian prepared an idea for the mid-1940s *New Naturalists* book series, together they explicitly sought to depict animals not as 'pinned, stuffed or pressed but as [they] occur in life, and in [their] natural habitat'.[69] In July 1945 Julian was appointed to a Committee on National Parks, chairing the sub-committee that considered the conservation of British wildlife specifically. Should there be national parks in England and Wales and, if so, where, and selected on what criteria?[70] Conservation was also part of Julian Huxley's internationalism, both institutionally and conceptually. In his Zoo years, he cooperated with the Society for the Preservation of the Wild Fauna of the Empire and the International Council for Bird Preservation. As director-general of Unesco, he was instrumental in bringing together conservation organizations and government representatives in Fontainebleau in October 1948, establishing the International Union for the Protection of Nature, later (1956) the International Union for Conservation of Nature and Natural Resources. The focus was on endangered species and habitat decline, very much Julian Huxley's mission. The use of the term 'protection' was a remnant of the nineteenth-century idea of 'protection of animals' (*from* humans). It turned soon enough to 'conservation' *for* humans and for other animals. Later, as Unesco representative, Julian undertook influential official journeys through Africa, spreading the word on wildlife conservation and the need for wildlife reserves. He was one of the few players in this early history of what became environmentalism to move regularly between Britain and Africa.

Julian's first visit to the continent was in August 1929, an official tour organized by the Colonial Office. His role was to advise first on the place of biological sciences in education, and second on 'the value of nature conservation' in East Africa.[71] Sailing from Southampton to Dar-es-Salaam, he used the journey on the boat to write poetry: 'A Freudian Faustulus'. He visited research stations in what is now Tanzania, travelled by train between Kenya and Uganda and later met Juliette in Jinja, at the source of the Nile, for a private journey. Their first nights were comfortable, staying with the governor, but it soon became adventurous, uncomfortable and occasionally dangerous. Julian loved it. They both did, taking an eight-week trip that included a long 'foot safari' into the Belgian Congo. With John Russell from the Uganda Education Department, they drove from Kampala to Kabale and then set off: Julian, Juliette, John, plus fifty local carriers. Julian thrived on the activity: 'we now were *walking* over the red earth, pacing the miles – it all added up to an inspiring experience'.[72]

All over Lake Edward they watched herds of antelope, zebra and water buffalo. They headed across country – 'an ancient track led us between foot-hills' – towards the administrative centre of one of Africa's oldest national parks, the Parc National Albert, the core of what is now Virunga National Park, where significant great ape research was undertaken by the primatologist George Schaller. They climbed the summits, Julian attempting Mount Karisimbi in search of gorillas. Camping at nearly 10,000 feet, he and John Russell walked to 12,000 feet but with no sightings of the animal, just a nest. The three of them got wildly lost during torrential storms, stayed variously with fathers and nuns of Catholic missions, camped on volcanoes and finally made it to their destination and home again.[73] Julian wrote *Africa View* (1931) from this adventure. It established his reputation as a three-in-one travel writer, nature writer and science writer.

Julian's engagement with wildness, animals and conservation in Africa was political all the way through. His 1929 journey was sponsored in a high colonial moment and through a formal colonial entity, the advisory committee on 'native education'. A second journey was undertaken in January 1944 with the Colonial Commission on Higher Education, also known as the Elliot Commission. Its mission was 'to consider the principles which should guide the promotion of higher education, learning and research and the development of universities in the Colonies'.[74] As always, he both attended to his duties and took the opportunity to watch birds, always with his binoculars and camera at hand. In Nigeria, he was spotted by the future novelist Chinua Achebe looking intently at the sky.[75] This 1944 African trip came hard on the heels of his Zoo downfall and just anticipated his appointment as director-general of Unesco. Conservation also became part of Julian Huxley's Unesco afterlife, continuing African work with a major 1960 tour, in which decolonization and development – 'the rapid emergence of African Governments' – was the new context.[76]

When Julian Huxley first went to Africa in 1929, the European-led League of Nations established itself as a trustee of people considered 'less developed'. Their wellbeing and political and economic status were 'entrusted to advanced nations who by reason of their resources, their experience or their geographical position can best undertake this responsibility', as the Covenant of the League of Nations put it.[77]

Later, himself on the edges of the United Nations' inner circle, Julian watched the world turn on its political axis. He heard Jawaharlal Nehru challenge the UN General Assembly with the demands of the new and non-aligned bloc: 'There are vast tracts of the world which may not in the past, for a few generations, have taken much part in world affairs. But they are awake; their people are moving and they have no intention whatever of being ignored or of being passed by.'[78] This was no longer the world of colonial commissions, where Julian could travel to parts of Africa, as he had in 1929 or even 1944, and expect to enjoy the hospitality of governors and colonial officers.

For Julian's later involvement in international conservation, the critique of the colonial past and present, the difficulties of decolonization and the protection of animals were part of one complex global political economy. These interests did not always align: perhaps they rarely did. He understood that and was alive to the post-war shift in global politics. He did not shy from the difficulties, even if he did not always get it right. And yet, notwithstanding an intellectual and political inclination towards a post-colonial world, Julian Huxley and early Unesco struggled to discard the old developmental framework of progressive stages of civilization that established (at best) some people as trustees for the wellbeing of others. Sometimes Unesco's commitment to the world's national parks and reserves signalled just this. As an indicator of development, wildlife reserves existed and should exist 'in practically every civilized country' and every country that aspired to civilization. At other times, wildlife reserves and national parks were part of a more internationalist idea of a global commons, part of the 'scientific and cultural heritage of all mankind'.[79] As far as Julian was concerned, conservationists needed to 'induce African Governments and the African public to understand and follow an ecological approach'.[80] Africans needed educational impetus and direction, in his view. The new governments needed persuading that it was now or never for wildlife conservation, not just game and nature reserves, but permanent national parks.[81] 'Internationalism' and 'decolonization' were delicately linked in the post-war years.

In June 1960, Unesco sponsored Julian on a survey journey to report specifically on 'the Conservation of Wild Life and Natural Habitats in Central and East Africa'. He valued this commission as the most interesting assignment of his life.[82] The Huxleys toured extensively for thirteen weeks, a schedule structured around visiting as many national parks and reserves as possible. They sailed to Cape Town, then on to Durban, Natal and the Kruger National Park in north-eastern Transvaal. They travelled through what was then the Central African Federation, including the Rhodesias and Nyasaland, then visited the Gorongoza Park in Mozambique, and on to East Africa, including the Serengeti and Manyara and Ngurdoto National Parks, and to Uganda and the Murchison Falls Parks. They ended their trip in Kenya, visiting the Tsavo, Aberdare and Lake Nakuru National Parks, game reserves at Amboseli and Isiolo, and the Mount Meru Reserve (figure 5. 4).

The Huxleys had planned to revisit the Congo, but a volatile transition from Belgian to Congolese government had just taken place on 30 June that year, and the new nation was in violent crisis. In January 1960, Congolese and Belgian leaders had met in Brussels, planning a 'hand-over' in Léopoldville, but there was no such smooth transition. There were large evacuations of Belgians coordinated by the United Nations, and an escalating crisis, including the assassination of Prime Minister Patrice Lumumba. On the dangerous border, the Huxleys met with the warden of the Congo National Parks in Kivu Province. This all signalled the great political shifts that had taken place since their first journey in 1929. Then young, they had walked as an adventure deep inside the Belgian Congo. Now they were older, the border itself was as far as they were willing to go on the cusp of independence, and sensibly so. Juliette compared their two trips: 'It had now become a sort of double journey, a counterpoint of inner and outer landscapes, of remembered and new visions. More and more landmarks fell into place, recalling our young selves walking like ghosts across our present'.[83]

In 1929, on their first visit, there were two national parks across sub-Saharan Africa: the Kruger National Park in South Africa, which had just been converted from an original (1898) game reserve, and the Parc National Albert, which had also just been established, in what was then the Belgian Congo. By their 1960 trip, there were almost forty parks and reserves of differing orders across the continent: national parks established by legislation and managed by a body of trustees with a dedicated

KENYA

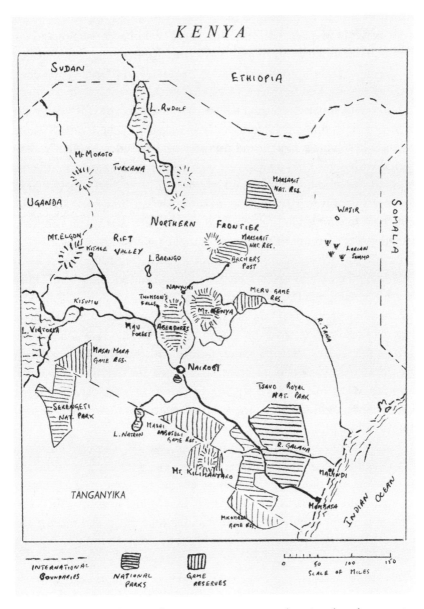

Figure 5.4: Map of East African game reserves and national parks *c.* 1963.

and salaried staff; wildlife sanctuaries and private game reserves of various kinds, including those established for managed hunting; and 'Tribal Parks' organized around Africans' needs and lands.[84] Julian and Juliette visited twenty-five of these in the Unesco-sponsored tour in 1960, as well as scoping potential sites for more wildlife protection zones. Throughout, Julian wrote, filmed and lectured about endangered species and habitats. His commissioned survey produced an 'alarming and challenging' report; precisely what Unesco wanted.[85]

Julian and Juliette returned to Africa quickly in 1961, this time for him to give lectures in Nigeria and Ghana; he lunched with Kwame Nkrumah, Ghana's first prime minister and president. Julian reconnected with dignitaries he had met on earlier journeys, for example the Alaki of Abeokuta, whom he had met first in 1944. And he lectured in some of the universities that the wartime Higher Education Commission had recommended be established. In East Africa, in 1963, he attended a conference of the International Union for the Conservation of Nature, which he had been instrumental in establishing nearly two decades earlier. That year he visited the Organization for African Unity in Ethiopia and met the emperor, who gave him every nominal support for the consideration of conservation strategies. Julian's final African journey was in 1971, over his eighty-fourth birthday. He and Juliette flew in planes this time, to watch hippos and ostriches from a bird's eye view. They saw new reserves and visited the Serengeti Research Institute, what Julian called the world's largest ecological laboratory. There they caught up with Jane Goodall, who was then studying wild hunting dogs.[86]

Most things were political for Julian Huxley. Wildlife certainly was. His message was animals in peril: the reduction of habitat, the reduction of numbers and thus the reduction of species. He publicized a political economy of conservation across Africa that linked it with world consumption. He was outspoken on poaching and the global market for animal parts, placing Africa in a long history of global capitalism and its worst manifestations: 'Like the slave trade, it is profitable, highly organized, extremely cruel, and quite ruthless.' He described the global trade in dead animals and parts of dead animals that had accompanied the natural history specimen trade of his grandfather's imperial milieu. Not just ivory and rhinoceros horn, but 'skins of leopards, colobes, monkeys and other species, antelope and buffalo horns, sinews, bones, and (on a surprisingly large scale) tails for fly-whisks also find a ready market'. Crucially he brought this all home, itemizing current London prices for these commodities. And then he pronounced on the trade's waste: 'From most of the tens of thousands of animals killed by poachers in eastern Africa today, the trophies are cut out, most of them to be used for trivial or ignoble purposes, and the meat simply left to rot.' Julian offered a slogan for the emotional and political economy of African wildlife in relation to sub-Saharan hunger, a local response to the global trade: managed properly, national parks could yield 'Profit, Protein, Pride, Prestige'.[87]

Organized poaching was one problem, but this was simply symptomatic of larger changes in global capitalism, according to Julian. He framed the dangers to wildlife in terms of a larger problem: human population growth on the planet as a whole. Overpopulation was a spatial and geographic question for his generation, as much as a biological one, and Julian Huxley was just then one of the problem's most outspoken analysts.[88] Humans invaded not only each other's land, but that of animals: 'the demands of modern civilization for more food and more space, have meant that the few untouched natural areas of our planet where animal herds can still roam freely are being steadily encroached upon'.[89] Even over his own lifetime, Julian wrote in 1961, the diminution in numbers was clearly observable, especially in East Africa, from his first visit thirty years previously.

The message about African wildlife was a challenging one, precisely because it raised an economy that implicated the rest of the world as traders and consumers. And yet the critique was received comfortably enough by the world's wealthy in the post-war boom. In the 1960s, the call for the conservation of African animals and their habitat morphed easily enough into the suddenly escalating nature tourism industry, as well as popular consumption of wild nature, from a distance – in a bookshop, a library or a cinema. The African wildlife park – and the books and films and *National Geographic* journalism about its animals – was on the cusp of a high moment.[90]

Julian thrived on the journalistic potential of his African travels: this was why he was such a valuable zoologist to sponsor. Writing as guardian of the wild and conscience of the world, he could combine words, pictures, adventure and politics for comfortable-enough consumption, despite the political economy message. Juliette Huxley also turned travel writer on their long 1960 adventure. She published *Wild Lives of Africa* (1963) to considerable success. The 'wild lives' of special interest to her were as likely to be human as other animal; and far more likely to be Europeans who called Africa home than Africans themselves. The wild lives of white folk who saved the animals were retold to eager American and European consumers on screen, as print journalism, and especially via the era's extravaganza that was *National Geographic*. While the Congo was in crisis, South Africa was enforcing apartheid, and the Mau Mau and British authorities were at war in Kenya, Africa and its wildlife came to the world via European mediators who lived exciting and dangerous lives with lions, gorillas, and chimpanzees.

George and Joy Adamson epitomized this combination of formal institutional roles in high modern wildlife conservation, and public roles in the booming industry that was post-war African wildlife journalism under the thinnest of 'conservation' lines. They were an extreme manifestation of the kind of trusteeship for animals that Julian admired, far more thoroughgoing, dramatic and devoted than his own passing responsibility for faux-wild animals at the London Zoo. The Huxleys joined the Adamsons as part of their long trip in 1960, visiting Isiolo Game Reserve in what was then called the Northern Frontier District. It was a politically precarious place and time. That year, British Somaliland became the Somali Republic. The Northern Frontier District – almost exclusively inhabited by Somalis – was carved out and administration over them granted to Kenyan nationalists.

George and Joy Adamson's larger stories accurately reveal the strange mix of game hunting, international zoo trade, animal protection and nature conservation that was mid-twentieth-century African animal politics. George Adamson was the experienced senior warden of the Northern Frontier District, having been in Kenya since 1924, for much of that time as a professional game hunter. Isiolo was in fact a game reserve, not a national park, and part of Adamson's job was to issue licences to shoot. The long history of game hunting became blended with, rather than obliterated by, high modern wildlife tourism. Juliette was disquieted by hunting but also relayed in her book that such regulated game hunting was not the real problem since it was known that animal populations could be sustained. Rather, 'non-selective' hunts and poaching in which hundreds of animals were killed were the real threat.[91] It is one of the least authentic lines in her otherwise rich account of Isiolo – seeming to come from George Adamson's mouth and mind, not her own. Julian himself disapproved of the great industry of game hunting in Africa, and very publicly so, although he had undertaken his own killing and stuffing for biological purposes. In 1915, between Rice terms, he holidayed in Wyoming and with Rice zoological responsibilities in mind shot a prairie dog, stuffed it and prepared its skeleton for mounting. He skinned a porcupine and sent its skeleton back to Rice biological collections as well, ready for teaching.[92] Perhaps, like vivisection, the purpose and intent made the difference.

Juliette liked the look of George Adamson, just then a minor wildlife celebrity. She had already seen him in the British press, with his 'short pointed beard, and bright blue eyes with the piercing vision of hunters and sailors'.[93] In 1956, as a game hunter, George Adamson had killed a lioness with three cubs, which he and Joy then rescued. The eldest two were sent to Rotterdam's zoo. The smallest they kept, raised and named 'Elsa'. The most famous lion ever, Elsa was their only child; the Adamson's 'wonder child', thought Juliette. Like so many others Juliette had consumed the massive publicity that surrounded the publication of *Born Free* that very year, 1960. Joy Adamson had written the book with stunning photographs that charmed the world. The drama and intimacy of the domestic–wild, animal–human family crossover story was caught in the stratospherically successful film released in 1966. It catapulted Joy Adamson to worldwide fame, the book and the film still appreciated as a landmark phenomenon that changed conservation's culture.

Given the immense and immediate popularity of the book *Born Free*, the Huxleys wanted to meet Elsa at least as much as they wanted to meet George and Joy Adamson. Joy was reluctant, however. There was a lot going on partly because David Attenborough was soon to arrive to film *Elsa the Lioness* for the BBC, and detailed preparations were underway. But there was another reason. William Collins, the publisher, had recently visited the Adamsons. Elsa had entered Collins's tent and had put her leonine jaws right over his head, leaving teeth marks: little wonder Joy Adamson was cautious. But eventually she agreed to a day trip to Elsa's territory, on the condition that Julian and Juliette stay inside the vehicles. That proved to be exciting enough. Though they initially thought they would be disappointed in their search, the lioness suddenly bounded out of the bush, 'Elsa's paws on Joy's shoulders, stood there like old friends meeting again'. Joy stroked Elsa, and Elsa nuzzled Joy, as Juliette recounted. The rest of them were stuck in Land Rovers 'and not allowed to do our own bit of loving'. But then their encounter came close enough. Majestic Elsa leapt onto the roof and then the bonnet. And there she lay down in front of the convoy, lazily stretched and comfortable. Nonchalant, observed Juliette.[94]

One of the twentieth century's most iconic species crossovers, the Adamsons' story is the strangest mix of loving and killing. It started with George Adamson shooting Elsa's mother, a lioness protecting three cubs. Elsa was raised by Joy and George as a couple, a kind of wildlife family in which parental responsibilities included teaching independence in the actual 'wild'. Then Elsa grew up, was successfully released and remained affectionately and willingly connected, especially to Joy, as the Huxleys had witnessed. Juliette called the George, Joy and Elsa triad a 'ménage-à-trois'. But really, and as it unfolded over time, Joy loved Elsa more than she loved George. Juliette Huxley certainly perceived the inter-species relationship between Elsa and Joy as a kind of love. She tried to work it out. Joy herself thought their relationship was about complete trust, 'lack of fear on both sides, total faith in each other', to the extent that when Elsa once took Joy by the throat, as she did her cubs, Joy 'accepted the gesture' and kept entirely still. Juliette described it as 'a moment . . . of perilous intimacy'.[95] For his part, Julian perceived Joy and Elsa's relationship to be a remarkable instance of inter-species connection. Or at least so reported the naturalist Gerald Durrell, who later visited Julian Huxley to drum up support for his conservation zoo on Jersey Island. As it happened, they sat down together in the Huxleys' Hampstead home to watch David Attenborough's *Elsa the Lioness* on the new television set. Gerald Durrell recounted: 'In silence we watched Joy Adamson chasing Elsa, Elsa chasing Joy Adamson, Joy Adamson lying on top of Elsa, Elsa on top of Joy Adamson, Elsa in bed with Joy Adamson, Joy Adamson in bed with Elsa, and so on, interminably.' Julian's response, as reported by Durrell? ' "It's the only case of lesbianism I have ever seen between a human being and a lioness," he said, quite seriously.'[96]

It probably *was* serious, even though Durrell twisted it all into a thin joke. Julian after all was the lifetime theorist of ritual, sexual selection and courtship; the inheritor of a century of Huxley thought on human animality and Darwinian animal–human emotion. He was also the unsurprised observer of the recent Kinsey reports on human sexual behaviour, which seemed to have shocked the rest of the world. 'I was not shocked or surprised,' he recalled. '[B]ut it interested me to see here proof that sexual morality in animals (of which I knew a good deal) is higher than in man.' [97] Juliette Huxley's thoughtful reflections on the nature of love between Joy and Elsa and Julian's perception of a kind of courtship behaviour make up a far deeper set of observations than Gerald Durrell's storytelling.

Joy Adamson continued to correspond regularly with the Huxleys, not least because a good friend, the Kenyan author and broadcaster Elspeth Grant, had married Gervas Huxley, Julian and Aldous's cousin, in 1931. Joy sent them updates on Elsa and her cubs, and snaps of their adventurous drive. 'Joy and Elsa', are in the Huxley family albums. It was Julian who wrote the foreword to the *Born Free* sequel, *Living Free: The Story of Elsa and Her Cubs* (1961), in which he raised the stakes of the Elsa–Joy relationship by analysing it in the context of Darwin's *Expression of the Emotions in Man and Animals*.[98]

Born free; living free; for ever free. All this animal freedom talk was the strange counterpart both to political freedoms of independence unrolling across the continent in those years and to the intimate human violence aimed at the Adamsons. Joy was murdered in 1980, and her now-estranged husband, the game shooter, was shot in Kora National Park, Kenya in 1998. There was animal love in death as well as in life: they were buried not with each other, but with their big cats. Joy's ashes were buried with Elsa the lion and Pippa the Cheetah. George lies next to his favoured lion, 'Boy'.

The kind of emotive protection of individual, named animals so popularly consumed in the 1960s and 1970s – especially those not-quite-wild and not-quite-domestic, like Elsa – was different from the high-level animal politics that Julian Huxley generally pursued. In many ways, the individualized 'protection' or 'rescue' of the likes of

Elsa recalled the 'prevention of cruelty' to individual animals, familiar from the RSPCA of T. H. Huxley's era, though perhaps stripped of the astute moral politics of the Victorians. This followed and drew from the long British history of guardianship of creatures, the prevention of cruelty to particular animals who cannot protect themselves, the innocents of the anti-vivisection movement. T. H. Huxley's generation *was* highly concerned about individual animal feelings and sensibility. He was less so, personally, even as he pursued and extended scientific understanding of what and how animals 'felt'. For the Victorian Huxley there was a presumed dominion over beasts, whether that was an ongoing scriptural dominion as some held from Genesis or a scientist's dominion over all nature, as he held himself. Twentieth-century Julian certainly eschewed biblical blueprints for human–animal relations; nonetheless the wildlife reserves he endorsed came to be conceptualized as a kind of mission, even with himself as a kind of institutional saviour of habitat, of final breeding pairs, of species from extinction.

Julian Huxley was thus less concerned with individual animals and more with whole populations, with species in an ecological context. It was neither 'cruelty' to, nor even 'protection' for, individual creatures that drove him forwards. Julian's animal politics was not quite 'animal rights', in either the original Benthamite model from the 1820s or Peter Singer's 1970s Benthamite 'liberation' model. True to his synthesizing inclinations and to his generation's refinement of ecological interactions between and within species in their environments, his conservation agenda was about just that: animals in complex environments, seeking and competing for food and land, sea and air. Animals were part of a political ecology that humans had come to dominate. Rather than protecting, or (as his grandfather would probably want to have it) sacrificing individual animals, Julian spoke for all animal populations as a trustee of evolution. It was foundational early conservation politics.

6

Primates: Apes and the Huxleys

When we enter the Natural History Museum, South Kensington, with five million other visitors each year, we step simultaneously into the age of the dinosaurs and into the Victorian age. We inhabit a grand Romanesque hall successfully campaigned for by Richard Owen, still awed by the terrible lizards that he named, and that he had carefully boxed over from the British Museum collection in the early 1880s to the new site. We are drawn to Charles Darwin's statue in the secular nave that invites cultural genuflection like Abraham in the Lincoln Memorial, like *David* in the Galleria del'Accademia, like a modest Jesus Christ in St Peter's. Richard Owen's own likeness used to be at the top of those stairs, but he was usurped and displaced in 2009, the Darwin anniversary, and now we have to search hard to find the old man to whom we owe a good deal, Huxley notwithstanding. There he is, up a level, in a nothing-place. The statue of Thomas Henry Huxley is nearer Owen than Darwin, and they still growl at each other.

And yet, for all this late-Victorian marbled hagiography, neither Darwin nor Owen nor Huxley is the real primate treasure of the Museum. Guy the Gorilla is. He now sits in his glass enclosure, taxidermied (actually latex-stretched, less dignified, but more technically correct). Guy is displayed like a simian Snow White in her glass coffin, suspended somewhere between life and death. Like a thousand stuffed beasts before him, but also like marble Darwin, Huxley and Owen, he is made to seem alive. In this strange, suspended afterlife, these four primates remain connected.

Julian Huxley adored Guy the Gorilla perhaps more than anyone, and in life, not death. His friend the extraordinary animal photographer Wolfgang Suschitzky, fresh from a Huxley family portrait session in Pond Street, Hampstead, caught Guy very much alive in what he singled out as the best shot of a lifetime, noting also how much Julian loved both the photograph and the animal.[1] This gaze is surely the steadiest and most commanding of centuries of intra-primate exchange, the ape's reflection (figure 6.1).[2]

Of all orders of animals, primates were core Huxley business. Their appreciation of simians stretching from the wild to the captive, the historical to the filmic. Thomas Henry Huxley wrote one of the most famous 'monkey books', as Charles Darwin called it, of all time, *Evidence as to Man's Place in Nature* (1863).[3] It was hardly the first study of 'man-like apes', yet, appearing at the height of the evolution controversy, the book focused attention freshly on the similarities and distinctions between humans and other primates, setting the shape of ideas for the rest of the century. Huxley's work was directly antecedent to the great boom in primatology, catapulting controversy over the human–simian link well into the twentieth century, where his grandson picked it up. Yet while T. H. Huxley preferred, or was forced, to study monkeys and apes dead, as skeletons or as soft anatomy specimens, his grandson looked at them living. Just as Julian watched the habits of the grebes as an animal ethnographer, so he came to value observational studies of primate behaviours. He was custodian of Guy's predecessors, Moina Mozissa and Mok, lowland gorillas kept at the London Zoo between 1932 and 1938, and then Meng, a mountain gorilla who arrived in 1938. He engaged closely with key primatologists of the twentieth century, first Solly Zuckerman, who studied captive apes, then George Schaller and Jane Goodall, who spent years in Rwanda, Uganda and Tanzania, their behavioural studies inspired by Julian's ethology.[4]

Thomas Henry Huxley occasionally observed the chimpanzee and orangutan in the London Zoo, yet he likely never saw a gorilla alive. Rather, he encountered them as rare and precious specimens, in anatomy collections or in libraries, in scientific reports and in a tradition of printed African travel accounts from the 1500s onwards. These great apes were quasi-mythical to him, even though he put so much store on them in his search for evidence as to man's place in nature. 'Africa' too was mythical to the elder Huxley as well as to his small-minded mid-Victorian critics who taunted at one point that he led a

Figure 6.1: Guy the Gorilla, by Wolfgang Suschitzky, London Zoo, 1958.

Gorilla Emancipation Society.[5] A century later, however, saving African gorillas and their native habitats from destruction was precisely, and seriously, Julian's mission.

MAN-LIKE APES

In the 1860s, man-like apes threw Thomas Henry Huxley into the public limelight and into a great controversy of his own making. He loved it, especially after the long, slow rise of his marine invertebrate work in the scientific community: few outside those circles were interested in tunicates. It was primates who sat at the core of his quickly escalating disagreement with Britain's leading anatomist, Richard Owen. Intellectual and professional, this fight was emotional and highly personal as well, and for both of them. It was a masculine battle between a silverback and a blackback asserting dominance in the family group that was British natural scientists.[6]

It all started with gentlemanly and professional patronage by Owen. In the years immediately after the return from the *Rattlesnake* voyage, Owen had eased Huxley's way into various advantages – into the Royal Society, into a position, into journals. Not quite a formal patron, nor was Owen yet a clear opponent. But once Huxley had a foothold, a salary and some key publications, his anti-establishment views mixed with his hot and adversarial temperament made him unstoppable. He was single-minded about his own way forwards, and Richard Owen was a singular block. It was a risky, even brave move on Huxley's part to take on the senior and well-positioned Owen. Every bit as low-born as Huxley, he was now without doubt the elder statesman of British anatomy with a long-established and well-earned international reputation, even if he was personally little liked.

Owen had worked as a comparative anatomist of humans and apes since the 1830s, when Huxley was still a boy. He considered the chimpanzee and the orangutan to be most similar anatomically to humans, yet he argued that the difference that remained signalled a difference in kind, not degree, between apes and humans. He was troubled that anatomists who thought otherwise were basing their claims on the dissection of juvenile, not mature, apes and monkeys, since so many died young in European menageries and zoos. Owen thought that the more mature the individual ape, the more developed their characters: thus juvenile specimens had not yet developed what most differentiated them from adult humans.[7] In charge of the natural history collection in the British Museum from 1856, his position meant that he could readily secure the few skeletons of chimpanzees and orangutans that were to be had, and he had first refusal to dissect the animals held by zoos who had died, and eventually even to the first gorilla specimens brought to Europe. Still, the number of specimens of any kind of ape was vanishingly small. His published anatomies from the 1830s to the 1850s were based on the dissection of an individual here and there, adding to the common pool of knowledge from comparable German, American or Swedish dissections, enabling small samples slowly to substantiate larger claims about primates.

And so, Owen studied in detail the jaws, teeth and hands of many kinds of simians, and of the greatest age range that he could find. It was their skulls and then brains that became the most important organs through which he insisted on an irreducible difference between humankind and great apes. He cross-sectioned orangutan, human and exceptionally rare gorilla crania, and, combined with data published by anatomists in other parts of Europe and in the United States, he drew conclusions about mean cranial capacities of different species. In one careful study published in 1851, based on thirteen chimpanzees and about 450 human skulls gathered from one source or another, he concluded that the gap between ape and human brain size was always larger than any variation in human brain size.[8] Later in the 1850s Owen set to work on soft anatomical differences: the cerebral hemispheres and the posterior growth of the cerebrum including what was to become the unlikely celebrity cranial organ of the mid-nineteenth century: the hippocampus. This became the specific point of disagreement between Owen and Huxley. Owen thought that the development – the expansion – of the hippocampus in humans was the categorical distinction from the apes, in whom it was absent.[9]

Owen was not opposed to the broad idea of the evolution of species over time, but he did reject claims about human descent from earlier kinds of apes or, as was sometimes claimed erroneously at the time, from living species of apes. Yet the fact that Owen pursued his human-ape comparative anatomy *when* he did meant that his work unavoidably confronted Darwin's new theory of evolution by natural selection. Or, put another way, it was partly because of the pace and intensity of Owen's claims about archetypes, apes and humans over the 1850s that Darwin began to consolidate his own material and test his ideas out carefully with a new inner circle, of whom Huxley was one. Owen's review of the *Origin of Species* in April 1860 was harsh and undiplomatic and included criticism of Darwin's 'disciples', Joseph Hooker and Huxley, and the battle lines were drawn.[10] For Darwin in the *Origin of Species* not much rested on chimpanzees or orangutans – that book was yet to come. Thomas Henry Huxley, however, launched himself straight into the hotspot – primates. The frisson of confronting Richard Owen and displacing the dominant male was irresistible.

Hardly an expert on mammals, let alone complex primates, Huxley set himself a new task. He launched an unremitting challenge to Owen's claims on the facts of ape and human anatomy, gathering information from every direct and indirect source he could. Huxley did not have Owen's immediate access to dead apes and monkeys, yet this was growing with his hostile takeover of the field. He dissected a young orangutan held at the College of Surgeons, and was able to observe one of the very few gorilla skeletons held in London collections. He carefully measured chimpanzees, gibbons, lemurs, spider monkeys and marmosets. He began with hands and feet and digits in soft anatomy specimens detailing the muscles, and in skeletal specimens the measurements and proportions of bones. It was all crude and basic anatomy, but carefully done. He showed twenty-seven muscles in the lower extremities of *Nyctipithecus* or Azara's night monkey from South America, for example. He took notes on the width of the hand of one gorilla specimen, to take another example, measuring it at 5¼ inches across and 9¾ inches in length.[11] And as any anatomist would in 1860, he looked at jawbones and dentition, at crania and brains, aiming to both link and distinguish chimp, gorilla and human. This was comparative anatomy.

The primate brain became key for the demolition of Owen. In January 1861 Huxley challenged Owen's hippocampus claim by arguing that the organ not only varied considerably among humans, but also that some humans had no hippocampus minor at all: how then could this structure signal human uniqueness?[12] He enjoyed the annihilation, but the larger point was less about the hippocampus than about Owen's remnant insistence on God's design in natural law, even evolutionary natural law, and Huxley's utter rejection of any divinely ordained nature. There was a political implication too. Owen's position bolstered an Anglican establishment, while Huxley sought to challenge it in every possible way. Theirs was not an argument for and against evolution (as were other battles that Huxley was soon to fight); it was an argument for and against the hand of a Creator, even the shadow of a hand, in a total evolutionary mechanism.

The next three years set the tone for a lifetime of bitter intellectual battle between these men in the extended single-sex family that was British natural sciences. When Huxley joined the council of the Zoological Society in 1861, Owen left it. It was all choreographed for maximum effect: Huxleyan theatre. He undertook public dissections and public lectures to prove his point on ape brains and human brains. Sometimes, for the extra impact of seeming impartiality, Huxley orchestrated others to dissect on his behalf, rallying his own supporters. The anatomist William Flower – emerging as an authority on primates – performed the dissection on ape brains at the 1862 meeting of the British Association. Huxley's strategy was to sidestep Owen's influence and embargoes, and build new networks altogether, bringing and keeping onside other men who were disinclined to the old views, invested in various new theories of evolution and happy for Huxley to both lead the charge and carry the cost, depending on how it all unfolded. When the Scottish anatomist Allen Thomson dissected a young female chimpanzee in 1860 who had died in Glasgow, for example, he was struck by the similarity between the juvenile ape brain and the human brain, comparing it to that of a child between two and three years. He offered the information freely to Huxley but retreated into polite Victorian gentlemanliness, begging his friend not to draw him into any controversy.[13] Slowly but surely, Huxley won out, occupying more and more of the public sphere and private networking that was mid-Victorian natural science. Intellectually for Huxley, marine invertebrates were out, and primate skeletons were in. Delivering his lectures and then publishing his book *Evidence as to Man's Place in Nature*, a defiant Huxley knew that it would cause an eruption and deliver further blows to Owen.[14]

In researching his book, Huxley found apes in three places: the zoo, the library and the anatomical museums. A few captured creatures lived in Regent's Park, but in the 1860s most great apes in London only came alive on the page. Early published travel accounts by mariners and missionaries from equatorial Africa, from the Indies and from the Americas were the repository for longstanding European fascination with 'man-like apes' and what came to be called (and are still called) old-world and new-world monkeys. The opening chapter, 'On the Natural History of the Man-like Apes', was a text-based history of these early modern travel accounts, especially of African encounters with a range of primates. Here Huxley researched and wrote more as an historian of science than as a naturalist; he found textual versions of man-like apes everywhere. It is a kind of literature review, only fascinating.

The earliest mention of African apes that Huxley noted was the Italian explorer Pigafetta's sixteenth-century account of the kingdom of Congo, in which he learned that apes imitated human gestures, to local amusement: 'For although they be unreasonable Creatures, yet will they notably counterfait the countenances, the fashions, & the actions of men.'[15] This information itself was second-hand, drawn from a Portuguese mariner's notes. The accompanying de Bry woodcut (1598) drew Huxley's attention, and he reproduced it, suggesting with his nineteenth-century knowledge that the long-armed, tail-less apes were in fact chimpanzees. He read an account of an Englishman who 'lived in the woodes' in the kingdom of Congo for nine months, and another who lived in Angola for nearly eighteen years in the early seventeenth century. The latter described animals 'hairie all over, otherwise altogether like men and women in their whole bodily shape'.[16] Huxley relied on Samuel Purchas's four-volume compilation of all-world knowledge, *Purchas his Pilgrimes, contayning a History of the World in Sea Voyages and Lande Travells, by Englishmen and others* (1625), which reported a dangerous 'monster' in Africa that locals called 'pongoes'. Purchas recounted that the 'pongoes' could not speak, but that much of their behaviour approximated human sociability. And Huxley concluded that this particular 'Pongo' had died out 'at least in its primitive form'.[17]

The main fact he took from all this richness was that these apes were occasionally bipedal. But this is not what his early modern predecessors were interested in. Their most pressing questions were not anatomical, but behavioural. How did apes and monkeys behave with each other, and towards their human observers or captors? What was their capacity for communication, even language, and therefore rationality? Were animals really dumb?[18] Samuel Pepys thought perhaps not. In 1661 he had described an animal brought from 'Guinny', which he thought was a cross, a monster, between a 'man and a she-baboone'. Yet in this case, the monster seemed to approach Englishness. 'I do believe it already understands much English,' he noted in his diary.[19] Right through the eighteenth-century Enlightenment, and into Huxley's Victorian and evolutionary era, men and women wondered their way through this profound possibility. Some evolutionists were beginning to ask in Huxley's time: was language solely human or might it have evolved from animal noises?[20]

In 1699 the English anatomist Edward Tyson described a live, but sick animal he sometimes called a 'Pygmie' from Angola, and at other times an orangutan.[21] Tyson's Royal Society treatise *Orang-outang, sive Homo Sylvestris; or the Anatomy of a Pygmie compared with that of a* Monkey, *an* Ape, *and a* Man classified the 'pygmie' as close to mankind, 'the Nexus of the Animal and Rational', in the great chain of being.[22] Studying Tyson's book closely, Huxley did not communicate in his own work just how the continuum offered by this biblically orthodox vision of all nature from minerals to angels prefaced the continuum of primates he was himself constructing. His seventeenth-century predecessor had written:

> from Minerals, to Plants; from Plants, to Animals; and from Animals, to Men; the Transition is so gradual, that there appears a very great Similitude, as well between the meanest Plant, and some Minerals; as between the lowest Rank of Men, and the highest kind of Animals. The Animal of which I have given the Anatomy, coming nearest to Mankind; seems the Nexus of the Animal and Rational.[23]

Evolutionary theories certainly 'unchained' the old great chain of being, precisely because, in strict Darwinian terms, there was no evolutionary direction or progress.[24] But as we shall see in the next chapter, Huxley was not thinking so much about past versions of humans and apes – the 'descent of man' – rather, he was simply aiming to show how humans and other primates were similar or different in their present form. Remarkably, the skeleton of the very animal which Tyson had observed and dissected (in fact a chimpanzee) came into Huxley's hands a century and a half later, the precious remains passed down through familial and museological connections. Tyson's grand-daughter brought the skeleton as the strangest of dowries to her marriage to a physician, who later presented it to his local Cheltenham Museum. It was that museum which lent the skeleton to Huxley in the mid-Victorian years. The apes once described as in-between – like this 'pygmie' – were precisely what Huxley sought, in order to fill out his continuum of primates.

In anticipation of Huxley's insistence on the continuum of humans and simians, the great eighteenth-century taxonomist Carl Linnaeus classified humans, apes and monkeys together under a new order, *Anthropomorpha* – man-like – later changed to *Primates*, the nomenclature that stuck. Huxley was less than impressed with Linnaeus's accounts of African or Asian man-like apes, but that the Swedish naturalist recognized a 'third species of man' interested him greatly.[25] It was useful for Huxley to be able to authorize his own anatomies with reference to Linnaeus's much earlier work. And yet, while 'ourang-outans', 'troglodytes' and 'pygmies' were classified alongside humans, the latter primate – humans – was the only species the famous Swede could actually study in eighteenth-century life. Linnaeus's French counterpart, the Comte de Buffon, by contrast, owned a live chimpanzee and was able to examine an actual gibbon. His two-year-old chimpanzee was well known, apparently trained to walk upright and to take tea.[26] Despite the animal's flair for human domesticity, Buffon's intellectual inclination was still categorically to distinguish humans and apes.[27] A gibbon, or 'long-armed ape', was gentle, Buffon noted, receiving its food 'calmly', but like so many exotic creatures 'it was very averse to cold and wet, and did not live long after being brought into a foreign climate'.[28] This simian tragedy was repeated over and over.

It was certainly hard to catch these animals and transport them safely. A large orangutan was killed in the Dutch East Indies in 1781, pickled in brandy and sent to the Netherlands, but was shipwrecked en route.[29] Those apes who did arrive safely were tragically short-lived. For naturalists this was an ongoing problem. In the supposed great age of taxonomy, Linnaeus, Geoffroy Saint-Hilaire, Buffon and Lamarck all spread as much confusion as clarity about pongoes, orangs, pygmies or gibbons, compounded by the problem of distinguishing juvenile or adult skeletons – and occasionally a live animal – from different species. Even for the early nineteenth-century naturalists – Georges Cuvier, Johann Friedrich Blumenbach and Richard Owen – far more knowledge was secured from examination of prized skeletons than from observation of the animal in life. It was not until 1835 that the first adult chimpanzee skeleton was analysed by Owen.[30] The provenance of these rare individual skeletons – their relation to the text-based and philological genealogies of names, and of a particular animal's often brutal capture and transportation – is partly what Huxley sought to clarify in his history.

While Huxley used 'man-like' generically, accounts of individual apes in these old records would almost always ascribe sex, and attach considerable cultural (more than anatomical) significance to the maleness or the femaleness of the animal. For male apes, their 'man-likeness' and human hybridity was often asserted through an alleged sexual desire for female humans. One way of searching for man-like apes, in other words, was by investigating or inventing how they treated women. In the late eighteenth century Buffon had made sexual behaviour one of the defining features of a new ape, reports just then reaching him of a beast 'as tall and as strong as man, and equally as ardent after a woman as its own females'.[31] This was 'the drama of interspecies rape'.[32] Stories of male-simian abductions of human females were old by Huxley's time, and he sided with his contemporary the American missionary and naturalist Thomas Savage, that they were unsubstantiated and ought to be repudiated.[33] Nonetheless Huxley couldn't resist repeating them and extensively cited an account of shipboard disputes over a 'she-cub' Mandrill, transported from Guinea. It was an account that relied on race, gender and species conflations and confusions. Given to the charge of a slave, one of the mariners 'told the slave he was very fond of his country-woman, and asked him if he should not like her for a wife? To which the African very readily replied, "No, this is no my wife; this is a white woman – this fit wife for you".' The following morning the animal was found dead.[34] In ways critical and uncritical, what Anglophone humans came to call 'gender' and 'race' have long accompanied the observation of primates. In wildly fanciful early modern accounts *and* in the most distinguished academic primatology, from scenes as misogynist as it is possible to be, *and* in canonical feminist studies through the twentieth century, human observation of other primates has been one of the richest sites (for better or worse) for the study of intersecting sexuality and gender and race. The ape's reflection is continually mirrored.[35]

THE GORILLA

By 1863, when Thomas Henry Huxley's *Evidence as to Man's Place in Nature* went on sale, the Victorian public were thoroughly primed with, as well as confused by, the cross-cultural and cross-species conceptual mess that constituted the 'man-like ape'. Chimpanzees and orangutans were familiar enough, and important in the scientific debates of the 1860s. But another great ape – the gorilla – had only recently come into European view, confounding anatomists all over again, as well as the simian-consuming public. This order of the animal kingdom quickly assumed a key position in evolutionary debate and in comparative anatomy, yet not a single live creature had been noted by any of the naturalists of the day. The one possibility was another 'Jenny', born in the Congo, captured as a juvenile and purchased by the Wombwell family, who operated travelling menageries through northern England in the mid-1850s. 'Lately kidnapped in the sunny regions of Africa', this Jenny was exhibited in rooms in chilly Scarborough, which is where Yorkshire naturalist and taxidermist Charles Waterton saw her in life, acquired her in death and preserved her.[36] She still sits in the Wakefield Museum. She lived her short English life as a 'chimpanzee', but Richard Owen and others later deemed her to have been a juvenile gorilla, based not on live observation, so far as we know, but on her crude taxidermy.[37]

Richard Owen had some of the earliest gorilla evidence and experience to go by. He was in direct communication with the American Thomas Savage, who in 1847 had stumbled on the skull of a massive ape by the Gaboon River, West Central Africa, and who later, with naturalist Jeffries Wyman, named it *Troglodytes gorilla* after the genus for the chimpanzee offered by Geoffroy St Hilaire in 1812. Owen was quick off the mark. He published an account of the skull in 1849, following Savage's initial troglodyte classification.[38] It was not until 1852 that the separation was made clear, and the genus *Gorilla* established.[39] The Americans thought it likely that this was the animal which had been called a 'Pongo' in the early 1600s. Huxley later agreed this was the case. By 1853, Owen had dissected a young male, preserved in spirits of wine.[40] And in 1855 and 1859, he detailed further accounts of the entire gorilla skeleton, 'and in these is always of the opinion that the gorilla is the nearest akin to man'.[41]

The gorilla became very famous very quickly. By the time Huxley published his book in 1863, he could already presume that his readers would be familiar with a creature called the gorilla, 'its present well-known appellation'.[42] However, people on the street knew about them less because of Owen's skeletonizing than because of the entrepreneurial visibility of someone who had seen gorillas in the flesh and in equatorial Africa. The French-American explorer Paul du Chaillu was sent on an African expedition by the Philadelphia Academy of Natural Sciences, observing gorillas, hunting them and eventually selling and shipping them, stuffed or pickled, all over the world. There was a great market. Everywhere, on the back of Owen, Huxley and Darwin, gorilla specimens were used to demonstrate likeness or dissimilarity between humans and apes. Du Chaillu had early on sent some skulls to Richard Owen and in 1861 brought a taxidermied gorilla to London. There was at least one attempt to ship a live gorilla to London from his second expedition in 1863–5, but it died en route.[43] In Scotland, three gorilla crania received from M. du Chaillu were exhibited to the Royal Society of Edinburgh.[44] In Melbourne, Australia, some of his stuffed gorillas were bought and displayed for public consumption. The lesson there was against Huxley and Darwin: the gorillas interpreted to show 'how infinitely remote the creature is from humanity'.[45]

Du Chaillu both created a gorilla market out of his expeditions and capitalized on the sudden ape-talk that surrounded the *Origin of Species*, publishing his *Explorations & Adventures in Equatorial Africa* in 1861. The gorilla was the book's unmistakable opener. A foldout frontispiece shows an oversized and upright beast, 'monstrous and ferocious', commanding the book as much as it commanded public attention (figure 6.2).[46] Du Chaillu detailed his confrontations with wild beasts, aggressive males standing on two feet that stared him in the eye, raising themselves up and beating their chests. But this was not all. His first hunt made him feel like a murderer, because the animal was like a hairy man, running away on two legs, with a cry that was too human.[47] For the Mbondemo people with whom he was travelling at this point, north of the Gaboon River, particular gorillas *were* men; or more specifically, they were possessed by the spirits of individualized deceased men. In these special possessed beasts, who by local tradition could not be killed, he relayed that 'the intelligence of man is united with the strength and ferocity of the beast'. With his later Mbenga companions, however, another gorilla was shot and then eaten. The reverse would not have been the case, for gorillas eat vegetables, roots, fruits and termites, not flesh. On the back of his dramatic and blow-by-blow, real-time account of gorilla confrontations and killings, du Chaillu's book offered anatomical information as well, and the benchmark of his primate knowledge was Richard Owen. He agreed with Owen that the gorilla is more human-like than the chimpanzee, walking more often on two legs. He repeated Owen's ordering of similarity to humans, receding thus: gorilla, chimpanzee, orang, gibbon.[48]

Huxley took exception to du Chaillu's work. We've heard all about gorillas for the last fifteen years, he wrote ungenerously. 'Once in a generation,' Huxley explained, 'a Wallace may be found physically, mentally, and morally qualified to wander unscathed through the tropical wilds,' referring to Alfred Russel Wallace's formative South-east Asian immersion.[49] But he was not going to accord this status to du Chaillu, an adventurer above his intellectual station. In short, he was not a scientist, and his information and methods cannot be credited as evidence.[50] Huxley was crudely flexing his professional muscles, making sure that, as he pushed out Owen, only he would become the guardian of the new primate business. Paul du Chaillu was an easy target, certainly not

Figure 6.2: Frontispiece to Paul du Chaillu, *Explorations & Adventures in Equatorial Africa* (1861).

one of the new professional scientists whom Huxley was championing. And yet du Chaillu knew more about gorillas than all the European and American naturalists put together. It proved simple enough to critique du Chaillu's florid account, but the fact is he had spent ten years travelling up and down rivers and mountains, without any support group (unlike, for example, the later Stanley), mostly befriending locals whose languages and customs, including hunting customs, he listened to, learned from and joined in with.

In West and Central Africa gorillas were alive and hunted and even possessed. In Europe they were, as yet, only dead. Nonetheless, with the creature's new fame, countless satirists, caricaturists and poets breathed life into the semi-mythical beast. The simian was made to speak. In fact, 'Gorilla' was a poet, nominally putting the famous evolutionary debates into verse.

> Am I satyr or man?
> Pray tell me who can,
> And settle my place in the scale.
> A man in ape's shape,
> An anthropoid ape,
> Or monkey deprived of his tail?

And later

> That apes have no nose,
> And thumbs for great toes,
> And a pelvis both narrow and slight;
> They can't stand upright,
> Unless to show fight,
> With 'Du Chaillu,' that chivalrous knight![51]

This first-person identity crisis was offered by 'Gorilla of Zoological Gardens'. This, too, was mythical. There were no gorillas in the Zoo in 1861. For the mid-Victorians, they were still to be found only in books and in anatomical collections. For the late Victorians, if they were very fortunate, a living gorilla might be observed. One was captured by a German expedition in 1874, arriving in the Berlin Zoo in 1876. He was a two-year old male, called Mpungu or, once again, Pongo, and, after being lent to London for a short time, died on the return journey to Germany in November 1877.[52] At the same time, a large female chimpanzee named Mafuca was kept in Dresden, but was she really a chimpanzee, some asked. The most learned naturalists of anthropoids at the time wondered whether she might have been a cross between a gorilla and chimpanzee, a species enigma.[53] Over the 1880s and 1890s there were a few gorillas brought successfully to Germany and France, all short-lived. One went to the Zoological Society in London in 1883 but survived only for a few months. A second, in 1897, also died quickly. By the early twentieth century, and despite these zoological failures – tragedies – gorillas in African *situ* were increasingly sought and prized. Hunting parties began to be commissioned by the world's great natural history museums, seeking more and more specimens to transport live if they possibly could, and if not, to kill, investigate and mount for display, in the manner of Du Chaillu's beasts. The Englishman Thomas Alexander Barns, for example, explorer, big game hunter, naturalist (it was easy enough then to be all three), travelled to the Virunga mountains in the early 1920s, not many years before Julian and Juliette Huxley, but unlike them he shot numerous gorillas: little wonder the Huxleys did not see any.[54] Others specialized more in live captures and transportations, and they were increasingly successful, the animals living longer in captivity. Often these were lowland gorillas, and this created one of primatology's research problems over the medium term: captive gorillas were often lowland, while those that went on to be studied extensively in the wild in the mid-twentieth century were mountain gorillas.

JULIAN HUXLEY'S SIMIANS

No primatologist, Julian Huxley nonetheless had unequalled opportunity for a few years to watch monkeys and apes quietly and closely, out of hours in the London Zoo. He was captivated by a juvenile gorilla, eighteen months old, who would trace the outline of its own shadow with its finger. Perhaps this was one of those excruciating zoo moments where a young isolated animal searched vainly for more of its kind, wishing or willing a shadow to become real. Julian took the action as a kind of art, or at least an artistic initiative, that he thought had not been observed in any other anthropoid.[55] Much later he wrote about the paintings produced by the London chimpanzee Congo, who drew with crayons and mixed paints. Congo had not reproduced anything 'remotely representational or symbolic', but Julian still interpreted this capacity as part of the primate evolutionary story. The 'sudden outburst of art' in the upper palaeolithic period must have developed from some prior base of skill or initiative. It all became more comprehensible, he argued, 'if our ape-like ancestors had these primitive aesthetic potentialities' to which could be added humans' unique symbol-making capacity.[56]

A century earlier Darwin had also watched the extraordinary animals kept at the Zoological Gardens, how they reacted and responded to their strange captive normal. The fascination lay in how the animals moved, sensed, responded and expressed themselves, not how their crania measured up. Both Richard Owen and T. H. Huxley had similar access to the animals kept by the London Zoological Society, but it seems they were too busy expressing anger and outrage to and about each other to think about how their dead apes might have acted and felt in life. Darwin, by contrast, climbed into Jenny the orangutan's cage. Sometimes he brought a mirror to London's orangutans, and he noted, with excitement: 'Both were astonished beyond measure at looking glass, looked at it every way, sideways, & with most steady surprise.'[57] He observed how the chimpanzee and the orang were frightened by thunderstorms; how a particular monkey wept tears, stricken by some grief; how angry baboons would strike the ground with one hand, 'like an angry man striking the table with "his fist"'.[58]

The anger that Darwin documented among baboons in the mid-nineteenth century was to become far more than that in Julian Huxley's time. One of the great leaps taken by the Zoo in the 1920s was to relocate some primates out of cages to new, open-air enclosures that simulated, if poorly, natural environments. One of the first of these was named Monkey Hill, designed and built over 1924 and 1925 and still a great drawcard over Julian's term as secretary, after 1935. Monkey Hill was promoted at the time as bringing baboons into 'natural surroundings'. But something grim in nature was also unleashed with the baboons' newfound freedom, or else had been contained in their former enclosures: they started to kill each other. This intra-species violence was triggered by the strange sex and age ratio of the artificial population crowded into Monkey Hill, vastly more adult males than adult females. It went on for years: the 'baboon massacre' some called it. Resident zoologist Solly Zuckerman, whose job it was from 1928 to dissect animals who died or were killed in the Zoo, suddenly had too many specimens on his hands. Owen and Huxley would have been delighted, were it 1860, and were comparative anatomy still the discipline *du jour*.

Solly Zuckerman was an intellectual and methodological hybrid of the Huxley grandfather and grandson when it came to these apes. On the one hand, by training and occupation he *was* an old-school anatomist, out of time, belonging to T. H. Huxley's world.[59] On the other hand, like Julian – in fact after Julian's ethological example – he was fascinated by behaviour, and even 'psychology'. And what extraordinary behaviour it was, playing out before them. Zuckerman observed and interpreted the violence and the slaughter between males, and between males and females, a lethal dominance and submission triggered, he concluded, by competition among males for reproductive females, and by hormonal changes in the latter. Zuckerman wrote up the simian war in *The Social Life of Monkeys and Apes* (1932), a classic of inter-war primatology.[60] This was not a social life that many wanted to accept or acknowledge, however, especially at a time when human males had also been killing each other on unsurpassed scales. Nonetheless this inter-war generation took the baboon massacre, and Zuckerman's interpretation, as indicative of both a latent and, recently, manifest human inclination towards intra-species killing.[61]

Although Solly Zuckerman had moved to Oxford by the time Julian Huxley took over as secretary of the Zoological Society, for decades they exchanged a great deal about primates – including humans – and about animals' social life. And they exchanged actual animals. 'Everything is arranged,' Huxley told him in the summer of 1936. 'We will house your four monkeys in a place where they can be kept together and studied by one of your pupils.' It was crude animal commerce: 'The exchange of the old Mandrill for the young Drill is agreed.' At another point two Huxley chimps were taken to Oxford to participate in a penny-in-the-slot-machine experiment.[62] Apes and monkeys were increasingly subject to zoologists' experimentation, and over the twentieth century this merged with an earlier tradition of domestication, animals captive not just in zoos, but also within households and families.

In 1917 a gorilla was bought in London and taken 'home' to a Gloucestershire village. Julian recalled seeing the famous 'John Daniel', who came to town regularly 'with his mistress' – his owner – 'in her company in the street, or in a cab'.[63] When living at 15 Sloane Square, John Daniel would regularly be taken to the London Zoo, where he was shown for the day and then went home again. This particular gorilla's life and death trajectory spanned every kind of being-in-the-world from a wild Gabon infancy, to a literal childhood in Gloucestershire schools, gardens and a bedroom where he reportedly made his own bed. Then, as he got too large and was sold on, he suffered the indignity of Barnum and Bailey's circus, then New York zoos and finally preservation and display in the American Museum of Natural History in New York. All this in a short gorilla life of four and a half years. John Daniel was a species crossover, and a crossover from nature to culture. He was, as one article put it, a gorilla 'in civilization'.[64]

Towards the end of Julian Huxley's life, this kind of familial domestication of apes was more common and more explicitly experimental than an idiosyncratic pet-curiosity. A new breed of psychobiologist and later sociobiologist inquired into anthropoid life. Among them were Yale-based Robert and Ada Yerkes, whose encyclopaedic study reproduced a photograph of an infant gorilla, one month old, being breastfed by a woman, a 'foster nurse'.[65] Over the 1950s, 1960s and 1970s, there was a great escalation of behavioural experimentation on chimpanzees especially, often in laboratory spaces. Attempts to teach them to talk and to paint were efforts to refine understandings of just where the human/animal and culture/nature line lay, and whether it could be shifted. Often enough the work was with malleable juveniles, their own parent replaced by the culture-shaping human family. This was the high era of behavioural studies, thanks in part to Julian Huxley's ethology, and so there was also a trend to research how various apes fared inside – and as part of – the quintessential institution of post-war civilization, the suburban, nuclear household. In the mid-1960s, for example, Julian was in contact with Julia Macdonald from Pasadena, who was writing a book about her baboon children. She had 'adopted' a young female 'as part of our family'. She later adopted a mate, but he only lived domestically for a short time.[66] When Aldous and Julian Huxley visited the San Diego Zoo together, they were pictured with a clinging baby chimpanzee, a strange version of a cross-species family, if a fleeting one (figure 6.3).

At the same time as such experiments in simian humanity proceeded, highly visible studies of primates in the wild were being pursued, ultimately far more influentially and even canonically. Pondering London Zoo's Guy, Julian expressed his desire to see a 'breeding colony where their strange natures could be carefully studied'.[67] George Schaller, then Dian Fossey, and then Jane Goodall improved on this idea and went to the lands of the great apes themselves, some of the same remote mountains through which Julian and Juliette had trekked in 1929.[68] Yet these were weirdly inverted studies – almost perversely so – sometimes infantilizing and sentimentalizing the animals as child-substitutes like Elsa. Knowing ourselves, through knowing apes, this was still 'simian orientalism'.[69] Yet what came through over time, in terms of ways of knowing, was not Richard Owen's and T. H. Huxley's observation of the dead animal, but increasingly refined cultural, behavioural and psychological study of the live one. It was a methodological victory of culture, of mind and of emotion over bones and brains.

Julian Huxley's ethology influenced this version of primatology directly. The mid-twentieth century's original primatologist-in-the-wild, George Schaller, attributed to him the key methodological breakthrough, to observe gorillas with full attention to their 'culture'. Schaller quoted Julian Huxley's observational philosophy in his book *The Year of the Gorilla* (1964): 'It is also both scientifically legitimate and operationally necessary to ascribe mind, in the sense of subjective awareness, to higher animals'; to investigate, in other words, their curiosity, anger, alertness, affection, jealousy, fear, pain and pleasures.[70] Julian's own influence in this matter was derived not from his grandfather's man-like apes, but directly from Darwin's observations of the London Zoo primates' expression of emotion. In nominating anger, jealousy, fear, pain and pleasure, Julian was recounting the table of contents of Darwin's book.

The following year, Jane Goodall went to Tanzania to work with Louis Leakey, already famous for his palaeoanthropology of human

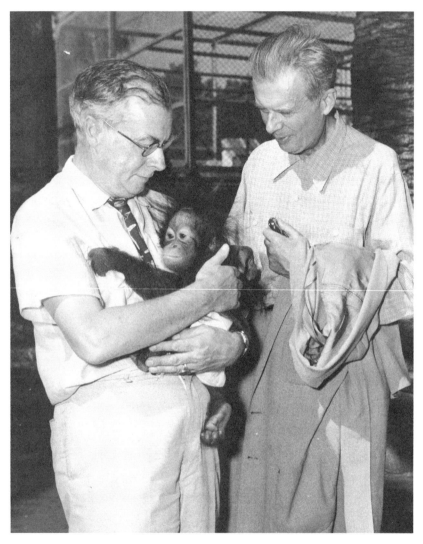

Figure 6.3: The fraternal primates Julian and Aldous Huxley, with
a young relative, at the San Diego Zoo, California, 1956.

origins, and already well acquainted with Julian and Juliette Huxley. Leakey was also thinking about human connections with chimps, directing Jane Goodall in that direction, a mid-twentieth century rendition of T. H. Huxley's search for evidence as to man's place in nature. Yet Goodall, like Schaller, was the very opposite of a palaeontologist. Her Cambridge PhD was to be in ethology – Julian's intellectual world – supervised by the primate ethologist turned psychological behaviourist Robert Hinde.

Jane Goodall and Julian Huxley corresponded regularly on the chimpanzees of Gombe. Unsurprisingly, he was fascinated with her work, and well before it became famous via *National Geographic*. For her part, in one letter she glowed to him over her 'baby's' first steps, as if Julian were a kindly uncle. She was also canny about the value of his international conservationist profile. By the 1960s, Julian's African visits were carefully orchestrated to maximize media impact, something on which Goodall capitalized from the beginning. She sought copies of scholarly work from Julian, and reported progress to him, sometimes sentimentally but more often scientifically. 'We have discovered that play between adults is far more common than was previously supposed. Especially between mature males and the old mother.'[71] She was making a film on chimp behaviour, after the genre that Julian had popularized. Julian was just as interested in return. He sought an extract of her film for screening at his Royal Society symposium on ritual in 1964. Some of her chimpanzees' activities, he wrote to her Cambridge supervisor Robert Hinde, 'really do seem to bridge the gap from primate to men'.[72]

If George Schaller attributed an ethnographic method directly to Julian Huxley, Jane Goodall was indebted too. She expressed that affectionately by naming one of her chimps 'Huxley', and certainly not after Thomas Henry. In her original observational group of fifty chimpanzees, she spotted a particular male: 'I became acquainted with Huxley, a tall slender male with a distinguished and slightly professional air, and named him for Sir Julian Huxley, the biologist.'[73] She followed 'Huxley' closely, like all the troupe, though he was never one of her central Gombe players. Julian was enormously pleased with, and flattered by, this latest addition to the Huxley family. In their conversations about bringing Julian and Juliette to Gombe, she very much wanted to introduce them to 'Huxley and co.'[74] But Huxley died, Goodall estimated at forty-one years, from a wound inflicted by another male.[75]

Julian routinely defended Jane Goodall against the mean-spirited criticisms from the primatology orthodoxy. Solly Zuckerman especially was intent on bringing her down, partly because he wanted to bring Louis Leakey down, partly because she was a woman-scientist then acquiring significant attention. Julian chastised him: 'she deserves every possible encouragement'.[76] Zuckerman chaired a panel on which she spoke at a Zoological Symposium in April 1962 and diminished her contribution in a letter to Julian: a 'ridiculous tendency' to generalize from limited observations, 'namely the meat-eating, of which she saw only three examples in some 600 continuous hours of observation'. Julian neither understood nor agreed. Indeed he thought her methodology perfectly sound. She reported simply what she observed, nothing more nor less. T. H. Huxley himself could not have faulted her empiricism. 'She was recording her actual observations on a particular colony of animals living in a wild state. Whether she saw meat-eating on one occasion or twenty occasions is to me quite immaterial: it is a new and highly surprising fact.' They made up over drinks at Julian's seventy-fifth birthday the following month.[77]

Julian was inclined towards Jane Goodall's work not just because it was behavioural, but also because she was partial to the observation of ritual, his grebe-inspired idea from so many decades before. 'She has also got some exciting new facts about "ritual dances" by adult male chimpanzees.' Indeed he couldn't stop writing about her to Solly Zuckerman. 'She says she has some wonderful photos this time, & also v interesting new results – notably having made real personal friends (!) with 3 adult Chimps, who take bananas from her hands.'[78] Zuckerman wouldn't have any of it and resorted to banal comments on gender and jealousy. 'If Miss Goodall has started making friends with adult male Chimpanzees we shall have to look at her observations on the social life of the species as representative of abnormal social life. What do the lady Chimpanzees think of all this?'[79] As a primatologist who specialized in gender and jealousy he should have known better and might, indeed, have looked rather more closely at himself.

It was both neat and incongruous that Goodall received the 2002

Huxley Memorial Medal, awarded annually by the Royal Anthropological Institute of Great Britain and Ireland since 1900.[80] It was neat and appropriate for the discipline of 'anthropology', since Goodall's ethology at Gombe *was* the ethnographic study of primate culture. It was ironic and incongruous because the medal honoured Thomas Henry Huxley, who would never have understood ethnography in this way, nor even the kind of animal behavioural study she favoured. It was his grandson who inspired and understood that.

PRIMATE HANDS

As Henrietta Huxley promised the young Julian, he did inherit much of his grandfather's library. One of the treasured books on his shelf in Pond Street, now sequestered in Rice University's archives, is Charles Bell's *The Hand, Its Mechanism and Vital Endowments as Evincing Design*, published in 1833 as the fourth of the Bridgewater Treatises. It's an important book, if a strange one to have ended up on Julian's shelves, since it epitomized the natural theology that T. H. Huxley spent his life annihilating. Following Paley's foundational natural theology, it presented evidence for the human hand's design as divinely purposeful. But the book started with monkeys not humans, those 'quadrumana' with not two but four simian hands, or so Bell thought. His chapter on the 'Comparative Anatomy of the Hand' opens with an illustration of a chimpanzee, demonstrating that the free motion of the simian hand is designed for swinging from trees. The human hand may be similar to its substitutes – fins, wings and paws – but in it there is 'perfection'. 'Man had hands given to him' by the Creator, as the wisest creature.[81]

It was the kind of statement that made T. H. Huxley recoil and then, in a split second, attack. At the same time, he was also a hand expert, well versed in Bell's work, and fully aware of the species significance of that celebrated opposable thumb. He described it in 1863: 'its extremity can with the greatest ease, be brought into contact with the extremities of any of the fingers; a property upon which the possibility of our carrying into effect the conceptions of the mind so largely depends'.[82] Julian followed through with this Huxley tradition, making the film *Fingers and Thumbs* with the Royal Zoological Society.[83]

As it happens, we have close records of the Huxleys' own opposable thumbs. One is a cast of Thomas Henry Huxley's right hand, stored away at Imperial College London (figure 6.4).[84] The other is a series of arresting photographs – portrait-like – of Julian's hands (figure 6.5). They were taken when he was middling old; it is not entirely clear when, and initially it is not entirely clear why.

It turns out that primatologists, evolutionary biologists and natural theologians are not the only experts on hands. So, too, are palm-readers. In the mid-1930s, when Julian was living in Regent's Park and Aldous was living in Sanary-sur-Mer, where he had recently written *Brave New World*, primate hands suddenly became fascinating and meaningful for the wider Huxley family via palmistry. In the strangest way, it all led back to Julian's London Zoo apes.

Over the 1930s Aldous and his first wife Maria were enough taken with palmistry to be studying it, Aldous especially intrigued with this particular somatic way into character or personality, if not the future. In the community of literary exiles in Sanary-sur-Mer, they befriended Charlotte Wolff, a Jewish refugee from Germany, physician and psychotherapist, analyst of hands as well as minds. Aldous cajoled his English and French circles to present their hands to Wolff, and he enabled her first book, *Studies in Hand-reading*, published in 1936 by Chatto & Windus, including some very famous handprints and palm-readings indeed: those of Romola Nijinsky, Man Ray, George Bernard Shaw, Maurice Ravel, Virginia Woolf and T. S. Eliot. It was as if Lady Ottoline Morrell's Garsington was being reunited. It was. There are Morrell's own hands for all to see, as bare and revealing as her Edwardian nudes. Aldous Huxley's own hands featured too, as did those of his brother.[85]

Wolff compared, clearly with some delight at the unique case, the hands of Aldous and Julian Huxley, nominally a man of letters and a man of science. Aldous's prominent Finger of Saturn, 'with its curiously large, spatulate upper phalange, reveal[ing] a predilection for rationalism', was a sign of scientific ability. 'The ascending lines of the Mount of Mercury show that it is the natural sciences which interest him, and that for these he has an inborn talent.' And yet, Aldous's

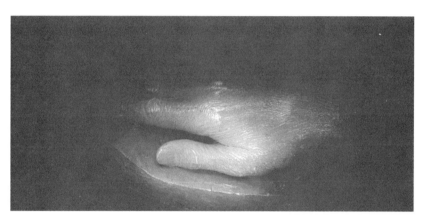

Figure 6.4: Cast of a hand said to be Thomas Henry Huxley's.

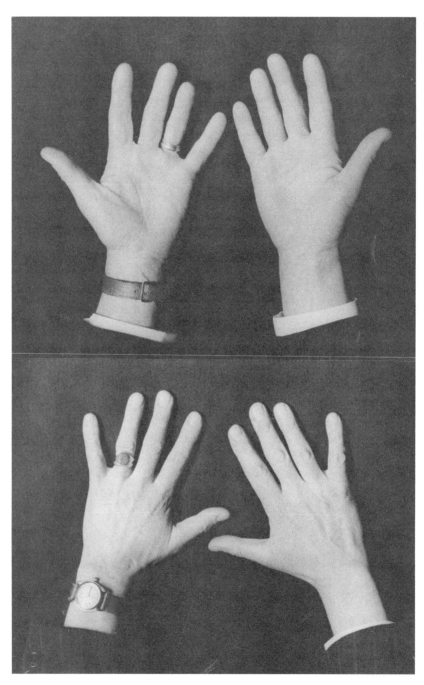

Figure 6.5: Julian Huxley, 'Hands'.

Zone of Imagination was where the concentration of emphasis lay in his hands. It was, she noted so closely connected with the Zone of Collectivity (family) that 'the two cannot be definitely separated'. Note the large number of lines in the Imaginative Zones, she guided her reader, with diagrams. And Julian?

> The upper part of the Mount of Venus shows sublimated eroticism, the need for affection or, to use the jargon of the psychologist, it signifies on the one hand the bond between mother and child, maternal affection, and on the other hand love of nature and tenderness for the defenceless.

Both brothers' hands show an 'enquiring mind, combined with particular sensibility of touch'. Julian's hands revealed 'unusual creative unrest and spiritual dissatisfaction. He is more than a scientist.'[86] Aldous was engrossed and fascinated, and any number of the extended Huxley family found themselves in Charlotte Wolff's room having their hands read and their palms printed for her records.[87] Julian indulged too, but for him the question of hands went in an altogether different direction: towards simians and evolutionary biology. For Charlotte Wolff, the medically trained palmist, that connection also made every sense.

Was Charlotte Wolff a charlatan, as T. H. Huxley might have put it? What would the famous Victorian anti-spiritualist and exposer of the fraudulent have thought? The answer is not so obvious. Indeed Wolff herself nominated her own intellectual genealogy in comparative anatomy of the nineteenth century, almost always citing Charles Bell and T. H. Huxley. The latter, indeed, 'may be called the father of these studies'.[88] In her books and articles for more than reputable scientific journals, it was Thomas Henry Huxley who ignited interest in the comparative anatomy of the extremities of humans vis-à-vis simians. And she sought to throw some of her own light on the study of evolution of primates. In 'A Comparative Study of the Form and Dermatogryphs of the Extremities of Primates' published in the *Proceedings of the Zoological Society*, now under Julian Huxley's remit, Wolff printed, read and compared the hands and feet of every kind of simian she could find: lemurs, baboons, gibbons, squirrel monkeys, orangutans, chimpanzees, gorillas and more.[89] And where did she find them, but in Julian's London Zoo?

Julian was Charlotte Wolff's entrée to the simian world. He secured her good access to his monkeys and apes in order to take handprints, footprints and in some cases fingerprints. He instructed the keeper of the primates to assist her, to advise on which chimpanzees were safe and which were not. As for the juvenile gorilla then in the Zoo, the keeper was firm: 'I do not allow anybody to go in with it for fear of infection, but if you will arrange a day, one of his keepers will take the prints for you under instruction' (figure 6.6).[90] Her disordered albums were preserved and still show the inky prints of highly individualized animals. No more 'Jenny', she took the hand-portrait of 'Daisy' and 'Girty', for example, both two years old, 'Fifi', fourteen years old, and 'Boobou', sixteen years old.[91] Wolff's simian palm- and finger-prints are a record as distinct and personal as her prints and readings of the Huxley brothers, their wives and their children. The commemoration of long-dead animals, these sequestered albums hold an intimate history of touch between species, and between unlikely ways of knowing.

Under Julian Huxley's watch, Charlotte Wolff engaged in simian palm-reading. She read the palm of Peter, a seven-year-old chimpanzee, in the same manner and with the same method as she read the hands of Julian and Aldous Huxley. Just like any human, 'the mentality and emotivity of chimpanzees have a correlation with the lineal composition of their hands.'[92]

> Peter, who is seven years and six months of age, has a hand with particularly few lines and an arrangement which is distinctly of the juvenile type. As the juvenile chimpanzee, which is not yet fully under the influence of the sexual emotions, is more intelligent that the adult one, the unusual intelligence of the chimpanzee Peter may perhaps be explained as a result of delayed emotional development.[93]

It was all a mélange of Darwin's expression of the emotions, T. H. Huxley's comparisons of old- and new-world monkeys, Aldous Huxley's other-worldliness and Julian Huxley's zoo-keeping, overlaid with early twentieth-century psychoanalysis. Again and again, Charlotte Wolff placed herself directly in a Huxleyan intellectual line. In a French article on Mok, London Zoo's then most famous gorilla, she began with reference to *Evidence as to Man's Place in Nature* and T. H. Huxley's study of the skeleton and musculature of the gorilla.[94]

Figure 6.6: Moina, a lowland gorilla, whose prints are being taken by a keeper, 29 June 1937, for Charlotte Wolff's research.

And while the comparative anatomist had worked mainly from skeletal material after an animal's death, when all those telling lines were gone, she noted that her own work was based on the sensory extremities of *living* animals. But not always.

When Congo-freeborn gorilla Mok died in a Regent's Park cage in 1938, Wolff pressed once more on Julian as zoological host. Together they obtained prints of Mok's hands, two days after his death (figure 6.7). Even the keeper had been unable to secure them safely from Mok in life. She took hand- and footprints, finger- and thumbprints from the dead beast and reprinted them, with analysis, in the *Proceedings of the Zoological Society of London*.[95] From dead Mok, nine years old, she confirmed that the form of the gorilla hand resembles more closely the human hand than that of any other anthropoid. She detailed her findings with as much anatomical diligence as T. H. Huxley ever did. The crease lines in the palm – her specialty – were analogous to those in the human hand, with the exception of two transverse lines that join and cross the entire hand.[96] Common in non-human primates, this configuration is rare but occasional in humans and is still sometimes called the simian line. Julian was as intoxicated by this research as he was by the gorillas themselves.

In 1947 the Zoo's most famous creature ever arrived to replace the dead Mok. Julian couldn't take his eyes off the great ape. 'When I saw Guy at the Zoo this summer, I could hardly tear myself away.' Guy was the finest animal he had ever seen, domesticated, captive or wild. But when Julian received that special primate return gaze, Guy managed to stare the ethologist down: 'Magnificent in his reserve and silent strength, he gives you a look of sombre dignity that makes you feel in some real sense his inferior.'[97] Julian, the great observer, felt watched. And yet eye contact alone was insufficient. He found himself wanting to talk with Guy. 'If we could only get through the barrier of non-communication set up by their inability to talk, what strange areas of ~~subhuman~~ mind we could explore.' Julian struck through 'subhuman' in his typescript for a BBC talk on Guy, acknowledging the primate's mind as equivalent to his own, not a greater and a lesser, a human and a subhuman. The expression of emotion projected by this particular man onto this particular animal was plain.

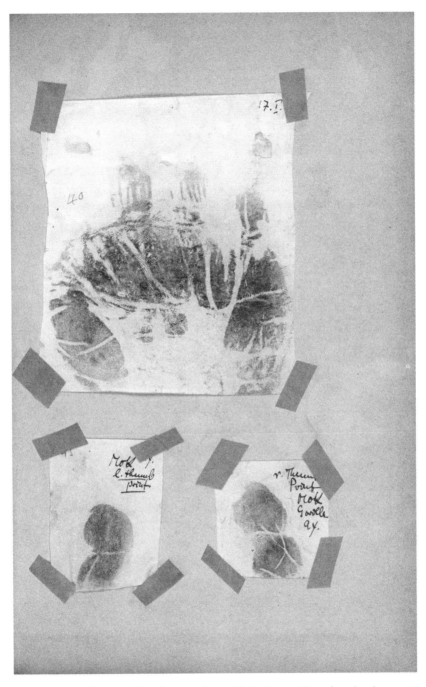

Figure 6.7: Palm- and thumbprints from Mok the gorilla, after death, 1938.

The photographs of gorillas kept in Julian Huxley's papers signal his captive attention both to individual creatures and to the species. He kept a copy of Suschitzky's magnificent portrait (figure 6.1) taken at the London Zoo in 1958. In many ways, it was framed by the convention of human portraits; in the convention, indeed, of the endless John Collier paintings of the Huxley primates. In fact, Suschitzky *was* the portraitist of this generation of Huxleys, over many decades photographing Julian, Juliette and their sons, Anthony and Francis.[98] The year that he captured Guy the gorilla, 1958, he also captured the brothers, Julian and Aldous. It is another intimate portrait of fraternal primates, Julian characteristically talking and Aldous's troubled eyes cast down, equally characteristically listening (figure 3.6). It's a moving image, especially when one knows the brothers' past. Yet it has none of the strength of the portrait of Guy. In this image, we feel emotion in man and animal with the same kind of power that Charles Darwin experienced when he climbed into Jenny the orangutan's cage, on occasion with a mirror. In this case, with a keeper's help, Suschitzky passed his own reflecting machine – his camera – through the bars, to catch not the enclosing iron itself, but, far more hauntingly, its shadows. Julian, like everyone else, was struck by the shadows of the bars across Guy's face, across his very self. Perhaps this signalled how prison bars and cages were disappearing from zoo design and aesthetics – Monkey Hill anticipated this. The sign of imprisonment and captivity was being replaced by devices of apparent clarity and freedom – the suspended disbelief of a clear-glass barrier, or of moats, zoological ha-ha walls, the kind of 'sunk fence' that had been built instead of high walls at James Huxley's Barming Heath asylum all those years before.[99] But it was all disingenuous, and for human benefit only. There was no escaping the fact that Guy was captive and contained.

Many years ago, the insightful art critic John Berger wrote an essay, 'Why Look at Animals?' He cut through mountains of scholarship on zoos and beasts with the clarity of one simple observation: in the wild animals stand and watch us, nervous, ready to flee or to fight, never letting us out of their sight. In captivity, however, despite being close up and proximate, the last thing animals do is hold our gaze: 'nowhere in a zoo can a stranger encounter the look of an animal. At the most, the animal's gaze flickers and passes on. They look sideways. They look blindly beyond.'[100] Yet there is Guy, staring down Julian Huxley and still holding our gaze with commanding power so brilliantly caught by Suschitzky. Defiant in the moment, 1958, Guy unknowingly defied Berger's insight. It is an excruciating image of life, especially when we try to hold Guy's gaze in death, now in his glass case in South Kensington. Neither on all fours nor standing proud bipedally as man-like ape, Guy sits on a fake rock, a kind of miniature Monkey Hill, and stares blankly. Berger was right even about Guy: he no longer looks at us, but past us into a beyond (figure 6.8).

Figure 6.8: Guy, a western lowland gorilla,
the Natural History Museum, London.

PART III

Humans

7
The Man Family: *Sapiens* of the Deep Past

The patriarch of one idiosyncratic family became not just the patriarch of science but also of what he called *Anthropini*, the Man Family. The 'science of man' was under construction over T. H. Huxley's lifetime, when human and natural histories morphed into the disciplines of anthropology and ethnology. This science of man was almost always about the past, but there were many kinds and chronologies of human history that Huxley's generation inherited, invented, challenged and discovered. An orthodox scriptural past offered a 6,000-year genealogy in which humans descended from Adam and from Noah after the flood. An unorthodox and potentially sinister idea resurfaced about a 'pre-Adamite' past, in which some humans existed before and separate from Adam's race, peopling continents other than Europe. A new idea of European 'prehistory' emerged in which classical Greek and Roman texts provided evidence of Europe's savage and barbarian indigenes and invaders. Orientalist pasts were deeply researched as well, by clever Oxford and Cambridge dons, and especially their German counterparts, reading ancient texts from South Asia and Central Asia in their vernaculars and providing translations for the less learned. Philological histories were pieced together to tell new stories of ancient human movement linking Europe and Asia: Indo-European or Aryan histories constructed homologies between current languages, just as Huxley or Owen found homologies between species' structures.

Perhaps most importantly, history was now in the ground and in rocks as well as in the Bible, in libraries and in remnant languages. This was the period in which human history took a palaeontological and an archaeological turn. The human past emerged in artefacts and skeletal remains within prehistoric archaeology and as fossils within the new palaeontology with which Huxley came to be so associated. Indeed the study of fossils was born more or less with him, palaeontology named in 1822.[1] Time was expanding as the geohistory of the earth receded ever deeper, and the 'antiquity of man' with it. Searching for the age of the Earth and the 'origins of man' drove Huxley's generation towards ever deeper periods, ages and epochs: even the classification of time was new.[2]

The mid-Victorian generation of natural scientists also pressed on with exciting ideas about how human bodies themselves – dead, alive, skeletal, fossilized and dissected – might tell the long past of their own species and perhaps that of ancestral others. The past was told – 'recapitulated' – by each human embryo, and for some evolutionary history was evident in 'rudiments', anatomies left over from earlier versions of *Homo*, the now-useless organs in each human body that were remnants or vestiges of almost impossibly long ago. Charles Darwin eventually put all this together to tell a genealogical history of humans and prehumans, of primate common ancestors and the related life forms that went before. We might imagine all of this analysis of the deep past of *Homo sapiens* as the slow emergence and uptake of an increasingly 'correct' evolutionary history. But the evolutionary history of *Homo sapiens* entered an already crowded field of knowledge and mass of activity about the deep human past, and was not at all essential even for the celebrity Darwinian of the 1860s, Thomas Henry Huxley. His *Evidence as to Man's Place in Nature* (1863) is almost always understood as antecedent to, as preparation for, Darwin's *Descent of Man*. Yet he barely mentioned Darwin's new theory of evolution by natural selection, and the genealogical image – the family tree of man – was absent altogether.[3]

Huxley and Darwin did share another system for thinking about *Homo sapiens*' deep past, however. An eighteenth-century secular 'universal history of man' underwrote and profoundly shaped their ideas, so deeply that it was almost presumed knowledge. During the French and Scottish Enlightenments of the previous century political philosophers as well as natural historians had theorized about human origins and human social change, proposing a staggered and staged social and economic development from primitive societies to civil states, in new worlds and old. In the nineteenth century, the natural history and then the evolutionary history of humans emerged around, with, and through these older ideas. This 'universal history' – known as stadial theory – slipped seamlessly into the science of man and lodged there for many generations to come. This idea of the long human past also had living evidence in the present. In its own logic, ancient human form, economy, and society were embodied by contemporary Indigenous people. Huxley, Darwin and countless others saw the past, even their own past, residing in 'aboriginal' people's bodies and cultures across the world.

By Julian Huxley's twentieth century, natural and human histories, natural and human sciences, tended to separate into specialties, into archaeological, linguistic, palaeo-anthropological or anatomical expertise and training. One line of inquiry was a straightforward continuation of T. H. Huxley's comparative anatomy between humans and other primates, and his analysis of skeletal and fossilized remains, what he called 'craniological ethnology'. Piece by slow piece a picture emerged out of the ground about human prehistory. Other methods of inquiry, however, were completely new in the twentieth century and almost unthinkable in T. H. Huxley's time. In a novel kind of bodily 'remnant', genes were starting to tell history too, and eventually chromosomal comparison between primates first supplemented and then overtook the crude comparisons of soft anatomy or the skeletal and fossilized skulls which T. H. Huxley held with such propriety.[4] Julian himself characteristically synthesized the new genetics with human history, telling it with all the intellectual grace and learning of his polymath grandfather, if with new knowledge. Yet he also found himself having to undo key aspects of his grandfather's work, the more he perceived the politics of insisting on 'man's place in nature'.

By the 1930s the science of man, long a science of race, was being put to new nationalist uses. European fascism was manufacturing and, worse, applying ideas about the species' past and about the racial past and present. With sharp political purpose and intent, Julian Huxley was one of a set of British scientists to name and shame a Nazi 'pseudo-science' of race, as he called it, and to declare its particular versions of human history a dangerous fiction. Through population genetics, a century of palaeo-anthropology, and with a new kind of critical and cultural anthropology at hand, Julian helped open up a deep history of human intermixture, not the purity of original race-couples. Julian Huxley killed off scriptural Adam and Eve in a manner that would have made his grandfather proud.

THE MAN FAMILY

Having set out his historical review of 'man-like apes' in his 1863 book, Thomas Henry Huxley reversed his analysis to consider the more difficult proposition: 'the relations of man to the lower animals'. How humans are animal or not-animal was hardly a new question, however. As Huxley himself noted, there had been constant attempts to answer this 'question of questions' within theology, philosophy, art and poetry from classical times onwards.[5] Unsurprisingly, for Huxley the evidence and the argument for connection between humans and brutes began not with these cultural inquiries, but with science and its special object of inquiry, 'nature'.

Huxley's opening strategy to show that humans had evolved – 'developed' – was not drawn from primatology, but from embryology. The lesson was how organisms, including humans, could and did develop over time from simple to complex, and he used his friend Herbert Spencer's analogy: the oak is a more complex thing than the acorn, and entirely unrecognizable from its origins. Far from starting with Darwin's new theory, he explained how his favoured von Baer had unravelled the complicated embryological development of mammals. The egg is 'animated by a new and mysterious activity' and divides and divides again. Many cells with nuclei are formed, and 'by and bye' a central thickening and groove creates what will become the vertebra, then its tail and limbs. So too with the development of humans: 'every living creature commences its existence under a form different from, and simpler than, that which it eventually attains'. And then came the implication of embryology for man's place in nature: 'there is a general law, that, the more closely any animals resemble one another in adult structure, the longer and the more intimately do their embryos resemble one another'. This was old scientific news, if still a stunning public claim. It has been absolutely certain for around thirty years, he told his readers, that the early embryological development of humans follows exactly the same process as the development of apes. It was only in the much later stages that the human embryo (and in fact, even 'the young human') is differentiated from the ape. This alone is sufficient evidence 'to place beyond all doubt the structural unity of man with the rest of the animal world, and more particularly and closely with the apes'.[6]

Does this mean that man is a mere brute? Nonsense, he said, dismissing those who were offended. And then, as usual, he doubled down, daring all doubters: that is an argument that any sensible child could refute. We were all once an insensible egg, he challenged in the plainest of terms, but that does not preclude the development of the arts and intelligence of the philosopher, the philanthropist or the poet grown from that egg, that 'yelk'. Observing the 'unity of origin of man and brutes' does not equate to 'brutalization and degradation'. Humans resemble different apes as they resemble each other. Likewise, humans differ from them as they differ from one another.[7]

Huxley's argument was based on documenting both similarities *and* differences between humans and other primates. On the one hand, any 'Saturnanian' who visited Earth pure and ignorant of theologies to the contrary, as he put it, would immediately see that the Man Family was more like apes than any other kind of animal. And thus he eased his way towards those newly famous gorillas that so few had actually seen, and which he regarded as the primate closest to humans in the present day. While others had distinguished humans from all apes with the claim that only humans had two hands and two feet (while apes had four 'hands'), Huxley disagreed: gorillas, like humans, have two hands that are anatomically quite different from their two feet.[8] This was evidence of the 'closeness of the ties'. And yet there were also clear anatomical distinctions between his own species of primate, *Anthropini* or the Man Family, and the gorilla: the comparative size of the brain-case, numbers of pairs of ribs, the length of vertebra in proportion to arms and legs. With regard to the skull, the differences between human and gorilla are 'truly immense', perhaps most importantly its 'cubic capacity'. Nothing less than 52 cubic inches had been measured in humans, while the largest gorilla skull yet measured was 34.5 cubic inches.[9] But there are even greater differences between other apes and monkeys, and they were all, rightly in his view, classified as 'primates' too. This undercut any argument for putting *Homo* and *Troglodyte* in different orders.

His core lesson was this: whatever differences could be documented between gorilla and human, the differences between the gorilla and other apes – the orangutan or the gibbon or the chimpanzee – were even greater. Primates ranged along a continuum that varied not just between but also within species. And so, there was no distinct line – on any criteria – that could be drawn between human and ape: neither powers of knowledge nor conscience of good and evil; neither feet nor brains; neither 'psychical distinction' nor 'the highest faculties of feeling and of intellect', could set humans fully apart from other animals in the primate order.[10] His consistent argument was based on the wide and continuous variations within and between *all* species of primates in the present: humans, apes and monkeys.

The evidence of 'man's place in nature', then, was in the first instance embryological and in the second instance drawn from comparative anatomy. Both of these sets of evidence *could* have been used by Huxley explicitly to support 'evolution', and possibly evolution by means of natural selection. But they were not. With regard to primates, it was all about present anatomy, not common ancestors. Only passingly did Huxley inquire into ancient or extinct species of 'ape-like men' in the deep past. And while we know that the term 'evolution' derived from just this tradition of embryology and recapitulation, in his foundational book Huxley did not establish this as stepping-stone to other and fuller discussions of the transmutation of species, still less to Darwin's particular theory.

Not only was Huxley's initial explanation of an 'origin' of man embryological, all his claims of humankind's similarities to and differences from other primates confirmed less Darwin's new ideas, by his own account, than eighteenth-century classificatory claims. He agreed enthusiastically with Linnaeus that 'man is a member of the same order as the Apes and Lemurs, "Primates"'. After all his own comparative anatomical work, his close study of prior scientific scholarship, his analysis of drawings and photographs of skeletal remains, as well as actual skeletal remains, he was clear that 'a century of anatomical research brings us back to his [Linneaus's] conclusion'. In a book we might expect to be about Charles Darwin, who had just presented a persuasive set of natural laws to the world, Huxley crowned Linnaeus, not his distinguished friend, 'the great lawgiver of systematic zoology'.[11]

Having established all that, *then* Huxley moved on to Mr Darwin, quite late in his book, and almost in passing. Darwin, he wrote, holds the one idea about causation that ('at the present moment') had evidence in its favour.[12] It was a modest enough acknowledgement, even a qualified one, that didn't chime with his far more positive published reviews of the *Origin of Species*. By 1863, Huxley could perfectly well understand and explain the 'Darwinian Hypothesis',[13] but very little of it – almost none – made its way into his explanation of 'the relations of man to the lower animals'. Perhaps Darwin was relieved not to be at the centre of such a discussion on man's place in nature; at least not yet. Indeed Huxley wrote rather more directly about Richard Owen than his distinguished friend, unable to resist another public assassination. This is further indication that Huxley's vocal support of Darwin was, at least initially, an expedient means of cutting Owen down to size. Archly, Huxley wrote that he had heard of another theory, another formula for evolution, Owen's 'ordained continuous becoming of organic forms'. Then the merciless thrust: 'it is the first duty of a hypothesis to be intelligible'. Owen's idea might be read 'backwards, or forwards, or sideways', but still makes no sense.[14] Owen might have been sly and obnoxious, but one does feel for him, slain over and over again.

If Huxley barely referred to Darwin, the reverse was not the case. Indeed, Darwin republished Huxley within the 1874 edition of *The Descent of Man*, a whole section transcribed as a 'Note on the Resemblances and Differences in the Structure and the Development of the Brain in Man and Apes'. Huxley's work – not on ancient skulls but on comparative soft anatomy – thus became part of Darwin's thesis on the evolution of humankind by means of natural and sexual selection.[15] 'Never mind where it goes,' Huxley demurred with entirely false modesty to 'my dear Darwin'. 'Put my contribution into the smallest type possible, for it will be read by none but anatomists.'[16]

And so, Huxley's *Man's Place in Nature* and Darwin's *Descent of Man* approached humankind and the past entirely differently when it came to evolution. Darwin's work was all about the process and mechanism of selection, and especially sexual selection over very long periods, the enormous lengths of time that geologists were still trying to fathom. For Huxley in *Man's Place in Nature*, variation within *Homo sapiens* was not documented to vindicate selection at all. Rather, it substantiated his comparative anatomy argument about the range of difference both within and between species across the primate order in the present. Darwin's book built towards an argument about genealogy: the descent of man, the tree of life. He was entirely clear about humans descended from 'a hairy, tailed quadruped, probably arboreal in its habits, and an inhabitant of the Old World.' This creature, like all the mammals, was itself descended probably from an ancient marsupial, and that from 'some amphibian-like creature' and 'through a long line of diversified forms' from some fish-like animal. Thus Charles Darwin wrote about 'the dim obscurity of the past'.[17] Huxley on the other hand not only had no conceptual interest in a family tree of man or a tree of life in this sense, he also discounted evidence before him for an 'intermediate' type; as we shall see, the newly famous skull found in the Neander valley in the German states in 1856 he *might* have assessed as ape-man or man-ape, but was disinclined to do so.

Little of Huxley's book, then, fed into an argument about genealogical descent. This becomes clear in one of its most enduring and recognizable images: the suite of primate skeletons that formed the frontispiece to *Man's Place in Nature*, drawn by natural history sculptor and illustrator Waterhouse Hawkins (figure 7.1). This much-reproduced and much-tweaked image of skeletal primates walking more and less upright – the gibbon, the orang, the chimpanzee, the gorilla and man – was not the evolutionary time sequence that it is often taken to be. It was, rather, a suite of present-day species' skeletons to be read comparatively.[18] Huxley's purpose was to show the evidence of relatedness in the present.

Similarly, Huxley presented a sequence of crania precisely to demonstrate anatomical differences (figure 7.2). Like the skeletal sequence frontispiece, this was comparative anatomy. It might seem to imply a time progression on a vertical axis, some kind of evolutionary ascent or descent, but it was not intended that way at all. It was not an evolutionary time sequence or a precocious phylogeny of man, elaborate versions of which Haeckel was soon to produce (figure 7.3). Rather, it

Figure 7.1: Skeletons of the gibbon, orangutan, chimpanzee, gorilla and man, drawn by Waterhouse Hawkins from specimens in the Museum of the Royal College of Surgeons. Frontispiece to Thomas Henry Huxley, *Evidence as to Man's Place in Nature* (1863).

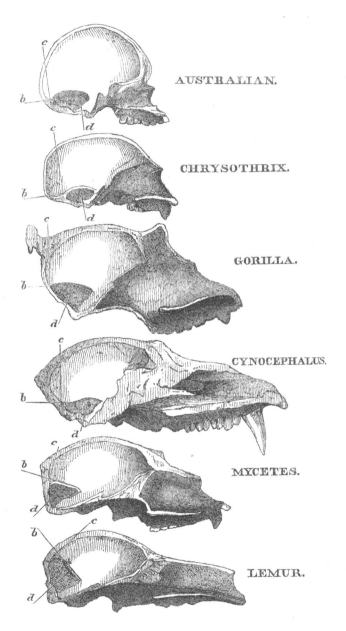

Figure 7.2: Sections of the skulls of man and various apes, from Thomas
Henry Huxley, *Evidence as to Man's Place in Nature* (1863).

PEDIGREE OF MAN.

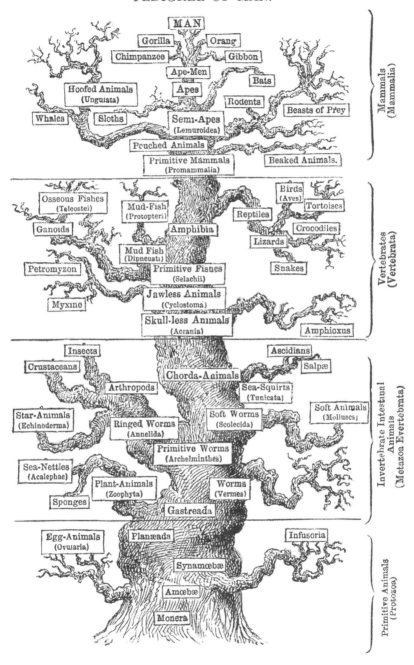

Figure 7.3: 'The Pedigree of Man', from Ernst Haeckel, *The Evolution of Man* (1879).

showed species' structures-in-the-present, extant humans, squirrel monkeys, gorillas and lemurs. The human cranium illustrated his key statement. 'The differences between a Gorilla's skull and a Man's are truly immense,' with the critical follow-up: as are the differences between the gorilla and other apes.[19] The age of the human skull was not relevant to Huxley because its purpose was not to show an intermediate type, but to represent the Man Family unambiguously.

In this illustration, Huxley represented humankind with the skull of an 'Australian', by which he meant an Aboriginal Australian. One thing Huxley was clear on was that all living humans fitted within the particular species *Homo sapiens* that Carl Linnaeus had named in the previous century. There was no question about this for him, though there was for some of his mid-nineteenth-century colleagues who were just then renewing various polygenist arguments of multiple origins, especially after the work of the Scottish anatomist Robert Knox. For Huxley, the Australian skull was in this comparative sequence precisely to represent 'the Man Family'. But clearly that's not all that is going on. His monogenist scientific sympathies did not mean that he saw all sub-groups of *Homo sapiens* as equivalent: 'Men differ more widely from one another than they do from the Apes.' Variation again, but not of Darwin's kind. The problem arose when Huxley explicitly applied a value to this variation, this difference: just as there were higher and lower apes so there were higher and lower humans. It is implicit, but not here stated, that the Australian cranium represented 'lowest Man'. And so, the illustration was to be read thus: 'even' compared to lowest Man, the differences are significant, let alone the differences between the gorilla and 'highest man'.[20] For us this is odious. For Huxley it underscored his greater point: even these large differences are exceeded by differences between gorillas and other species of ape and monkey. We are all in the primate order together.

The racism within Huxley's comprehension of the Man Family is clear. Almost without thinking (precisely because of the centuries of European thought behind him), human variation was arranged for him on a hierarchical axis. Supposedly higher and lower humans were mapped onto time and history in complicated and mixed ways. Images like this borrowed from, but were neither as static nor as doctrinal as, the old great-chain-of-being axis that traversed from the simplest organisms traditionally to the angels and to God: this was an 'ascent of man', to be sure, but not over time.[21] Huxley obviously set himself apart from such scriptural modelling of higher and lower natural, human and supernatural worlds. Yet nor was Huxley offering the evolutionary time-based axis of 'transmutation' (of different humans) that we might expect, that Darwin was to explain a decade or so later, and that Haeckel was to illustrate in his beautiful 'Pedigree of Man'. How then, *was* Huxley's understanding of variation within *Homo sapiens* about time and history and change?

UNIVERSAL HISTORY AND
THE MAN FAMILY

In T. H. Huxley's 1863 'Man' book, humans in the past and in the present were conceptualized on an axis of civilization. The continuum from savage to civilized also worked – in its own logic – from 'lower' to 'higher', from 'simple' to 'more complex', and from older to more recent (though not in all societies). It was a schema for thinking about humans across temporal and geographical scales that pre-dated 'evolution' in the embryological recapitulation sense, and 'transmutation' or 'evolution' as Darwin was starting to use these terms. All societies had once been savage was the simple claim. Some still were.

This 'stadial theory', the progression of human history by stages, underwrote almost every theory of human development over time, not just in Huxley and Darwin's intellectual era, but in Julian Huxley's twentieth-century world as well. It emerged less from a natural history tradition than from the pens of social thinkers: economists, lawyers, political philosophers. Stadial theory posited that human societies progressed from savage states to civilized states through (usually) four stages of all-human history. Commercial and civilized societies had all once been 'savage' (hunter-gatherers) or 'barbaric' (nomadic herders), then became cultivators of land. Social, political and economic organization became increasingly complex as societies passed through these stages, or as they stalled in the early or earliest stage. In some versions, societies might regress.[22] Thus, for European observers and thinkers like Huxley, human societies and individual humans in the present were almost always differentiated as higher (more developed, more complex) and lower (less developed, simpler).

For Huxley, all these ideas had a usefully secular origin and had long been applied to natural laws. This was a thesis that (in its own flawed logic) traversed the globe, nominally taking in all time and all place. It was thus apparently universal. A time axis always operated in this universal history, useful for later evolutionary versions of human development. It was progressive and allowed for change. And it aligned with a thesis of one species, *Homo sapiens*, while recognizing difference within. For these reasons, eighteenth-century stadial theory offered Huxley the clearest way to conceptualize a deep human past. Yet 'universal history' was profoundly about the present too, since many Indigenous societies in Huxley's world were understood *not* to have developed along the four stages, and thus even to have no history. And so, with stunning longevity and influence, the thesis of historical stages of human social development presumed and presented the belief that Indigenous people in the nineteenth-century present lived as they had millennia before, but also, importantly, as *European* ancestors had lived in the long distant past. The savages and barbarians of hunter gatherer and nomadic stages were thus sometimes placed in the deep European past and sometimes identified in the present, in Tierra del Fuego or Tasmania, in the Andaman Islands or sometimes in the Pacific Islands. These were not simply humans from other parts of the world, but humans suspended in an early stage of development, so it was argued.

There is no question about it: Huxley saw some people in his present as living ancients. It was a key part of the logic of his comparative anatomy. 'All the Polynesian, Australian, and central Asiatic peoples ... were, at the dawn of history, substantially what they are now.'[23] That explained the how and why of his cranial and skeletal comparisons. After sending a copy of *Man's Place in Nature* to Charles Kingsley, he wrote from the Suffolk coast in May 1863 about the absence of progression in certain human societies: 'I suspect that the modern Patagonian is as nearly as possible the unimproved representative of the makers of the flint implements of Abbeyville.'[24] The universal history tradition also brought into Huxley's modernity an economic and cultural version of human *unity* that had previously operated within a scriptural great chain of being. The paradox is that objectionable stadial theory was also the intellectual tradition that yielded the history of the species-as-one. More often than not, a single human nature was presumed: societies that had not 'developed' *could* still do so. Connection and relatedness across time and space was more often the point than incommensurable difference. That was about all that redeemed the idea, and made it useful, eventually, for 'evolution'.

With little self-awareness, then, the nineteenth century's most famous fact-finder and lover-of-evidence fell in with longstanding presumptions about higher and lower humans. He sounded fair and 'evidence-based' as we would now say: one specimen of an 'ape-like' brain among non-white races does not mean that that prevails universally, he declared. But then his generation's racism tumbled out: it does not mean that it prevails 'among the lower races of mankind *however probable that conclusion may be*'.[25] His valuation of higher and lower, savage and civilized humans kicked in rhetorically and relentlessly when he tried to explain various arguments. When commentators objected that an original common ancestor reduced humans to apes, for example, he responded thus: even if all humans were once man-like apes or 'some naked and bestial savage', this does not diminish the difference now between 'civilized man' and the brutes.[26] For all his scientific argument that we are all inside the Man Family together, his recourse to the image of the bestial savage showed how thin that claim was. And for all his insistence that there was currently no 'intermediate type' – the Aboriginal skull categorically represented *Homo sapiens* – such statements made people suspect otherwise. It was the thinnest of attempts to be evidence-based, his work reinscribing all the presumption and bigotry that had long paraded as the philosophy and the universal history of man, and that was turning into biological anthropology.

The universal history tradition helped evolution by natural selection make sense when it came to the descent of man. Darwin linked the evolutionary history of all organisms with deep human history: 'there can hardly be a doubt that we are descended from barbarians.' The evidence he put forward was neither archaeological nor palaeontological but instead from his own, very personal and recent history: 'The astonishment which I felt on first seeing a party of Feugians [sic] on a wild and broken shore will never be forgotten by me, for the reflection at once rushed into my mind – such were our ancestors.' For Darwin, these people were not just living representatives of some kind of early man, they were animal-like: 'their mouths frothed with excitement, and their expression was wild, startled, and distrustful'. He couldn't be more explicit: 'like wild animals, [they] lived on what they could catch'. He used this image to redeem his core proposition: that humans had a lowly origin, including common descent with apes. Shamelessly, and entirely for effect, he stated that he would rather be descended from a monkey and a baboon who had demonstrated small heroic acts than 'a savage who delights to torture his enemies, offers up bloody sacrifices, practises infanticide without remorse, treats his wives like slaves, knows no decency, and is haunted by the grossest superstitions'. In the final paragraph of the book, Darwin offered that Man might feel pride at having 'risen, though not through his own exertions, to the very summit of the organic scale'.[27] This was evident in various noble qualities that included 'sympathy which feels for the most debased, with benevolence which extends not only to other men but to the humblest living creature'.[28] What he had in fact just demonstrated is the power of intellectual and cultural inheritance, the way the most learned and experimentally experienced scientist incorporated a set of invented ideas about other humans, and frankly, the way in which Darwin felt closer to Jenny the orangutan in Regent's Park than he did to his fellow humans in Tierra del Fuego. When he concluded his book with a statement of natural history humility – 'Man still bears in his bodily frame the indelible stamp of his lowly origin' – he missed his own low political culture, which failed to be able to consider human difference in any way except progressively from simple to complex and hierarchically from low value to high.[29] Huxley was no better.

HUXLEY'S SKULLS: ETHNOLOGICAL CRANIOLOGY

Huxley's mid-nineteenth-century generation reinvented 'the science of man' by bringing the universal history of so-called primitive humans into conversation with comparative anatomy on the one hand, and on the other a new study of fossils – palaeontology. What he called ethnological craniology was to become one of his own specialties, catapulting his reputation worldwide. Since Blumenbach, he noted, skull collecting and skull measuring 'has been a zealously pursued branch of Natural History'. But as things stood in the 1860s, he found the research field wanting. Huxley regretted the lack of specimens available. This was despite huge collections all over Europe. On his London doorstep was the eighteenth-century Hunterian Collection, which had been government-bought from John Hunter in 1799 and kept by the Royal College of Surgeons. The collection was conserved in Huxley's time by the most eminent of anatomists, by his enemy Richard Owen and later by his friend William Flower. From that collection, Huxley could study and most importantly compare not just the gorilla skeleton on which he based most of his findings, but also human skeletons. From the Hunterian collection, for example, he analysed the skeleton of a male Bosjesman or 'Bushman', and compared it to that of 'a woman of the same race'.[30]

Huxley was comprehensively up to date on the studies undertaken by his natural history colleagues across Europe, with a deep learning in German studies and collections especially. One of his illustrations was drawn from von Baer – the skull of a 'Calmuck' – a Mongolian. Huxley was working up his own classificatory system, sorting out variation in human skulls. The Mongolian was an instance of a round or 'brachycephalic' skull. He compared this to a 'Negro' skull, which was long or oblong, 'dolichocephalic'. But he typically took issue with the research to date within ethnological craniology, which measured skulls in different ways. A baseline or standard was sorely needed for the field to legitimately compare specimens, and (of course) that measuring method should be his own. Comparative anatomists should measure from the base of the skull, which is fixed first in developmental terms, in the human embryo and infant: all the other bones grow and reshape. This would offer a standard measurement, what he called the 'basi-cranial axis'.[31] Male skulls were the templates from which measurements were standardized, from which evidence as to 'man's' place in nature was to be adjudicated. While sometimes he took pains to define his skull specimens by sex, at other times that they were male was presumed. Unless it was the sole specimen, female skulls were particular, not to be taken as representative, as type specimens, or as any kind of anthropometric baseline for *Homo sapiens*. In this sense, and in many others, the Man Family was just that.[32]

In such ways, Huxley insisted that he was working only from the evidence before him and often made a show of his apparent impartiality. Because the range of variation of human structures was 'imperfectly studied', and often only single skull specimens were available, caution needed to be exercised in asserting that this or that feature was characteristic of this or that type or race or stock of human. He often offered counter-intuitive evidence to press home this point. One comparison that ended up in *Man's Place in Nature* was of a 'Negro' and an 'Australian' skull, a 'Tartar' skull held by the Royal College of Surgeons, and a skull from a cemetery in Constantinople that was part of his personal collection, 'of an uncertain race'. In this case, the 'prognathous skulls' – long-jawed skulls – of the Australian and Negro he considered 'less ape-like', on at least one measure, than the Tartar and Constantinople skull. His conclusion demonstrated, he thought, his evidence-based reliability, precisely because the conclusion was 'singular' – unexpected. Along similar lines, some muscles in the white man were arranged more like apes, he declared, than the same muscles 'among Negroes or Australians'. Huxley also made a show of dismissing unverifiable results, the claims about a 'Hindoo' cranium for example, the capacity of which had been measured by amateurs to 46 cubic inches. Nonsense, and not to be entered. Still, far from setting aside the larger inquiry into human skull capacity – the long-established method of filling crania with millet grain or with water – he instead sought larger and more rigorous studies of precisely that. In this instance, he drew instead on the 900 human brains recently weighed by Professor Rudolph Wagner.[33]

This was comparative anatomy at its crudest, even if Huxley thought it was comparative anatomy at its most persuasive. Yet he thought it could and should all be improved methodologically. Until there was a much larger collection of human skulls from across the world, all measured in the same way (his way), there would be no safe basis to make any conclusions about the 'crania of the different Races of Mankind'.[34]

Huxley was only occasionally interested in dating skeletal remains.

Indeed, as yet it was impossible to do so except in terms of the geological strata in which they were found: this offered broad ages and epochs, not a chronological dating. After the stadial model he was likely to nominate a European skull as generically 'ancient': an ancient Danish skull, for example. But when he compared this with, for instance, an 'Australian skull' or 'the skull of a Bushman', it was more or less irrelevant to him whether it was ten or ten thousand years old. They were all crania of *Homo sapiens*, and it was the variation in the present, rather than age or antiquity, that was his point.

In his chapter on 'some fossil remains of man', Huxley brought palaeontological evidence to the discussion. Such remains were exceedingly rare, but had recently taken a decisive turn with the discovery of a skeleton, including a fragmented skull, in the Neander valley in the German states. It was this skull, along with another discovered in caves in Belgium in 1829, on which he focused. Might these be evidence of a transitional form or a connecting link between the Man Family and the higher apes in the deep past? He approached the question 'with anxiety', yet even this evidence he decided not to enter as anything but part of the Man Family. His question was this: did they or did they not depart from the normal wide variation in structure amongst present-day humans? Was the variation similar to the current variation of skulls in contemporary man? He thought so, though conceded the difficulty of assessing just this matter without better evidence on 'the range of variation exhibited by human structure in general'.[35]

Both the skull from the caves of Engis in Belgium and the skull from the Neander valley near Düsseldorf had been analysed closely by geologist Charles Lyell, who was far more interested in the antiquity of fossils. Lyell was unsure of the antiquity of the Neanderthal skull but dated the Belgian skull according to the mammoth and woolly rhinoceros remains with which it was found in the strata. The discoverer of the Engis cranium, Philippe-Charles Schmerling, had published his assessment back in 1833, which Huxley had translated and reprinted within *Man's Place in Nature*. At that point, Schmerling had suggested that its shape approximated an 'Ethiopian' more than a 'European'. Huxley disagreed, working from a plaster cast made of the Engis skull, and camera lucida drawings created by an antiquarian friend, the surgeon and palaeontologist George Busk. Huxley assessed the Engis skull to be like some Australian skulls and unlike others. It was also like some European skulls and unlike others. He decided that he could not definitively align it to any present 'race', and in the end nominated it to be 'a fair average human skull' that fitted easily enough into the current known variations. This might seem like Huxley redeemed of his racism, but he undid himself again with fancy prose and supposed wit: it was a cranium that aligned with so many specimens then available that it 'might have belonged to a philosopher, or might have contained the thoughtless brains of a savage'.[36]

The Neanderthal skull was different, however. It was a game changer, and Huxley knew it. Discovered in a cavern only recently, in 1857, it was still being analysed when Huxley was writing his lectures and gathering them together into *Man's Place in Nature*. Professor Hermann Schaaffhausen of Bonn, the anatomist who first assessed the skull, found it to be unlike any other human cranium known to exist 'even in the most barbarous races'.[37] Huxley agreed: 'the most pithecoid of human crania yet discovered'.[38] Not everyone thought the Neanderthal skull was special, however, and there was the widest of assessments of its age and significance. Professor Mayer of Bonn thought it was a 'rickety "Mongolian Cossack"', driven by Russians through Germany and into France in wartime, 1814. Irish geologist William King, at the other end of possibilities, thought it so old as to be a distinct species, *Homo Neanderthalensis*. Huxley concluded that although it was old, it was similar to contemporary human skulls and thus part of a series 'which shall lead by insensible gradations from the Neanderthal skull up to the most ordinary forms'.[39] It was a variation of a human skull, still within 'the Man Family'. Huxley speculated that its capacity was about 75 cubic inches, comparable to 'Polynesian' or 'Hottentot' skulls, he noted, as if we should be impressed, although in his own logic this comment did not make the 'lower races of mankind' more ape-like, it made the Neanderthal skull more human-like. The rest of the Neanderthal skeleton suggested to Huxley a similarity to 'Patagonians', who live unsheltered in a climate 'not very dissimilar from that of Europe at the time during which the Neanderthal man lived'. Really, he was saying that the Neanderthal was the most extreme of a human series, 'from it to the highest and best developed of human crania'. It was like some Australian skulls (which he put in the present) and it was even more like some Danish skulls, which importantly he put in the past tense, 'ancient people who inhabited Denmark during the "stone period"', drawing from the work of his friends George Busk and John Lubbock on 'the Stone Age' – what was beginning to be called prehistory.[40]

Neither the Engis cave skull nor the Neanderthal skull convinced Huxley that a non-human or pre-human form had been found. More palaeontology was needed, and in one of very few explicit statements, he posed the question of age:

> Was the oldest *Homo sapiens* pliocene or miocene, or yet more ancient? In still older strata do the fossilized bones of an Ape more anthropoid, or a Man more pithecoid, than any yet known await the researches of some unborn palaeontologist?

Time will tell, he told his reader. And it did, in Julian's era. Still, he conceded that the Neanderthal skull especially indicated that the antiquity of Man must be extended 'by long epochs'.[41]

Partly because he was so upfront in *Man's Place in Nature* about lack of palaeontological evidence, colleagues and antiquarians began to send skulls and photographs of skulls to Huxley, dispatched by private individuals or museum officers, missionary or military personnel. By the mid-1860s, he was well known as an authority on the comparative anatomy of human remains, central to the new ethnological craniology that capitalized on imperial connections. He came to be surrounded by skulls: plaster casts, drawings, paintings with skulls, as well as actual crania sent to him from all over the world. This was all a long way from marine invertebrates. Many photographs of crania – some of them strangely beautiful – are hidden away in Huxley's papers in Imperial College London. One series captures the patriarch's world: the skull, the brain, the mind propped up by books of learning (figure 7.4).

Up and down the social and natural history scale, the likes of Huxley presumed they could and even should collect, steal, exchange or buy human remains. Some held on to them as antiquarian treasures, sending Huxley casts or photographs. Others released them to metropolitan authorities, to add to the Hunterian Collection or those in the British Museum, soon to move to the new site in South Kensington. In 1866, for example Huxley learned that a skull secured by military

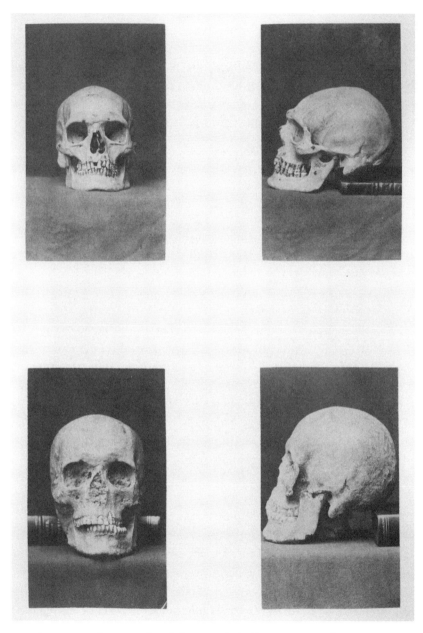

Figure 7.4: Photographs of the 'Bondi skull', sent to Huxley in 1864.

officers in Patagonia was being offered for exhibition. In this case, the human remains were stolen from graves. Everyone involved knew that this was plainly wrong, but were troubled only insofar as the Patagonians were concerned: that is, only if they found out: 'The Skull was handed to Capn Watson as privately as possible in order that the Natives might not know that any of the Graves of their ancestors had been disturbed.'[42] Likewise, many Australian Aboriginal skulls, or records of them, were sent to him. Huxley felt he had a particular expertise in Australian ethnological craniology. He knew of an important cranium found in Bondi, on display at the Australian Museum, Sydney's natural history museum (figure 7.4). In 1864 this was described as approximating 'the famous Neanderthal skull', famous not least because of its publicity in Huxley's book. The trustees ordered casts to be made and 'transmitted to the European museums'.[43] In this display, like so many others, ancient humans from widely dispersed parts of the world were starting to be classified and displayed together, apparent remnants of a global 'Stone Age'.

From a different continent and another decade, the explorer Henry Morton Stanley asked Huxley to adjudicate skulls from the Congo River. On his 1875–6 journey, Stanley had reached a village where he saw two long rows of skulls. When he asked whether they were human, the response he received from the 'chiefs and Arabs' with whom he travelled was 'soko', or chimpanzee. The local men described '*nyama*', an ambiguous animal who 'walks like a man and goes about with a stick, with which he beats the trees in the forest'.[44] Stanley shipped two skulls to Huxley, by then the undisputed expert: were these in fact human or chimpanzee remains? Huxley was clear that they were human – one a man, one a woman – and both the skulls of 'the ordinary African negro'. Queen Victoria, for one, was both interested and apparently pleased that Huxley had disposed of the *soko* affair, which had hit the press.[45]

Long after Huxley's expertise was exceeded by a generation of more specialist palaeo-anthropologists and pre-historical archaeologists, he continued to be sent skulls and photographs of skulls. As late as 1894 he received yet another photograph from a rural New South Wales surgeon: 'As it appears to me to approach the Neanderthal type more than the average Australian's skull it might be interesting to you, as step towards the missing link!'[46] The same year, Huxley received three photographs of skulls found near the channel of the Fraser River in British Columbia. They were discovered in a 'very ancient shell mound', as skeletons, the knees of which were drawn up to the chin, and assessed locally to be differently shaped to 'common Indian types now prevailing', taken as a sign of their possible great antiquity. But what did the great Huxley think? 'My desire to know whether there is anything remarkable about the skull photographed must be my excuse for troubling you,' asked his correspondent.[47] Here, still, the key point was to differentiate or identify the cranium vis-à-vis Indigenous people, specifically to see whether the remains were to be assessed as human or something else. Huxley, the marine invertebrate specialist, had turned into a leading palaeo-anthropologist, a comparative analyst of vertebrate skeletons and of primate skulls. He was now the global master of dead humans.

THE ANTIQUITY OF MAN

When Huxley, like others, considered the antiquity and history of *Homo sapiens*, he did so inside a heady but confused mix of palaeontology, prehistory and stadial theories of primitive humans. These ways of knowing were not for some time to become conceptually distinct. The terms 'old', 'ancient', 'prehistoric', 'post-Pliocene', 'pre-Adamite', 'primitive' and 'fossilized' were all imprecise, waiting to be fixed and standardized, possibly even dated. The geologist Charles Lyell did some of that initial fixing, and he did so in 1863, just when Huxley's book came out.

The publication of *Man's Place in Nature* coincided with Lyell's *Geological Evidences of the Antiquity of Man* (1863), although Huxley's lectures had already been published in journals. Lyell used these extensively, especially Huxley's anatomical work on hands and feet.[48] Yet Lyell's book was entirely different, because he was far more upfront about the new transmutationist theories. There it was in the subtitle, 'with Remarks on Theories of the Origin of Species by Variation'.[49] *The Antiquity of Man* was Lyell's cautious declaration of his turnaround on evolution, having strongly criticized Lamarck's transmutation in his much earlier *Principles of Geology*. Yet even after longstanding conversation with Darwin, Lyell's statement on human evolution was cautious at best. And far from endorsing natural selection as a mechanism for human evolution, he indicated that the gulf between human and animal was still a mysterious matter. Human capacity for forethought, including the idea of an afterlife, quite possibly set humans apart entirely.[50] Thus even though Lyell was still inclined towards a classificatory difference between human and animal, it was his own geological study that propelled the very idea of the (evolutionary) antiquity of humankind.

The Antiquity of Man periodized human prehistory into two 'Stone Ages', the Palaeolithic and the Neolithic. And the geological periods that most natural scientists worked with in the 1860s were established by Lyell as the Pliocene, the Miocene and the Eocene of the Tertiary Epoch, defined in *Principles of Geology* as periods of pre-human history, before 'the earth was tenanted by man'.[51] Many thus looked for human remains in the 'post-Pliocene period', but perhaps humans had tenanted the earth in the Pliocene period itself, or even earlier? For some this was the most exciting of possibilities. For Lyell himself, however, this deep earth past had to be negotiated very carefully indeed, given the trend towards gradualist development ideas, the insensible transmutation of species. Especially with regard to the implications for a simian origin of humans, Lyell was profoundly uncomfortable. More than that, he was privately repelled.[52] Lyell must have been shocked and discomfited, then, by Huxley's brash confrontation with those who (like him) were alarmed at the prospect of a Man Family within a primate order. Huxley's signature response was: what of it? 'What matters it whether a new link is or is not added to the mighty chain which indissolubly binds us to the rest of the universe?' And for the theologically concerned: 'Of what part of the glorious fabric of the world has man a right to be ashamed – that he is so desirous to disconnect himself with it?'[53]

Lyell was hardly alone in trying to reconcile an ancient earth, evidence of transmutation of a human species and Christian learning and conscience. Across the world natural historians, theologians and theologically inclined geologists were trying to make it all one story. Huxley's old Sydney friend the Reverend Robert King was one. 'The Bible's account of our race, Adam's race, will stand even after the presence of a Homo priscus 180,000 years old is established.' He seemed blissfully unaware of how such a combination of polygenist and scriptural argument would infuriate Huxley, signing off his letter most affectionately with fond memories, and warmly referring to a portrait still treasured in his album.[54] The antipodean geologist Rev. W. B. Clarke was another. He also sought Huxley's views on 'the new theory of the "Antiquity of Man"', trying to make sense of the evidence before him in Australian strata. 'Here we have all the general facts relied on by the theorists. We have the teeth of animals slain in hunting . . . the *Australian* skull, which is the assumed type of the pre-Adamitic man. We have the *stone* implements of the human race, and geological evidence of deposits of comparatively old date.'[55] Clarke presented as uncontroversial fact that Aboriginal remains were 'pre-Adamite', not necessarily of a different species to *Homo sapiens*, but not descended from Adam.

Notwithstanding this imperial research, archaeological and palaeontological 'prehistory' was under construction in the first instance as a *European* object of inquiry: the uncovering and piecing-together of the cultures and remains of ancient Britons, ancient Danes, ancient Germans. As president of the Ethnological Society, Huxley brought prehistory home in his presidential address 'On the Ethnology of Britain' (1870). He drew partly on Roman texts – by Caesar and Tacitus – that mixed ancient history and prehistory in scintillating ways and speculated on two physical types of original ancient Britons: one tall, fair and blue-eyed; the other short, dark with black eyes, rather like his own and his mother's. Huxley's thesis was that no 'new invasions' changed this fundamental mix, with fair and dark 'types' evident within contemporary Britain. He aligned these 'ancient Races of men', their flint axes and bone skewers with 'the lowest savages at the present day', who were themselves unchanged from the 'time of the Mammoth'. At least, he demurred, we have every reason to believe so.[56] Huxley learned of Stone Age humans across Britain: a cranium from Sudbury, another from a tumulus at Keiss in Caithness.[57] And one summer's day he found Stone Age man at 107 Pall Mall. At the Athenaeum Club he sketched a pre-human Huxley complete with remnant tail, the sign of 'nature', but also stone tool, the sign of 'culture' (figure 7.5).

Figure 7.5: Huxleyan Stone Age Man: embodying nature with his
vestige-tail, and culture with his axe. 19 July 1864.

ARYAN HISTORIES

Language was another axis through which the antiquity of humans and relationships between different humans and their ancient migrations were assessed. Huxley generally thought it sensible to keep the study of language and of physical characters separate. Like prehistory and palaeontology, philological study of human antiquity involved as much European introspection as pronouncement about colonized and racialized others. The hot debate over Aryan languages was a story that successive eighteenth-, nineteenth- and twentieth-century Europeans told about *themselves* and *their* past. Both Thomas Henry and Julian Huxley dealt with the strange history of the 'Aryan', sometimes a language, sometimes a society, sometimes a race. It was an intergenerational Huxley concern over an era in which the application of science to race-based nationalism intensified.

So-called 'Aryan' languages had been extensively researched, especially in the British orientalist tradition, after Sir William Jones's late eighteenth-century work on Sanskrit and its ancient connection to Armenian, Greek, Latin, Lithuanian, German and Celtic. The German-born professor of comparative philology at Oxford Max Müller vastly extended this research in the mid-nineteenth century, one theory locating original or primitive Aryans in the Ganges valley, whence the western world was peopled. Later, a theory emerged about an Indo-Iranian original language spoken by a people of whom the Aryans of India and Persia were descendants. These linguistic theses had 'Aryan' languages and therefore people spreading over time and space from somewhere in Central Asia or North India. From the 1880s, however, both German and English philologists and anthropologists began presenting a wholly new thesis that reversed the Asia-to-Europe axis. Perhaps Aryans originated not in Central Asia or north India, but in Russia or Scandinavia or the Baltic? The northern origin – what was to morph into the 'Nordic' idea – was suggested,[58] inspiring an English version that Celts were the original Aryans.[59]

In 1888, poor Oxford-based Max Müller tried to clarify – and distance himself from – his earlier and loose use of the term 'Aryan race', which he agreed had thoughtlessly linked philology and biology.[60] But it was too late. The idea of a northern Aryan race was a tantalizing one, already picked up by anthropologists and biologists and soon to be connected to a stark new generation of race-nationalists, first American and then German. In an avalanche of Aryan scholarship, a competition for the aboriginal Aryan unfolded, substantiated by the strangest mix of evidence based on comparative language, comparative mythology and comparative head shape. Advocates of the northern Aryan thesis were delighted to base their modern craniology on authoritative Huxley's work, especially his thesis on the 'great race of fair-haired, broad-headed Xanthochroi'.[61] The capacity to just invent a new word for a 'race' in this period was an essential tool in an ethnological craniologist's box.

Head shape was Huxley business if anything was, and he couldn't resist joining in. He published 'The Aryan Question and Pre-historic Man' in *The Nineteenth Century* (1890) as he was moving into his retirement in Eastbourne. Past language, 'peoples' and 'races' were becoming hopelessly mixed in these theses, Huxley was quick to point out. Yet he applied biological anthropology to the old linguistic idea, thinking his scientific and agnostic credentials rendered him personally immune from further myth-making.[62] He announced that he could only possibly look at the Aryan question as a biologist, but in truth it was a stretch. Signalling Müller's early work, Huxley did declare that the purely biological conception of 'race' (which he himself puts in inverted commas) 'illegitimately mixed itself up with the ideas derived from pure philology'. 'Aryan' can never be a race, that is, a sub-division of 'species', in the proper zoological usage. Nonetheless he was going to enter his views and weigh them against various kinds of historical and archaeological evidence. 'From the purely anthropological side', by which he meant speaking as a biologist not a linguist, he was inclined to favour the recent proposition: that the original seat and distribution of Aryans lay in Europe, rather than in Asia 'in primaeval times'. What was his evidence? Bodily characters. A type still found in Scandinavia and the Baltic – tall, fair, blue-eyed, 'yellow or reddish hair' that covered typically long skulls – he thought had moved out of the regions bordering the North Sea and the Baltic. And as far back as archaeology takes us 'into the so-called Neolithic age, the great majority of the inhabitants had the same stature and cranial peculiarities as at present'. That is, they were original – aboriginal – to the region. Partly using archaeology, partly using Roman sources, partly using ancient scholarship from India and Persia, and partly using his own ethnological craniology, Huxley mapped the distribution of long-skulled and broad-skulled, tall and short, blond and brunette types across Europe, and all that onto the distribution of languages. He made a highly qualified statement, but one that nonetheless opened up a blond, blue-eyed, Nordic Aryan origin for a generation of race theorists early in the twentieth century, not a Central Asian or north Indian one. He even reintroduced and reauthorized the idea, the term, 'Aryan race'.[63]

Huxley thus lent his name to the reverse story of Aryan geography, migration and history, speculating that India and Persia were the final not the original 'seats' of the Aryans and of the so-called Indo-European languages. He linked this to the much-debated Neanderthal skull, found fifty years before, his 'old friend'. The skull was 'a pretty sure indication of the physiological continuity of the blond long-heads with the Pleistocene Neanderthaloid men'. In other words, the northern Aryans had been *in situ* a very long time and constituted an origin of the Aryan type, who migrated south and eventually into Asia, not the other way around. It was an idea that his sometime student H. G. Wells would utilize in his own long history of the world, including a chapter on 'The Aryan-Speaking Peoples in Prehistoric Times'.[64] By the end of the nineteenth century, then, scholars and journalists announced that the 'tyranny of the Sanskritists' – Jones then Müller – had been checked, and that up-to-date intellectual forces had linked 'prehistoric archaeology, craniology, anthropology, geology, and common sense', and that included the ageing and retired but not retiring Huxley.[65]

Thereafter, a new generation of polymath scholars and journalists lifted and shifted 'Aryan' from the deep past to the political present, putting the idea to work for the explicitly racialized states that were emerging in the new century.[66] The Aryan and the Nordic deeply shaped American race-based nationalism in this era, tied to manifest destiny, to American empire and unequivocally to white supremacy. This was the era of unrivalled scientific racism, and the Aryan idea swiftly amplified to fascist versions. By the 1920s Adolf Hitler was tying 'the Aryan' to his strident German nationalism. His was a cautionary tale about the mighty prehistoric Aryans and the loss of their power and purity through racial intermixture. As he spread across Europe and Asia,

> [t]he Aryan surrendered the purity of his blood, and thus lost the right to the Paradise which he had made for himself. He went down in the mixture of races, and gradually lost more and more of his cultural capacities until finally he began to resemble the aborigine more than his own forefathers, not only mentally but physically. For a time he could still live upon existing cultural substance, but then ossification set in, and finally oblivion claimed him.[67]

There was some official use of the term 'Aryan' by the Nazi government after 1933. Yet historians have also suggested that few scientists in Germany actually afforded the idea of an Aryan race much credibility. In any case the term proved too difficult and imprecise for Nazi legal purposes of identifying, separating and segregating people by 'race', by genealogy. By 1935 'German-blooded' (*deustchblütig*) was preferred.[68] It has been argued that part of the ongoing 'myth of the Aryan' was a 1930s Anglophone overstatement of Nazi deployment of it. Certainly Nazi recourse to the Aryan was strongly criticized by English anthropologists and biologists that decade; Julian Huxley foremost among them. And in order to attack Nazi pseudo-science he had to undo his own grandfather's anthropology of the previous century.

The critique of scientific racism, including the spurious history of pure European pasts, was one of Julian Huxley's prime communication agendas in the mid to late 1930s.[69] Putting nature and culture together in a new way, he joined forces with Cambridge cultural anthropologist Alfred Cort Haddon to write the short, cheap and influential book *We Europeans: A Survey of 'Racial' Problems* (1935). Together, they took on the new fascist nationalism sweeping across Europe – in Spain, in Italy, in Germany – seeking to undermine the racial ideas that were being deployed so starkly. Theirs was a fundamental critique: what is race anyway? It was the vaguest of ideas, they said dismissively. Like 'blood' it was based on the most elemental biological error and the most facile misunderstanding of basic genetics.[70]

Huxley and Haddon walked readers through various truths and fictions of histories of the species, its unity, and its variations, and in the process they exposed the fallacy of the Aryan.

Julian explained that language does not establish common ancestry, and there is no such thing as an Aryan race.[71] But it had nonetheless been pursued, he recounted, by such luminaries as Thomas Carlyle, J. A. Froude and Charles Kingsley. This sequence kindly but disingenuously overlooked his own grandfather's intervention, Julian even claiming that the idea of the Aryan race 'ceased to be used by writers with scientific knowledge' in England and America, but not, alas, in Germany.[72] It was a Huxley sleight of hand. The English story that Julian recounted was not his grandfather's late-in-life dalliance with Aryanism, but Max Müller's 1888 correction of the mistake, the linguist who saw 'the enormity of his error'.[73] Julian's message was that the English had rightly self-corrected, but the Germans had not. The Nazis had run with the idea of the Aryan and the Nordic, and unforgivably with the idea of a historic and genealogical pure race that can be recovered through contemporary reproductive practice and policy. 'This concept of an original ideal type, later obscured by the admixture of alien elements, has provided the intellectual foundation for all the arguments of the "Nordic" and "Aryan" theories and the attitude these involve towards "racial" and political problems.'[74] He insisted on placing 'race' in inverted commas throughout, but then so had his grandfather for different reasons. After the war, Julian tried to explain again the Aryan nonsense in the film *Man – One Family* (1946), a documentary rendition of *We Europeans*, updated for the ideological pivoting of the post-war, post-Holocaust world (figure 7.6).

The 'Aryan Question' saw successive Huxleys in the thick of race science, and then in the thick of the critique of race science. Its intellectual history captures the crossover of human and life sciences, studies of human culture and studies of human nature. It is the story of anthropology, and its own crossover from bodies to culture, skeletons to ritual. The strange history of Aryanism thus fast-tracked from the ancient past to the most pressing of contemporary European politics. It was part of Europe's myth-making, about itself and its own prehistory, a European origin story put to horrendous political uses.

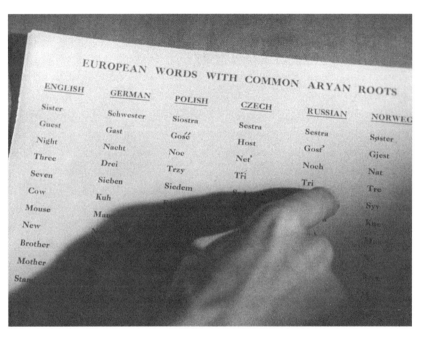

Figure 7.6: Julian Huxley points out that 'Aryan' is simply a linguistic categorization, not a biology, race or a people. *Man – One Family* (1946).

JULIAN HUXLEY AND
AFRICAN ORIGINS

At the same time as Julian Huxley and others were debunking Aryanism, another origin story – one based on clearer and different evidence – was under construction. A nineteenth-century scientific consensus about a Central Asian origin for all humankind had by the 1950s shifted to a consensus that the earliest humans had emerged within, and then moved out of, Africa.[75] Methodologically, this involved a straightforward con-tinuation of Huxley's 'ethnological craniology' well into the twentieth century. For Africa-loving Julian Huxley, the geographic shift in emphasis was fascinating, and he joined a chorus of Europeans who granted Africa a history, as if it were theirs to bestow. Indeed, Julian barely did that, bestowing a *pre*history only, and declaring *that* remarkable. 'Africa is a continent almost without a history,' he announced in *Africa View*, 'but it is rapidly acquiring a surprising pre-history.'[76]

Partly because hominid fossils were so few and far between, the twentieth-century shift towards Africa as a site for the origin of *Homo sapiens* was incremental at first. In 1924 Raymond Dart made a for-tuitous South African discovery of what he called a 'man-ape', and he published the fossil news in *Nature* the following year.[77] Like Huxley, Dart was an anatomist, and had recently moved from University Col-lege London to Johannesburg's University of the Witwatersrand. At Taung in the northern Cape there was considerable mining activity, and a concerned local sent Dart several boxes of fossils that had been blasted out of the ground. There was the skull, and Dart knew at a glance it was significant. It was too big to be a baboon or chimpanzee, but it was not big enough to be human. Dart well knew that Darwin had been one of the few who had early speculated on an African ori-gin of humans: 'Was I to be the instrument by which his "missing link" was found?' he later immodestly wondered in his autobiog-raphy, *Adventures with the Missing Link*.[78]

Dart named the specimen *Australopithecus africanus*, the southern ape of Africa. But at the time all kinds of theses existed about 'origins', and few palaeo-anthropologists were interested let alone convinced. Many still looked to Central Asia, the traditional nineteenth-century idea. Others followed the discovery of 'Java Man' by Eugène Dubois in the early 1890s, inspired by Alfred Russel Wallace's and Charles Lyell's separate hunches that South-east Asia was a most likely site for the origins of humankind. And over the 1920s, while Raymond Dart pressed his case, global contenders multiplied. Some influential anatomists and palaeontologists sought evidence for Europe itself as the origin site, an alternative and older hominid story than the Aryan variation. This extended the significance of the Neanderthal discovery as well as the Cro-Magnon skull found in France in 1868. The anatomist Grafton Elliot Smith, strongly guided by T. H. Huxley's work, studied all kinds of brains and skulls over the 1920s, including mummified Egyptian remains. He argued that humankind had diffused outwards from Egypt.[79] And around the same time teeth were found near Beijing, determined to be those of an ancient human, and so *Sinanthropus pekinensis*, or Peking Man, was reborn.

All this unfolded over the middle 1920s, when Julian was between Oxford and Kings College London, and starting to imagine both *The Science of Life* and scientific life beyond a professorship. And so, when he sat down with the Wells father and son to consider how the story of the evolution of humans was to be told in their book, any number of theories and sites presented themselves, but on a thin set of specimens. With characteristic authorial charisma, they explained that the palaeontological evidence of 80,000 years of hominid history could be 'packed carefully in paper and put into a small suitcase'.[80]

Dart's South African discovery was news enough to find its way into *The Science of Life*, but it was not accompanied by any suggestion that therefore Africa was the site of human origins. Indeed it was a fairly unexcited and cautious reportage. What Dart had so pointedly called 'a man-ape', Julian described as an 'ape'.[81] This was still not the 'true missing link', the ancient common ancestor. Indeed so remote, as yet, was the possibility of African origins that Julian turned with far more interest to a cranium found in East Sussex, in a village called Piltdown, not far from his Surrey childhood home. The British Museum proclaimed *this* as a key 'missing link', and it drew experts' interest from across Europe, including Julian's soon-to-be favourite, the French Jesuit palaeontologist and geologist Pierre Teilhard de Chardin. 'Discovered' in 1912, dated to 500,000 years old and alluringly both 'British' and close to London, the Piltdown skull turned out to be neither ancient nor even from one species. Its authenticity was quietly questioned from the beginning, and much later, in 1953, the bombshell hit: the Piltdown skull was confirmed as fraudulent, a composite of a medieval human skull, an old orangutan's jaw and likely chimpanzee's teeth.[82]

Many British prehistorians and palaeo-anthropologists had wished and willed the Piltdown skull to signal a local British origin for humankind, for the 'dawn of man' to rise like the summer sun over the home counties. And for a while it did. Piltdown Man featured in *The Science of Life* far more prominently than Raymond Dart's South African skull, precisely because it seemed closer to the Man Family. It was 'obviously a man and not an ape, but so different from ourselves as to demand being put in a new genus, Eoanthropus or Dawn-Man'.[83] By 1935 Julian and Alfred Cort Haddon announced in *We Europeans* that there was now a balance of evidence between Eurasia and Africa. In their anti-fascist book, Huxley and Haddon were quite certain that Europe was not humankind's 'original home', by then seeking to de-escalate Aryan and Nordic theses. 'At the moment the balance of probability is more or less equal as between Africa and Central Asia.' And they put a chronology to it all in a way that T. H. Huxley never had. By then the timeline of Earth and human history was broadly familiar to countless natural history museum visitors in the new vogue for human origin exhibitions and dioramas. Perhaps a million years earlier both Piltdown Man and Peking Man had roamed the Earth, they explained. Neanderthal man existed 50–100,000 years ago, while Cro-Magnon and varieties of *Homo sapiens* were perhaps 20–40,000 years old.[84] And then prehistory changed to 'history' around 3400 BC in Egypt with cereal agriculture, and when Neolithic – 'Stone Age' – folk across Europe changed into their Bronze Age and Iron Age descendants. That was now the conventional manner of telling the history of *Homo sapiens*. But the old trope of another chronology, of living primitives, sat alongside it still. Julian and Haddon's diagram of climate, period and human type included 'Neolithic: lasting in Oceania to present day'.[85]

By the time that Julian Huxley ascended the pulpit at the 1959 Darwin anniversary in Chicago, much of this had changed. Piltdown had been thrown out, now unquestionably a hoax, to the embarrassment of many, including Julian. And enough hominid discoveries had been made in East Africa to more or less secure the African origins thesis. In large part this was the work of the Leakeys, whom Julian knew well. With Juliette he crossed paths with the palaeo-pair in a sequence of African journeys. In 1929 'Mr and Mrs Leakey' – at this point first wife Frida Leakey not Mary Leakey – showed him the Elmenteita caves in Kenya. They talked over 'the genesis of the Rift, the correlation of geological events here with the Ice Age' and Leakey's 'fossil man'. Julian the neo-Romantic poet and scientist was stirred. Coleridge helped him: 'But oh! that deep romantic chasm which slanted / Down the green hill athwart a cedarn cover! / A savage place!'[86] This was an African Xanadu.[87]

Decades later the most important Leakey discoveries were made. In 1959 Mary Leakey uncovered the quickly famous (care of *National Geographic*) *Zinjanthropus boisei* living some 2 million years earlier. Julian and Juliette visited soon after, in 1960, to see the place where the latest ape-man or man-ape had been found. The Huxleys met Louis and Mary Leakey at the remote Olduvai camp 5, made up of their caravan, their tables of fossils, books and notes and their five Dalmatians. '*Zinjanthropus* does not quite deserve the title of man,' Juliette recorded in her travel memoir, explaining African prehistory to British readers: he did not use fire; he was barely yet a hunter; he still lived on vegetables. Julian and Juliette were guided down into the famous gorge together: 'scrambling about was a treasure-hunt, for at every step rough stone axes and scrapers litter the ground'.[88] She marvelled at the colossal *Dinotherium giganteum*, with 4-foot tusks, partially and painstakingly being uncovered. The near-relative to the elephant was 'not yet ready to emerge from its stony sleep'. Juliette nominated Africa's 'obscure pre-adamic past', and it affected her deeply, a kind of intellectual vertigo. 'I am losing ground; I am falling down those cliffs at a thousand years a second . . . I humbly reflect on the relentless evolution which has linked me, across this million-years gulf, with the white bones under the cliff.'[89] On their next visit in 1963, the year of Kenyan Independence, the Huxleys stayed in the colonial government house that was on the cusp of becoming the State House. Louis Leakey this time showed Julian his latest finds from Olduvai Gorge and his newer site on the shores of Lake Rudolf.[90] By then, the Africa thesis, if not the Leakeys themselves, had been largely vindicated.

All of the Leakeys were idiosyncratic twentieth-century inheritors of T. H. Huxley's craniology, photographed with their skulls as often as Huxley was painted with his. Louis Leakey was chronically controversial, an academic outsider, but Julian was both a friend and supporter, and was inclined towards – likely envious of – the generous off-grid support Leakey received from *National Geographic*. For years, he tried, but failed, to get Louis Leakey a fellowship of the Royal Society, defending him against criticism, especially from the influential Solly Zuckerman, who always doubted anything Leakey said or did regarding primate anatomy.[91]

Zuckerman was down on the Leakeys, on Raymond Dart and on Africa in general, and tended towards South-east Asia as the most important site for human origins. In setting the Leakeys' work aside, Zuckerman channelled T. H. Huxley's more vicious public assassinations. He could have *been* Huxley standing there at the 1947 meeting of the British Association in Dundee, scientifically humiliating the celebrity couple with his statement of comparative anatomy fact. Much more of what Zuckerman called 'biometric study', and what Huxley would have called 'comparative anatomy', was needed before any phyletic claims could or should be made.[92] Solly Zuckerman, primate anatomist, was a far truer latter-day Huxley than his own grandson. Zuckerman wanted to work from the facts. Much in the field had been deduced 'not by the hard facts one would prefer, but mainly by speculation'. He recommended far more careful anatomical study and even used Huxley's 'basicranial axis' as one of his key measures, reproducing Huxley's sketch of this axis from nearly a century before. He used all this evidence in the mid-1950s to puncture the claims about what he called the *Australopithecinae*, 'predominantly ape-like, and not man-like creatures'. The South African finds were probably an ancestor of the modern gorilla and chimpanzee, not of humans at all.[93]

When Julian considered palaeo-anthropology, human origins, and the study of anatomical affinity between *Homo sapiens* and other primates, he was inclined to refer and defer to his grandfather's scholarship, not always his pattern on other subjects. All the twentieth-century anatomical work on cranial and other skeletal specimens that had engaged the likes of anatomists Arthur Keith, Grafton Elliot Smith and later the primatologist Zuckerman had certainly extended T. H. Huxley's evidence, but Julian rightly considered that this had not 'radically changed the general principles of study which Huxley formulated'.[94] He dutifully recorded and published the progress of palaeo-anthropology, the clearest twentieth-century extension of his grandfather's work, but he was not excited by it. It was too dead. Other kinds of bio-histories excited him far more, especially genetic ideas about how the distant past might be alive in every human body.

THE DEEP HISTORIES
OF JULIAN HUXLEY

Like most twentieth-century biologists Julian Huxley knew and repeated Charles Darwin's observations that any and every human carried around the species' history within them, what Darwin had called 'rudimentary organs', leftover vestiges of an evolutionary past perpetuated in human bodies of the present, in the ear, in various muscles, in human body hair.[95] Every living body was a museum of evolution, *The Science of Life* explained, landing on the perfect analogy again: 'the private development of each one of us is an affidavit swearing to the evolutionary history of our race'.[96] Remnants of our early gilled-selves and our early tailed-selves are evident in every human embryo, Julian wrote, although unlike his grandfather he was now able to source and reproduce a photograph of a four-week-old human embryo, a quarter of an inch long, complete with tail.[97] Julian and his generation were beginning to see much deeper into the evolutionary past and how that past was reborn with each act of human reproduction. Soon enough, in the 1950s, photographs not just of embryos but of human chromosomes would show that the human karyotype has forty-six chromosomes, while other primates have forty-eight.[98]

Julian Huxley may not have known this when he considered and reconsidered the evolutionary history of humans over the 1920s to 1930s, but he did know Mendelian genetics, and that signalled a very different method for assessing population-level evolutionary histories of humans. He knew that genes both mattered and mutated. With Darwin, Mendel and Ronald Fisher to help him, Julian was keen to revise the deep history of humans and their origin stories, especially with regard to human migration and reproductive habits, and in the process he questioned the applicability of the model of genealogy itself, of the 'family tree of man'. In the 1930s especially, this all underscored his argument about the impossibility of purity, even the impossibility of 'race'.

One peculiarity was 'man's migratory propensities'. There is a great deal of evidence that humans have always moved. He thought this depended on *Homo sapiens'* particular lack of interest in 'large differences of colour and appearance' when choosing mates, differences that would deter 'more instinctive and less plastic animals'. Thus, in human evolution, mutually infertile groups were never produced: 'man remained a single species'. Furthermore, while crossing between distinct types was a rare phenomenon in other animals, in humans this was early normalized. 'Man is thus more variable than other species for two reasons,' he explained.

> First, because migration has recaptured for the single interbreeding group divergences of a magnitude that in animals would escape into the isolation of separate species; and secondly because the resultant crossing has generated recombinations which both quantitatively and qualitatively are on a far bigger scale than is supplied by the internal variability of even the numerically most abundant animal species.[99]

Humans are thus unique animals because of the degree to which 'crossing' has taken place. When considering early humans and prehistory, Julian often put it this way: convergence was as frequent as divergence. In human history, a species with great variability regularly moved and crossed between 'types', far more so than any other animal.

In a 1953 version, Julian Huxley's evolutionary history of humans looked like this: 'Proto-men' flourished in the early Pleistocene and are all now extinct, including the Heidelberg and Neanderthal man. Then came the fully human phase. Perhaps a quarter of a million years ago arrives *Homo sapiens*. But these *Sapiens* had to wait for their evolutionary and environmental opportunity. 'It dragged on as a numerically small group for perhaps four-fifths of the time since its origin; and then, towards the very end of the last Ice Age, the human deployment really began.'[100] Human dominance included internal variation, but never beyond the species itself. Of course, he wrote, there are still geographical sub-species produced among *Sapiens*, as among other animals. Differences, 'largely adaptive in references to the area of their evolution', shaped the characters of the 'Negro', the 'Mongol' and of his own 'race', the 'typical Leucoderm'.[101] The latter was one of scores of attempts to fix a 'race' term for the 'light-skinned', for 'Caucasians', for 'whites'. Initial divergence then gave way to continual convergence of this species, 'the recurrent migrations and irruptions of prehistoric and historic times'. This was and is continually driven by a 'migratory urge' and a 'psychological adaptability'. In evolutionary terms this gave rise to 'a single interbreeding unit, instead of an increasing number of non-interbreeding units'.[102] Fisher-style population genetics was the only method that could adequately capture and study the patterns of variation across human types.

The first implication was that there were never pure races within the human species. The crossing of different human types had always been widespread, and human groups are all of mixed origin. Searching for pure races is an entirely misplaced enterprise on biological grounds, let alone on political grounds. Searching for 'common ancestry' is also misplaced among humans. A second implication was the disposal of 'the' missing link. While T. H. Huxley had been clear that no 'intermediate type' or 'missing link' had yet been unearthed, Julian Huxley rejected it for different reasons. If a genealogical tree of hominids were put together by any competent anatomist 'it will be seen that none of these early men stand in the direct line of ascent to modern man. We must indeed abandon the hope of finding the "missing link".' It was hard news for Raymond Dart, who continually traded on the idea. But Julian explained that there is no single missing link joining humans to the apes. Rather, 'innumerable forms converge as we go back in time upon the common anthropoid ancestor'.[103]

The third implication was that Darwin's favoured 'tree' image simply did not work for humans. The conventional ancestral tree may have some advantages for representing the descent of animal types, however 'it is wholly unsuitable and misleading for man'. The idea of a race descending from an original couple

> seems to spring largely from the family trees beloved of genealogists, in which a family is traced back to a single founder and his wife. Such family trees in reality trace the descent of a name and have little to do with biological heredity: they are social, not genetic, documents.

Human ancestry, human pedigree, diverges as well as converges, with constant inter-crossing. It does not descend from a single pair.[104]

Ultimately, Julian's vision and history of a deep human past was not based on a T. H. Huxley-style accumulation of evidence regarding homologies or anatomical comparisons, nor on the fragment-by-fragment piecing together of ape-man fossils. He was conceptualizing a past that was far more alive. It was about the breeding, moving and thinking inclinations of early humans. It was even about their courting habits. Julian Huxley was imagining ancient humans with rituals, at one level just like his grebes. At another level, he considered humans of a different order altogether to grebes or any other animal, including great apes. One of the key distinctions between humans and other primates was matters of sex and reproduction. He pointed to non-seasonal sex and to a long post-maturity.[105] Similarly, while there still exist physiological differences in sexual desire at different phases of the female sexual cycle, these may readily be overridden by psychological factors in humans but not in other animals. Humans, he summarized, are continuously sexual; other animals are discontinuously sexual.

Julian was trying to give a modern twist to his grandfather's question: what exactly distinguished humans from other species; what was man's place in nature? His answer was quite different, however. Far from seeking to naturalize humans, Julian argued for the uniqueness of man.[106] Why? Because, as he would have it, at some point in evolutionary history early humans evolved the capacity consciously to transmit ideas from one generation to another. Natural selection had randomly produced an organism that was unlike any other, symbolized by its capacity to understand its own evolutionary past and to pass that information on to future generations. The deep human past as well as possible human futures were fundamentally shaped, for Julian Huxley, by the transfer of knowledge and therefore ultimately of human culture. He comprehended culture itself within an evolutionary framework: human habits, language, customs, even religion could and should be considered in terms of natural selection.

This delight in folding culture into nature, and nature into culture, made Julian Huxley an inspired choice to lead the new United Nations Education, Science and Cultural Organization after the Second World War. He immediately set to work to have the multi-volume deep history of human nature/culture connection written, a twentieth-century 'universal history'. A total history of humankind would bring scientists and humanists together, and it would be based on the phenomenon of cultural transmission in evolutionary terms. Julian proclaimed that as biologists studied the mechanism of genetic transmission to understand biological evolution, so humanists can study the mechanism of cultural transmission 'for understanding the process of human history'.[107] This included the transmission of ideas, rituals, symbols, skills, beliefs and even works of art – 'the self-reproduction and self-variation of mental activities, operating through the various media of cultural inheritance'. It started, rightly and predictably enough in the universal history tradition, with a volume on prehistory leading to the beginning of civilization, with cultivation and with written records. The 'history of mankind' project was commendable in that it was not just European and neo-European pasts that were being told. Huxley himself, aided by Joseph Needham and their Unesco staff, well knew that an East and South Asian history needed recording, and from its own perspective; that a long Middle Eastern history, a Central Asian history, a North, Southern, East and West African history and an Oceanic history all had separate and linked ancient and modern pasts. But universal history ruled still, and Huxley continued to organize the world's people in evolutionary stages: primitive, barbaric, intermediate and advanced.[108]

Thinking through 'man's place in nature' was the great Huxley drive. As the 'Inheritor', Julian Huxley *should* have reaffirmed the connection between humanity and our brutal companions, his grandfather's foremost intellectual ambition. Yet Julian put a twist on the idea: of course man was part of nature; evolution by natural selection was the great, the only, law in operation across all time and all place. But culture was part of that nature, uniquely developing in humans over evolutionary time.

We might consider this intergenerational Huxley shift from nature to culture through the science and arts of craniology. Thomas Henry Huxley held, collected and was constantly portrayed with his skull-orbs. These were his key to the literal nature of humans: 'evidence as to man's place in nature'. Julian Huxley's 'skulls' were more likely to be cultural than fossilized, art not artefacts. Two were interior décor in his Hampstead home.

Dating to around 1300 CE, the Ife sculptures had been excavated by the German archaeologist Leo Frobenius in the early twentieth century. Some were removed to Berlin, one or two to the British Museum, but many were left at Ife, Nigeria, as Julian Huxley saw for himself when he visited in 1948. The ancient life-size heads – some bronze or copper alloy, some terracotta – were ill-kept, and, by his own account, Julian was personally responsible for instigating sufficient care and an appropriate museum for the precious heritage.[109] One seems to have become Huxley décor at their Pond Street home (figure 7.7). It sat in the company of the sculptor Dora Clarke's bronze head 'A Kikuyu Girl' – 'Negro Art', she called it – modernist African primitivism that served, apparently without irony, as the frontispiece to Huxley's *Africa View* (figure 7.8).[110]

The remarkable Ife heads offer a counter-story to Thomas Henry Huxley's ethno-craniology, to Raymond Dart and the Leakeys' precision search for ape-men in Africa and to universal history's presumption that non-European 'primitives' had no history. Julian might have imagined himself as a cosmopolitan adjudicator of culture, appreciative of African art (and Dora Clarke's African primitivism) and self-appreciative of his own influence in saving African heritage. But he simply couldn't imagine the Ife heads as solely West African, insisting that they were evidence of contact with Europe, with the Mediterranean. As he wrote his memoirs in 1965, he was aware that various art authorities styled the Ife heads as autochthonous, independent of Europe. He disagreed, insisting that the method of casting 'is a very complicated business, and I find it hard to believe that it was independently developed by African Negroes, however, cultured'.[111] Still, Julian stares into the royal African face from 1300, his appreciation deep, if flawed. He helped bring prehistory, anthropology and evolutionary ideas together, despite these shortcomings, making science and humanism talk to one another across nature and culture. The biological Man Family by the mid-twentieth century, and partly by Julian's hand, was becoming the cultural 'Family of Man'.

Figure 7.7: Julian Huxley's culture of man. 'JSH and cast of Ife head (W. Africa)', *c.* 1948.

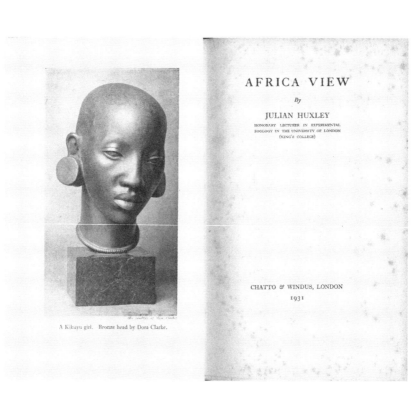

AFRICA VIEW

By

JULIAN HUXLEY

HONORARY LECTURER IN EXPERIMENTAL
ZOOLOGY IN THE UNIVERSITY OF LONDON
(KING'S COLLEGE)

CHATTO & WINDUS, LONDON
1931

A Kikuyu girl.   Bronze head by Dora Clarke.

Figure 7.8: Dora Clarke's bronze head 'A Kikuyu Girl'.
Frontispiece to Julian Huxley, *Africa View* (1931).

8

Human Difference:
Sapiens in the Political Present

'The Family of Man' was a precarious idea of mid-twentieth-century high modernity: 'The unity of all mankind within its varied diversity', as Unesco put it in 1956.[1] The continuities and subtle distinctions between T. H. Huxley's Man Family and Julian Huxley's Family of Man are themselves a journey over the nineteenth and twentieth centuries from biologically defined to culturally defined humankind. Herein lies the fascinating history of biological and cultural anthropology, in which we find these remarkable Huxleys at play again and again, and from its founding professional moment in Britain. The emerging discipline that studied humans – not jellyfish or monkeys or crayfish – was not just about the distant past, it was also about all kinds of people in the here and now, in the nineteenth- and twentieth-century present.

The Huxleys were resolute about a singular humankind, yet they were still theorists of variation, experts in changing modern systems of biological thought on human difference. Both Huxleys asked: if humans are not all the same, how do we value their difference at individual and population and political levels? They gave all kinds of answers, pondering the relationship between an ethics of cultural and biological difference and the in/equality of their fellow men and women. The human identity-difference problematic was as foundational to political philosophy as it was to natural history and biology. Political systems and legal systems recognized and placed different values on different kinds of humans all the time, in ways implicit, quotidian and unnoticed, as well as explicit, regulated and authorized. Part of the Huxley story is active participation in that politics across the nineteenth and twentieth centuries, an era of great manifestos and doctrines, of civic protests and liberation movements, of wars and revolutions. What was the best relation between individuals and states? How do we adjudicate differences between men and women? Just what are 'races' of humans, and how is that variation valued? And how do some humans get to govern other humans, with what justification or explanation, lawfully or unlawfully, ethically or unethically?

Over the nineteenth and twentieth centuries, lives were put on the line over these matters, over race and colonial rule, over slavery and emancipation, over women's political place vis-à-vis men's, over workers and owners of capital. Between 1825 and 1975 these Huxleys lived through colonial wars, civil wars, class wars, world wars and a cold war. And they were influential actors in a science that engaged openly with developments in democracy and human rights, with republican, liberal, socialist, communist and fascist polities. This family witnessed, thought about and acted both within and against ideas of 'liberation' and 'freedom', from the rights of man to the declaration of universal human rights. Theirs was a political biology, and generations of this family both knew it and owned it, for good and ill.

IMPERIALISM, ANTHROPOLOGY AND INDIGENOUS WORLDS

T. H. Huxley the anthropologist was created out of his early years in the Pacific. Like Darwin on the *Beagle* voyage, he arrived in the South Sea with all kinds of pre-conceived ideas about the people he met on those coasts. These were as much social and economic ideas – about agriculture and food, fertility and family, custom and beliefs – as they were physical or biological or comparative anatomical ideas. In the 1840s Huxley was not yet fascinated with the skeletal remains that came to preoccupy him in the 1860s. If anything, he was more drawn to bones as adornments. In Papua, for example, he took particular note of a 'peculiar necklace, to wit, a piece of their rude string on which several pieces of bark and a human odontoid vertebra were strung'.[2] And like his fellow naturalist on the *Rattlesnake,* John MacGillivray, he was fascinated by another set of human bones turned into a bracelet. This repurposing of a jawbone plus two human clavicle bones lashed together with twine was collected by Huxley and in 1869 donated to the British Museum, where it still lies.[3] For the moment, bones spoke to Huxley about humans' place in culture, not nature. He was immersed in living encounters in the present, not a palaeontology of the past.

Huxley caught some of this life in watercolour (figure 8.1). *Natives on the New Guinea coast 1849* is not a scene motivated by comparative anatomy or a Blumenbach-style drive to classify human difference, but nor is it, of course, unmediated by other aspects of Huxley's own culture. His scene fits particular philosophies about 'original man' that had been strongly argued for centuries: we might view this delicate watercolour less as mature Huxley on 'man's place *in* nature', and more as younger Huxley reflecting on 'man in a *state of* nature', recalling seventeenth- and eighteenth-century political philosophy that imagined societies before civilization. His was not a scene of Hobbesian pre-civil primitive violence. Here, instead, was familial sociability more like that theorized by his favoured David Hume in *A Treatise of Human Nature* (1739).[4] In this instance, at least, Huxley eschewed the idea of 'savage violence' and put pleasing familial order, peaceful abundance and lack of want into watercolour.

Yet as he painted in 1849 there *was* violence in and around the Louisiade Archipelago, categorically so. Huxley and his countrymen themselves crashed through this philosophical mythology of noble Islanders and peaceful and fertile sociability, however mythic or misplaced that no doubt was. Other sketches from the *Rattlesnake* that year show a 'pinnace and boat firing on native canoes' (figure 8.2).

This incident took place near what the English had just named 'Joannet Island' – Pana Tinani – in July 1849. Huxley recorded the event in detail. At dawn three canoes with about thirty men approached the English vessels, and one of the English sailors was struck with a stone axe in the head, with more strikes and spearing following as the Islanders tried to overturn the vessel. Two muskets were fired, and then the 'long gun from the pinnace opened fire upon them, first giving them a round shot ... and afterwards sending a charge of grape among them'.[5] Across the bottom of the sketch is scored 'Ye natives finding ye Grapes are soure', the kind of jocose comment Huxley might well make, though this particular making-light of imperial violence is not in his hand. Huxley assessed it all as 'a very treacherous business altogether', but he was not talking of

Figure 8.1: *Natives on the New Guinea coast, 1849.*
Thomas Henry Huxley, watercolour.

Figure 8.2: Pinnace and boat firing on native canoes with figure studies, *c.* 1849, attributed to O. W. Brierly and/or Thomas Henry Huxley.

the Englishmen. The Islanders, he thought, were insufficiently pun-ished for their disturbance of peaceful English visitation, and the naturalist John MacGillivray, for his part, considered that the vio-lence of the episode confirmed the savage disposition of the New Guineans.[6] In twisted imperial logic, Huxley later understood the English weapons themselves to signal civilization and imperial right and might, romanticizing the Islanders' lack of firearms. For him, that was the intellectual frisson of the cross-cultural encounter:

> we had the interest of being about the last voyagers, I suppose, to whom it could be possible to meet with people who knew nothing of fire-arms – as we did on the south coast of New Guinea – and of making acquaintance with a variety of interesting savage and semi-civilised people.[7]

This for him was a key marker of civilizational difference.

The violent encounter took place after the *Rattlesnake* had sailed up the Australian coast, leaving Sydney in late April 1848. The coast and hinterland were the sites of many calm and comparatively civil exchanges between the British and Aboriginal people, as well as robust but not necessarily violent or one-sided negotiations for traded com-modities. Other meetings were perceived by the British as important to the locals; they may well have been. At Cape York, the northern tip of Australia on the Torres Strait, Huxley travelled inland and was wel-comed by a local elder who recognized Huxley as 'the returning spirit of his dead brother', or at least so Huxley interpreted the situation. Some Aboriginal people joined the British men on various inland par-ties. One, whom they took the liberty of naming 'Jacky', accompanied the land-party that trekked from Rockingham Bay on the Coral Sea, south of Cairns, and was to be met by the *Rattlesnake* at Cape York. He was the only person who survived, the rest of the party perishing by starvation or speared and killed.[8] This affected Huxley deeply, not least since he might well have been one of them, having joined the leader, Kennedy, on several challenging early survey trips inland.

Another encounter in Rockingham Bay was with a Girramay elder who approached in a canoe and came aboard the *Rattlesnake*, 'very much disposed to make himself at home'. He and Huxley shared no language, but they appraised each other closely. The Indigenous man inspected Huxley's shift and black neck-ribbon. And Huxley noted the local adornment, especially 'a bar of red paint across the bridge of the nose and under the eyes'.[9] Huxley painted and sketched the portraits and figures of many local people on the *Rattlsenake*'s cruises. But in this encounter, he himself was 'painted' after the local fashion, the Girramay man applying red marks across Huxley's nose, behind his eyes and on his forehead. Huxley sketched the two of them in this intimate encounter. 'Hal being painted', Henrietta later wrote, and Julian included it in the 1935 publication of his grandfather's diary (figure 8.3).[10]

Such intimate cross-cultural encounters with Indigenous people were never to be repeated for T. H. Huxley, whose later travels were confined to Europe and North America. As he became one of the foremost authorities on the biology of human variation, in fact he rarely if ever again engaged with living, breathing, feeling, thinking non-Europeans at all. Frankly, his closest contact thereafter was with skeletal remains, which he assessed alongside ancient remains from the Scottish Islands or Cornish estuaries or German valleys. Perhaps the nearest he came was a peculiar 1869 project to anatomize and taxonomize various Indigenous people not via their remains, nor in person, but via photographs of their bodies-in-life. Would the Colonial Office oblige by circulating a request to scientists, missionaries and officials across the Empire for photographs of natives? Could they be arranged just so – see instructions – and can they please be entirely naked? Huxley received some compliant responses: full-length side, frontal and naked portraits, with ruler held by the person, to their left. However, other photographs arrived on Huxley's desk of people fully clothed. From New Zealand, Bermuda and North America came photographs and letters in which people refused to conform to the demand. Some missionaries and 'protectors' declined in no uncertain terms, arguing that their civilizing work in 'dressing' people would be jeopardized by Huxley's anthropological demand for nakedness.[11] Moreover, while comparative anatomist Huxley sought to pin down 'types' anthropometrically, some of what he received were rather more portraits of individuals.

Photographs arrived from Tierra del Fuego, conventional Victorian studio portraits of family groups or men standing next to dining

Hal. being painted

HUXLEY BEING PAINTED BY A NATIVE ACQUAINTANCE

Pencil, with red paint on the forehead, the bridge of the nose, and behind the eye, in both figures. Unsigned, undated : the legend is in his wife's handwriting. The scene is probably Rockingham Bay.

Figure 8.3: 'Huxley Being Painted by a Native Acquaintance', 1849.

Figure 8.4: Photographs of people from 'Terra del Fuego', *c.* 1870, sent for Huxley's ethnographic study.

tables or chairs (figure 8.4). The photographs sent of Maori were also of particular people: they were named. In all but one, either traditional or European dress was worn. The photographs that arrived on Huxley's desk from the Cape were mainly taken in Breakwater prison, and they were also identified. They were not generic, not just a 'type' or a 'race' or in Huxley's preferred term a 'stock' of humankind, but instead were highly individualized. The Cape photographs were even accompanied by elaborate genealogies that circled outwards generationally and temporally (figure 8.5). These individuals were thus placed in a familial context as well, providing Dutch or English names as well as their Khoisan name.[12] Not a tree-image genealogy that Darwin had imagined, and that Ernst Haeckel had developed, in a strange way this patterning anticipated the circular graphics of late twentieth-century genetic genealogies. Beyond names, it detailed key events and causes of death, identifying multiple generational relations in concentric patterns around the central boy, in this instance. One woman, for example, was 'married after her 1st husband's death ... who died of starvation'. And we learn that another relative was 'killed by a lion'.[13] Of course, all of these people were subjects of their own lives, far more than they were ever objects of observation and inquiry for Huxley's emerging science of man. Like the Huxley family, they lived through joys and tragedies, births, marriages and deaths.

There was another disciplinary use of family trees that this so-called 'Bushman Genealogy' from the 1860s anticipated: the kinship genealogies that were starting to turn anthropology towards culture, and away from biology. By the late nineteenth century, all kinds of new cultural and social methods began to challenge biological anthropology, Huxley-style, including the tracking and documenting of kinship through family trees, not in order to trace the inheritance of Mendelian characters, but in order to examine relationships of marriage, kinship, power and ritual.

One of the most important turning points for the discipline of anthropology in this regard was the 1898 Cambridge expedition led by Alfred Cort Haddon to the Torres Strait islands, around which the *Rattlesnake* had sailed fifty years earlier. Young Haddon had been early mentored by T. H. Huxley, and their early careers were similar. A zoologist who began by researching embryology and then marine

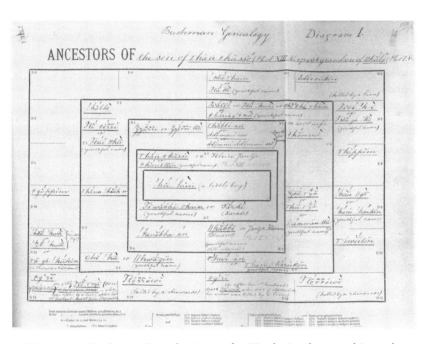

Figure 8.5: 'Bushman Genealogy' sent for Huxley's ethnographic study.

biology, Haddon first went to the Torres Strait to study coral reefs (after Darwin) and travelled around the islands and coasts of what became German New Guinea and British Papua in the early 1880s. There, and in the Torres Strait islands, he was 'anthropologized' and acculturated, much more fully than Huxley ever was.

On these journeys, Haddon began to wonder just what the British imperial mission was, and then he saw it with stark clarity. 'Between ourselves and our dependent races there is a great gulf of tradition, language & religion,' he wrote to Huxley, who knew as much. For Haddon, however, this great gulf was the starting point for his emerging anti-imperialism. In 1891 he was in conversation with Huxley about the new Imperial Institute, which was being built between the Albert Hall and the new South Kensington Natural History Museum. Now demolished except for the Queen's Tower of Imperial College, the Imperial Institute was intended to collect and display industrial and commercial products and processes from across the Empire. Haddon wondered aloud to Huxley about its purpose. Might it be redirected away from commerce and industry and instead become a cultural repository for anthropological knowledge of peoples of the Empire? Old Huxley's eyes narrowed suspiciously as young Haddon explained himself, in his self-edited plan for the Imperial Institute. 'The Anglo-Indian & the Colonial have too frequently a ~~(insolent?)~~ contempt for "niggers", a term of reproach which implies ~~the same~~ a hatred & superciliousness similar to that with which the Jews regard Gentiles.' And when Haddon appealed to their shared first-hand experience of the colonies, he lost his mentor altogether. Haddon's anti-imperialism was undisguised.

> It is hardly possible for those who have not visited the Colonies to realize how very deep, bitter & undisguised this contempt is. The Anglo Saxon ruthlessly forces every native into his own ~~narrow?~~ Procrustean bed of usage & belief, & in this is supported by the resources of the Imperial army & navy

Huxley, of course, had been part of that very navy. It got worse: 'The result of this policy is that we exterminate slowly or rapidly, unintentionally or by force, the inhabitants of the countries we annex,' challenged Haddon. Not mincing words, Huxley threw 'a bucket of cold water' on the idea. He told Haddon that a programme for converting the heathen to Christianity would have more chances of success than one to convert the new Imperial Institute to anthropological purpose of this sort.[14]

It is an important exchange that tells us that critique of British rule over Indigenous people was already alive and well in the late nineteenth century. Anti-imperialism is not just retrospective, but is itself part of the history of Empire, and indeed, of anthropology as a discipline. It also tells us how unwilling Huxley was to even begin considering an anti-imperial position. By then a firm critic of British imperial rule, Alfred Cort Haddon had fully misread his Huxley mentor. He presumed sympathy and alignment of ideas, perhaps by Huxley's boasting that he was sociologically qualified by his early journeys,[15] perhaps by their shared experiences of Torres Strait and Melanesian societies. Perhaps, indeed, Haddon's guard was down because of the substantial institutional and career support that Huxley had consistently, even generously, offered him. At the end of the day, Haddon's anti-imperial accusations were too close to the bone, even fifty years after Huxley's own South Sea journey, with its intermittent violence.

Haddon was turning the discipline of anthropology in a new direction, with a new method and purpose. Just after T. H. Huxley died, he published *The Study of Man*.[16] It was an evolutionary treatise, Huxley-style in one sense, but in another altogether different. In it, Haddon explained how the so-called 'civilized' people of the world carried vestiges of the past in their minds and cultures and behaviours as well as in their bodies. Colonial violence was one such vestige. There had been a slow shift in imperial anthropology from a time when expedition artists could depict naval fire on men in a canoe, and then with impunity quip about it, to Haddon's anthropological critique of colonial violence on Indigenous people, and his not unrelated turn away from biological anthropology. It was a delayed vindication, but for Haddon there must have been real satisfaction in later writing *We Europeans* with Huxley's grandson in the mid-1930s, an anti-racist critique of anti-Semitism, of fascism's misplaced use of 'biology'. The bucket of cold water was returned.

In the 1930s Julian was inclined to diminish his grandfather's anthropological credentials. When he read and edited the *Rattlesnake* voyage diary, he was upfront about its shortcomings: 'He had none of the trained anthropologist's insight into the black man's mind, little conception of the alien ways of thought and feeling in which a primitive savage is enmeshed.'[17] Even by that point, and even for Julian the zoologist, minds, thoughts and feelings were in the anthropological ascendant, over bodies, anatomies and physiologies.

Julian was never a field anthropologist, but he was certainly a traveller and an observer, and in his Unesco role was offered a kind of charge over scientific, educational and cultural analysis of 'peoples of the world', of 'the family of man'. He and his camera relished the journeys. In 1953, when he travelled to India to receive the Unesco Kalinga Prize for the Popularization of Science, he characteristically took the opportunity to visit as many countries as possible on the way. After crossing the American continent, visiting Aldous and Maria in California, he went on his own airborne South Sea voyage. He was delighted by Hawaii. Too delighted. Leaving any critical skills to one side, Julian joined the then-standard chorus about the new American state's lack of 'race prejudice', buying into the myth of all 'living together in these enchanted isles'.[18] He flew to Fiji, where he was impressed that rugby was played so well, but the island was ecologically disastrous, he noted, with so many introduced species. In Sydney, he thought of his grandfather and grandmother meeting and falling in love, and he followed the *Rattlesnake* journey in rapid time, flying up the east coast of Australia to the Great Barrier Reef, which left him awestruck. And then to Cape York, where in a precarious three-seater plane he flew over the forest that Huxley had slashed through with poor Kennedy and his party, who had later perished. Having worked so carefully through his grandfather's diaries and sketches, Julian knew that story better than anyone then alive. And then he experienced Torres Strait Islanders for himself. Shamefully for a recipient of the Anthropological Institute's highest honour (or perhaps appropriately because it was the 'Huxley Medal'), in his 1970s memoir Julian wrote offensively, but he thought colloquially: 'I met some cheerful abos on Melville and Goulburn Islands.'[19]

Partly because he was a believer in progress, including evolutionary progress, Julian sustained the idea of a worldwide 'development' of all people and all societies. Yet for him, unlike for most cultural anthropologists by the 1950s, differences were still hierarchical, not relative. Indeed the intellectual trend to the latter irritated him. 'One school of anthropologists', he complained in 1953, 'is never tired of proclaiming the doctrines of cultural relativity – that different cultures are not higher nor lower, but merely adjusted in different ways.' He appealed to his core discipline and to his Huxley roots:

> This is to neglect the lessons of biology . . . there can be no doubt that the social and cultural organization of the Australian black-fellows is a survival of a very low and primitive type of culture, even though some features of it, like the intricate details of its totems and its marriage system, are clearly a recent specialization.[20]

Analysis of culture – 'totems' and 'marriage systems' and 'kinship' – was what most anthropologists were now doing, and Julian only appreciated this knowledge as an amateur, as a traveller-observer, in many ways not that differently to his grandfather, despite his leadership of the cosmopolitan Unesco.

The whole point of social and cultural anthropology was to be as closely enmeshed as possible with local societies. This was not the distant shipboard observations of Huxley's *Rattlesnake* voyage, with the occasional close encounter where one's face was painted in traditional style, or where a local could be persuaded to be sketched. Professional anthropologists now spent long field seasons with their subjects, a new ethnographic method. Oxford's Africanist Edward Evans-Pritchard described the purpose of fieldwork thus in 1951: the anthropologist 'must live as far as possible in their village and camps . . . physically and morally part of their community. This is not merely a matter of physical proximity. There is also a psychological side to it.'[21] Kinship studies had unfolded quickly, especially in the Pacific and in Australia after Bronislaw Malinowski's work, and across Africa by many of his students, not least Evans-Pritchard.

In 1951, when Evans-Pritchard pitched a psychological side to ethnography as president of the Royal Anthropological Institute of Great Britain and Ireland, yet another Huxley was at Oxford. Fittingly, unsurprisingly, Julian and Juliette's son Francis Huxley had just completed an Oxford degree in zoology at Balliol College and was pursuing, after his father's example, ornithological expeditions to the

Norwegian Arctic Island of Jan Mayen, and to Gambia in 1948, after initiation into anthropology by Evans-Pritchard. Then Francis discovered Brazil. In 1951 and again in 1953 – just when Julian was making dismissive comments about cultural relativism – Francis Huxley set off into the Amazon jungle, to Urubu villages between the Pindaré and Gurupi rivers. Julian and Juliette watched with considerable pride, though also with anxiety when their second-born was 'lost' in the Amazon, incommunicado for many months. As Evans-Pritchard instructed, he was becoming physically, morally and psychologically as much part of the community as he could. T. H. Huxley would likely have dismissed the method as highly suspicious not least with regard to the impartiality required of the science, but he would have praised his great-grandson's decision to write a popular travel book from those years, in the family tradition. Its title – *Affable Savages* – if not its content, would be comprehensible to the Victorian Huxley. Francis used the term more knowingly than his father and great-grandfather, but prurient interest in 'the savage' still sold books in 1956, and a cultural anthropologist could still use the term with impunity, probably the last generation to do so. Cruel and vindictive in war, Francis wrote, Urubu people could kill and eat a prisoner. Yet such acts do not therefore indicate lack of principle or order, he explained. 'Savages in fact have morals,' and nothing is trivial: superstitions, customs, day-to-day rituals. Francis Huxley published a genealogical chart to show how the members of two villages were related, how the marriages and intergenerational movement between the villages worked.[22] Culture helped him understand the Amazonian's own comprehension of 'man's place in nature': 'Taboos are especially interesting to the anthropologist, for they arise wherever man sees the supernatural entering – or perhaps emerging from is the better word – the world of nature.' The distinction between the natural and the supernatural he thought clumsy, but still useful.[23] For Julian and Juliette's son, Huxley and Henrietta's great-grandson, anthropology was all about understanding thought and feeling that was initially alien, and only through immersion familiar.

Francis Huxley's own idiosyncratic career was as unprofessional as it was possible to be, while still occupying the identity of 'anthropologist'. T. H. Huxley's effort to make salary, position and status available to men of science radically attenuated in this line of the dynasty – Francis would say for the better. At the same time, it was strengthened in parallel family lines, most especially with the physiologist Andrew Huxley, Julian's much younger half-brother. While Andrew Huxley followed the most stellar professional career possible – master of Trinity College, Cambridge, Nobel prize-winner, president of the Royal Society – Francis Huxley moved further and further away, certainly from natural sciences and then even from conventional human sciences. Led, perhaps drawn, by uncle Aldous and trained by R. D. Laing, Francis moved towards analysis of interior cultures, of minds, of psychology and psychoanalysis. He eventually found more of a home, like Aldous, on the west coast of the United States than in the home counties or in London. The eccentric desert-dwelling, psycho-analytically inclined, LSD-experimenting expert on shamanism and voodoo experienced and pursued anthropology as a profoundly cultural and psychological inquiry.

Francis represented one endpoint of the long and strong connection between the Huxley family, anthropology, Indigenous people and imperialism. In 1969 the author of *Affable Savages* co-founded the organization Survival International, originally the Primitive People's Fund. Unlike his father or great-grandfather, Francis was by all accounts no organizer but is remembered as the intellectual force of the group, cautioning against Rousseau-like romanticization of Indigenous people. He travelled back to Brazil in 1972 on a government-sponsored trip and co-authored the report on Amazonians for the Aborigines Protection Society, the same society that had launched his great-grandfather's Ethnological Society in the 1840s.[24] Survival International still defends the land and rights of Indigenous peoples: 'for tribes, for nature, for all humanity'.[25] Alfred Cort Haddon would have been delighted, a strong version of anti-imperialism finally embraced by a Huxley.

These Huxleys embodied the generational fortunes of British anthropology, its methods, its politics and its strange dance with knowledge of bodies and minds, with nature and culture. Yet it is not quite sufficient to see this simply as a shift over time from 'biological anthropology' to 'cultural anthropology'. T. H. Huxley himself inherited knowledge of human difference that was fundamentally economic, not biological.[26] And as we have seen, Darwin began to consider culture and even psychology and morality within the framework of natural selection, something that piqued Julian's intellect especially strongly. By the same token, by the time Julian Huxley died in 1975, a new kind of biological anthropology drawing on molecular biology and genetic studies was in the ascendant, sitting uncomfortably alongside the cultural and psychological work in which his son Francis was immersed.

Perhaps 'biology' and 'culture' intersected most forcefully and mutually with, and for, Julian. It did so through his fascination with the concept and practice of 'ritual'. Fifty years and two world wars after his observations on the grebes that magical spring with Trev, he curated a symposium for the Royal Society on Ritualisation of Behaviour in Animals and Man. The aim was to think through ethology, psychology and ritual across animal and human domains, extending the Huxley traditional understanding of 'man's place and role in nature' towards a new era of sociobiology. As an evolutionary biologist, Julian was convinced that the purpose and result of ritualization in higher animals and 'man' are analogous, a key mechanism in his particular post-war specialty, 'psychosocial evolution'.

What are such rituals? Julian drew from his life as much as his science: birth and name-giving, initiation into adult life, marriage and death. He nominated 'scapegoat' rituals, from the Jews' expulsion of a single symbolic goat to the Nazi scapegoating of the entire Jewish race. He spoke of rituals of atonement for sin and military rituals of discipline and rank-hierarchy; rituals of national bonding like Russian May Day celebrations; psychiatric and moral rituals, including their relation to tabu and religion.[27] Ritualization of Behaviour in Animals and Man was a landmark conference and then publication for ethologists, evolutionists, primatologists and anthropologists, a testimony to, and culmination of, Julian Huxley's twentieth-century reach across evolution and behaviour, across bodies and minds, across nature and culture. It was yet another intergenerational Huxley affair. Julian's own son was there, extending some of the work on ritual and religion and minds on which they had been collaborating. In Julian's book *The Humanist Frame*, Francis contributed 'The marginal lands of the mind'.[28] And in his own introduction to the significance of considering ritual and human psycho-social evolution together, Julian drew on Francis Huxley's insights into voodoo gods in Haiti, analysed in his book *The Invisibles*.[29]

Each year, the Royal Anthropological Institute still awards the Huxley Memorial Medal, its highest honour, and through that measure alone we can witness anthropology as a discipline shift over time from the study of nature to the study of culture. In the early years the Huxley Memorial Medal was awarded to classifiers of race and to biometricians such as Francis Galton (1901), Karl Pearson (1903), Joseph Deniker (1904) and William Z. Ripley (1908). In 1920 the Huxley Memorial Medal was awarded to Alfred Cort Haddon, a key acknowledgement that post-war British anthropologists were confirming the ascendance of cultural over biological anthropology, mirroring the shift in the United States first to the influence of Franz Boas and then of his students, including Ruth Benedict and Margaret Mead. Partly due to Haddon's work in the Torres Strait, anthropology was less about bodies and more about what other people thought and felt, acted and believed, about their kinship systems, their rituals, their symbols, their families and communities. Everything was changing with and for anthropology. By the middle of the century and onwards, the Huxley Memorial Medal marked a different disciplinary tradition altogether, awarded to Alfred Radcliffe-Brown (1951), Edward Evans-Pritchard (1963), Claude Lévi-Strauss (1965), Clifford Geertz (1983), Mary Douglas (1992) and Marilyn Strathern (2004). If we imagine T. H. Huxley as a very long-lived man, we can be sure he would have read all their work carefully, but would he have recognized it as anthropology? And would he have endorsed the new fieldwork methods in which immersion and identification with other cultures were valued and expected? Would he have awarded them his own memorial medal? It is safer to think of his inclination towards a minor line of medallists, the anatomical evolutionists and primatologists whose work he would have recognized more readily: anatomists Arthur Keith (1928), Grafton Elliot Smith (1935), Wilfrid Le Gros Clark (1958). We can also be safe speculating that the patriarch would have endorsed the award of the Huxley Memorial Medal to Julian in 1950.

ANTHROPOLOGY, SLAVERY AND
EMANCIPATION IN THE 1860s

Thomas Henry Huxley did more than authorize biological anthropology in Britain, he was instrumental to the shape, nature and the very title of its professional association. This was all the patriarch's personal handiwork.[30] What is now the Royal Anthropological Institute of Great Britain and Ireland was initiated, named and structured by him.

The 1860s was a decade in which disciplinary distinctions in the science of man were being worked out. 'Ethnology' involved intra-human study, according to Huxley. It could be zoologically oriented – that is, the study of anatomy and physiology between different humans, his own specialty. But it could also be culturally oriented – the study of language, for example, 'the most human manifestation of humanity'. Ethnology in the 1860s *could* also entail the study of manners and customs and beliefs, anticipating cultural anthropology. Together, these might offer 'a clue to the origin of the resemblances and differences of nations'. Anthropology, by contrast, was then the study of humans' place in larger nature, in relation to other animals. Like a disciplinary Russian doll, ethnology sat inside anthropology, which sat inside zoology. And zoology, Huxley explained, was the 'animal half of biology – the science of life and living things'.[31]

The Ethnological Society of London had not emerged organizationally out of natural history, however. It had been founded in 1843 as an offshoot of the Aborigines Protection Society, the humanitarian organization that was to send Francis Huxley to report on Amazon tribes 130 years later. The Ethnological Society broadly maintained its liberal, humanitarian and anti-slavery ethos, but in 1863 some members disagreed enough to form the breakaway Anthropological Society, led by James Hunt, an idiosyncratic speech therapist turned student of human difference. Unlike Huxley's group – 'the Ethnologicals' – Hunt's 'Anthropologicals' were in the main polygenist: Hunt himself was explicit about his belief in more than one 'species' of human. Indeed, he positioned himself and his new breakaway society thus precisely by contrasting himself to Huxley.[32] Hunt read his own paper on man's place in nature in 1863 at the British Association meeting in Newcastle.

It was titled 'On the Physical and Mental Characters of the Negro' and was ill-received, to say the least. When Hunt delivered the same ideas to his own smaller and internal audience at the Anthropological Society of London, his message was entirely clear: 'fully considering the white man and the Negro distinct species'.[33] Hunt's proclamation was foundational to the new Anthropological Society, constituting the first article in its *Memoirs*, with the title pointedly changed: 'On the Negro's Place in Nature', an unmistakable reference to Huxley's recently published book, and a dispute with Huxley's monogenism.[34]

Just when British intellectual elites were warring over man's place in nature, Americans were in civil war, 1861–5. Bloody and brutal battles were fought between those who supported the expansion of enslaved labour across the continent and those who did not. If the Ethnological Society had an undergirding affiliation with abolitionism and humanitarianism through the Aborigines Protection Society, Hunt's Anthropological Society was, through its secession, the opposite. It was connected, if not quite to slavery (or as Hunt put it, the 'so-called "slavery" of the Confederate States of America'), certainly to the South. Hunt's *Negro's Place in Nature* was republished in the United States in 1864 as an anti-abolitionist tract.[35] It was one of the crudest justifications of slavery to emerge from the British natural sciences. Huxley's group knew it and responded. When Berthold Seemann, a vice-president of the Anthropological Society, suggested an 'anthropological garden' along the lines of the Zoological Gardens in London, 'where living specimens of the principal varieties of the human race might be seen and compared', Huxley's friend and X Club member George Busk hit back with what shouldn't have needed saying, but in mid-Victorian London apparently did.[36]

Despite the extreme positions of the Anthropological Society, Huxley sought to solve the organizational rift that had emerged within the science of man, negotiating with the Anthropological Society rather than casting it out. In fact, his own highly qualified abolitionism, in the end his disbelief in the equality of all humans, as we shall see, made negotiation with them personally feasible. That Huxley came to be cast *against* the Anthropologicals, as a liberal and progressive anti-racist monogenist, and sometime emancipationist, is if anything a measure of the extremity of racism of the high-Victorian era. He attempted several times to merge the two societies and was finally successful in 1871, his elegant resolution a new name altogether, the Anthropological Institute of Great Britain and Ireland.[37]

Huxley stood against the mid-nineteenth century revival of polygeny in a way that sometimes served his patriotic posturing more than any principled position, shoring up the idea of a particularly British commitment to anti-slavery. For him, opposition to irrational polygeny often associated with American biology, and justified by twists on Genesis, was the opportunity to parade his British approach. 'On these islands,' Huxley wrote in his characteristically multi-claused prose,

> we are in the habit of regarding mankind as of one species, but a fortnight's steam will land us in a country where divines and savants, for once in agreement, vie with one another in loudness of assertion, if not in cogency of proof, that men are of different species; and, more particularly, that the species negro is so distinct from our own that the Ten Commandments have actually no reference to him.[38]

There *is* a biogeography of human difference, but he warned that only a careless zoologist would therefore pronounce a *species* difference. The evidence against that idea was to be located in continuous reproduction between all kinds of humans: the singular mark of different species was infertility.[39] And yet belonging to the same species, *Homo sapiens*, hardly meant equivalence for Huxley and certainly did not confer anything like equal political rights.[40] One problem was Huxley's arrogant certainty about his own position. Pursuing the doctrines and discourse of emancipation in the 1860s, he asserted 'nature' to be the final arbiter. With fair field and no favour, nature would reassert the hierarchical status quo. It was arrogant because 'nature' was his own special knowledge; Huxley was nature's spokesman.

And yet, as the Civil War unfolded, Huxley's stance on species unity was useful enough for emancipationists to bring him onside, this notwithstanding his thinnest of abolitionist positions, certainly not based on a political philosophy of equality.[41] With apparent magnanimity, but in fact with a deep presumption of superiority, he argued that the emancipation of the enslaved was warranted less on a political than a moral ground: arbitrary domination damages the master. And so, with abolition of slavery in the United States, 'the master will benefit by freedom more than the freed-man.'[42]

Just what influencers like Huxley thought about the men and women of the large African diaspora was hardly an American question alone in the 1860s. Huxley was born into a British world that still engaged in slavery if not the slave trade, sugar from the West Indies grown, harvested and milled by the labour of enslaved men, women and children. That was not ended until 1834, and in some places, 1838. His own Caribbean-born wife Henrietta could just remember hiding during slave rebellions in the late 1820s. A generation later, in 1865, there was a rebellion in Jamaica that changed everything. Protests over poverty and exclusions from the franchise escalated violently. Governor Edward John Eyre came down hard, declaring martial law. The Morant Bay rebellion, as it came to be known, was suppressed with Eyre's troops killing perhaps 400 people, followed swiftly – too swiftly – by mass arrests, trials, corporal punishments and executions. One of the men executed under this much-contested martial law was George William Gordon, businessman and mixed-race member of the Jamaican House of Assembly. The entire tragic sequence immediately sparked controversy in Britain, and a 'Jamaica Committee' was established and eventually chaired by John Stuart Mill, with Huxley on the executive committee. They called for the criminal prosecution of Eyre for murder, specifically for the murder of Gordon after a summary trial. Huxley's liberal friends lent their names to the Jamaica Committee as well: Charles Darwin, Herbert Spencer, Charles Lyell, Henry Fawcett, each perceiving British legal process to have been fundamentally undermined. It was a highly public and divisive matter that set friends against each other. A counter-committee was established to support Eyre, driven by Thomas Carlyle. Especially challenging for Huxley was the active and public defence of Eyre by his closest friends, John Tyndall and Charles Kingsley. Huxley wrote them careful and warm letters, assuring them both of his own best efforts to recognize profound disagreement of principle, while maintaining and sustaining precious friendship, even, he wrote to Tyndall, their love. 'If you and I are strong enough and wise enough, we shall be able to do this, and yet preserve that love for one another which I value as one of the good things of my life.'[43]

Despite the caricature of reserve, mid-Victorian affection between men was often honest and warm and forthcoming.

Mill, Huxley and the Jamaica Committee insisted that their mobilization concerned the blatant disregard of British legal process and principle. But it was disingenuous to think that the whole Morant Bay episode was not about race. It brought old British pro- and anti-slavery arguments, as well as current American racial politics on slavery, to the fore. Mill and Carlyle had long been pitched against one another, Carlyle arguing not for slavery but for a kind of serfdom, for him an admirable feudalist ideal in which everyone worked. Mill's 'the Negro Question' was the response in which abolition of slavery was part of a long-term programme for the extinguishment of all tyranny.[44] Even if the actual circumstances in Jamaica are set aside, longstanding British debate, as well as the immediate context of the American Civil War, meant that being for or against Governor Eyre implied – at the very least – being for or against chattel slavery.

Unsurprisingly, then, Huxley's work with the Jamaica Committee was popularly read as a liberal statement on race, equality and 'man's place in nature'. *The Pall Mall Gazette* saw it thus, with a 'would-be smart allusion', as Leonard Huxley later dismissed it, to Huxley's 'peculiar views on the development of the species'. The editor considered that defence of a mixed-race Jamaican politician amounted to 'bestowing on the negro that sympathetic recognition which they are willing to extend even to the ape as "a man and a brother"'. Huxley responded quickly, writing from the Athenaeum Club that his executive work with the Jamaica Committee had absolutely nothing to do with his views on species, nor with his work on man's place in nature. More precisely (and for us more tellingly), it was neither motivated 'by any particular love for, or admiration of the negro', nor by a base vengeance against Governor Eyre. Rather, with Mill, he saw the Morant Bay episode as a question of British legal order of the most fundamental kind: 'Does the killing [of] a man the way Mr Gordon was killed constitute murder in the eye of the law, or does it not.'[45]

The Pall Mall Gazette's 'would-be smart allusion' was at one level nonsense. At another level it correctly signalled the dire mixing of comparative anatomy Huxley-style, evolutionary ideas about common descent and the crude racism of the 1860s. It was reasonable to make the link between Huxley, 'the Negro Question' and the Jamaica Committee, not least since Huxley had the year before gone on record with a highly positioned essay on emancipation.

'Emancipation – Black and White' (1865) was prompted by the Civil War, not by the Jamaica rebellion.[46] The terrible war in the United States was very much on Huxley's mind because his much-loved and much-missed sister, Lizzie, lived in Tennessee. More than that, his nephew was serving with the Confederates. He wrote to her with sympathy regarding her position in the South, 'my heart goes with the south, and my head with the north'. Still, he lay down his position on slavery before her: 'whether you beat the Yankees or not you are struggling to uphold a system which must, sooner or later, break down'. In a long letter written in May 1864, he declared privately what he would soon say publicly.[47] His opposition to slavery did not derive from a conventional humanitarian politics. He was not swayed by the Wedgwood credo that was in Darwin's blood and soul, 'Am I not a Man and a Brother?'[48] In fact he confessed to Lizzie no sympathy for the 'black man'; 'don't believe in him at all, in short'. His position derived from what 'slavery means for the white man': bad political economy, bad social morality and bad influence 'upon free labour and freedom all over the world'. He wished slavery to be ended 'for the sake of the white man'.[49] So much for Huxley's idea of black 'emancipation'.

Closer to his own home was what he called 'white emancipation'. The Ladies' London Emancipation Society met regularly at Aubrey House in Kensington. This was not an association for the emancipation of ladies, it was a women's association for the emancipation of American slaves, although over time the ambiguity became part of the political point; the political lesson. It was established in 1863 after women found they could not join the London Anti-Slavery Society, and Harriet Beecher Stowe was among the well-known American abolitionists calling for English women to create their own society in support of the North. The Ladies' London Emancipation Society was alive to Huxley's work. They borrowed his name, his authority and his expertise for their cause, notwithstanding his highly qualified argument against slavery, and not to mention his positive reluctance to concede the equality of all men, let alone all humans. They published a pamphlet in 1864, *Professor Huxley on the Negro Question*.[50]

It set out Huxley on his usual material: the comparison of skulls, feet, teeth, brains and his usual conclusions, regarding the singularity of species and the lack of any intermediate type. This was hardly the most persuasive argument against slavery. Indeed the Society didn't have much to work with, but deploying Huxley was expedient because he had so publicly debunked Hunt's absurd claims in *The Negro's Place in Nature*, and the supposed science of various slavery advocates. And yet, as if in a show of impartiality and fairness, Huxley also called the idea that 'the negro is the equal of the white man' the view of 'fanatical abolitionists', and a proposition so 'hopelessly absurd as to be unworthy of serious discussion'. Still this did not stop the Ladies' London Emancipation Society, and others, from praising Huxley and claiming him for the anti-slavery cause:

> He believes in the doctrine of freedom, or equal personal rights for all men, and he pronounces the system of slavery to be root and branch an abomination – thus making his physiological definition of the Negro's place among men equivalent to an earnest plea for Negro emancipation.[51]

It was a misrepresentation – at the very least an expedient stretching – of Huxley's position.

WOMAN'S PLACE IN NATURE: GENDER AND THE MAN FAMILY

Emancipation black and white was a more proximate enterprise than Huxley imagined when he came up with his arch title. It was Clementia Taylor of Aubrey House, Campden Hill, who published 'Professor Huxley on the Negro Question' under the Ladies London Emancipation Society's instruction. Representing perhaps 200 members and subscribers, the society included Charlotte Manning, Harriet Martineau, Frances Power Cobbe, Emily Faithfull and Helen Taylor. Most of these women were involved in their own emancipation too, ultimately with far more energy and focus. As with an earlier generation of anti-slavery societies, the Ladies London Emancipation Society was an abolitionist group in purpose, and a women's group in practice and in effect. In 1860s London, and with this personnel, the former quickly morphed into the latter.

London's west was criss-crossed in a new feminist geography in the 1860s. Aubrey House was just a mile away from the Huxleys' home. A mile further was the office of the *English Woman's Journal* at 19 Langham Place, Westminster. The Langham Place group had been meeting since 1859, developing initially cautious positions on matters like girls' education, legal restrictions on married women's property rights, women's employment and eventually women's right to vote.[52] From 1865, many of them joined the Kensington Society, which met to debate and refine the issues and arguments at Charlotte Manning's house, 44 Phillimore Gardens. And from that base, Emily Davies and Barbara Bodichon prepared a petition in consultation with John Stuart Mill and his stepdaughter Helen Taylor, the outliers who lived over in Greenwich. It was a petition that Mill brought to the political centre of Britain and the British Empire, the parliament in Westminster, to no avail. And so the first committee to campaign for women's suffrage was formalized back in Aubrey House, in October 1866, and the Committee of the London National Society for Women's Suffrage held its first of many meetings in July 1867.

A key platform of London women's organizations in the 1860s was educational: women chipping away at the restrictions arbitrarily placed on their entry to spaces of learning and exchange. Education, as well as science, was Huxley's core business, especially his well-known dedication to the broadening of education for working-class men. His involvement with women's education had different class implications, however, and he was regularly approached for support by women of his own circle. His pattern of response was to support women, but as different and subordinate, not as equals. The journalist and novelist Eliza Lynn Linton wrote to him in 1868: might women be admitted to meetings of the Ethnological Society?[53] No, he replied, but at least he was consistent. As secretary of the Geological Society, he also opposed the admission of women, reserving it for professionals, that is to say, men. With more success, Maria Georgina Grey sought Huxley's support for Girton College, Cambridge. This he would agree to. Separate education was tolerable to him, possibly even desirable. Women, Grey argued, want education because:

they, like men, were created in the image of God, because to develope
[sic] and cultivate and perfect that Divine element within them, is their
right and their duty, and no man has the right to disbar them from
entering on this glorious inheritance of the children of God.[54]

It was hardly an argument to convince agnostic Huxley, although
Moravian-educated Henrietta was more persuaded by this theology
of sex equality.[55]

Huxley was more inclined to ventures that operated in secular terms.
Sophia Jex-Blake, a force behind the London School of Medicine for
Women, asked him in 1872 to support separate anatomy classes for
women at Edinburgh University. He would not hesitate to teach anything to a class composed solely of women, he replied, but certainly
not to mixed student groups.[56] True to his word, Huxley became a
governor of the London School of Medicine for Women, founded in
1874. He had for some time taught physiology separately to women in
South Kensington, his 'S.K. Lectures to womankind' he called them,
writing to his physiologist friend Michael Foster, who kindly took
them over in January 1872, when Huxley was on the cusp of another
breakdown.[57] His own pupils assisted over time. Margaret Macomish
taught physiology to women in London.[58] She also illustrated *Boys
and Girls in Biology; or, Simple Studies of the Lower Forms of Life*, a
series of Huxley's lectures edited and published with his permission by
another of his South Kensington students in the 1870s, the American
Sarah Hackett Stevenson.[59] She went on to become professor of obstet-
rics at the Northwestern University Woman's School, and the first
woman member of the American Medical Association.[60]

Huxley himself was a beneficiary of the ideas and talents of well-
educated women. He relied on them often enough to translate his own
work, one of the few ways in which women could earn money independently. Henrietta expertly assisted his German translations, though he
received that work for free. The French Londoner Marie Halboister
sought his permission to translate and republish in Paris his essay on
the 'Natural Inequality of Men'. She had her own politics in mind, concerned that Rousseau's eighteenth-century principles put into practice
were bringing 'calamities upon my country' all over again. Huxley was
more than happy for his own wisdom to be thus translated and applied.[61]

But it was a different story when women communicated their own philosophical or political positions in the public sphere. His responses were dismissive, to say the least. '[A]bsurd epicene splutter', he described one of Helen Taylor's pieces in the *Fortnightly Review*.[62]

Huxley was surrounded by women every bit as clever as himself: his neighbour, George Eliot, for one, his daughter-in-law Julia Arnold Huxley for another, and her sister, Mary Ward, for a third. By the time Huxley died in Eastbourne, on 29 June 1895, mourned deeply by the devoted and conventional Henrietta, a different species of woman had become part of the Man Family. The 'new woman' was educated, independent and political. She was the beneficiary of the successful agitation over the 1860s and 1870s. She was likely to support not just women's capacity but also therefore their right to vote, to represent themselves in parliament, to be full subjects and citizens. But she *lived* her new version of femininity rather more than she created committees or organized petitions. The new woman exercised as much liberty over her own person, education and place in the world as she possibly could and so was as often disappointed as she was vindicated. The new woman looked just like Huxley's daughter-in-law, Julia.

The Arnold sisters – Julia, Mary the anti-suffragist and Ethel the suffragist and lecturer on women's political rights – lived through the second great moment of British feminism, which became focused on parliamentary representation.[63] Mary and Julia talked often about political matters, including women's and men's roles therein. Yet the surprise was that Mary Augusta Ward, philosopher, novelist, translator, activist, so well known publicly as Mrs Humphry Ward, was president of the Women's National Anti-Suffrage League from 1908, in that role editing the *Anti-Suffrage Review*. The civic world should mirror nature and natural human differences, Mary Ward was convinced. The premier argument against women voting was '[b]ecause the spheres of men and women, owing to natural causes, are essentially different, and therefore their share in the public management of the State should be different'.[64] Huxley would have agreed, but he did not live long enough to see the resurgence of campaigning in the Edwardian era.

'It certainly isn't a matter that must come between you & me!' Julia Huxley reassured her sister in 1908, just when Mary had taken up the new anti-suffrage role. She did not feel especially strongly about it, she wrote, a little disingenuously, 'except when I have just been reading Mill!', and then she argued that humanity, rather than gender, was surely the point:

> we are all human beings before we are men & women – that if it is shown to be, on the whole, right & wise & for the improvement of humanity that people should learn to govern themselves, it is just as much so for women as for men – because before all else they are just human beings.

But Julia Arnold Huxley, educator of girls, still held a qualified view:

> I don't want adult suffrage yet, for either men or women (but I am quite certain the woman of the lower classes is much more fit to have it than the man!) but it seems to me it ought not to pass the wit of man to devise something between that & 'the same terms as men' enfranchisement, which so obviously doesn't meet the case well.[65]

Ailing Julia just lived long enough to witness the mass London rallies and the suffrage campaigners turn militant, with the imprisonment of women from 1905, but she did not live long enough to witness the prison hunger strikes and forced feeding that began in 1909. Living until 1920, Mary, on the other hand, saw the Parliament (Qualification of Women) Act passed in 1918, to which she was so opposed, but not the Representation of the People (Equal Franchise) Act of 1928, which extended votes to women on the same term as men, that is, as adult humans.

Aunt Mary Ward and young Julian Huxley occasionally discussed women and men and voting. She offered one version of her arguments to him in 1912, just as he was setting sail for the new Rice Institute opening ceremonies. It was what we would now call a biologically determinist proposition, and so she bravely entered Huxley territory. Julian was trying to argue that women's and men's brains were the same and it was opportunities for education that shaped their capacities one way or another. But Aunt Mary would have none of it:

> The functions of sex determine by their use the leading characters & activities of the brain. If women were not mothers their brain might work in the same way & the same ends as that of men. But they are

mothers, actual & potential, & that ~~determines~~ develops the brain power of the woman along channels of her own – determined by the facts of her own special environment, home, children, etc. while the brain power of men turns naturally to political uses owing to the man's different environment. This surely amounts to a fruitful differentiation of functions with the human race as a whole.

Mary Ward's 1912 position accorded more or less with Thomas Henry Huxley's, circa 1860, and Henrietta Huxley's for that matter. 'It only leads to waste of power for both sexes, if their division of labour is not recognised both socially & politically,' Mary declared to Julian.[66] She was smarter, more learned and more successful than most of the men in the Huxley and Arnold families, and earned more money from her writing than Leonard Huxley or Julian Huxley or even Aldous Huxley ever did. And it was she, not they, whom Theodore Roosevelt invited into wartime cultural diplomacy.[67] But such capacity, success and influence was not to interrupt or curtail a gendered – masculine – public sphere. 'Complementary powers, correspondingly but not identical functions – to each sex its own', Aunt Mary finished. And she signed off poignantly to Julian in 1912, unaware of the family tragedies just around the corner, 'much love to Kathleen and my Trev'.[68]

POLITICAL BIOLOGY

Over the 1920s and 1930s, Julian Huxley was part of an increasingly active and vocal group of biologists, lending their names to a suite of liberal, progressive and, for some, socialist and communist agendas. The decades were extreme. Stalin's communism was embedded, the National Fascist Party had been governing Italy since 1922, German fascism followed in 1933, and the Spanish Popular Front called its communist international brigade to fight General Franco in the Civil War from 1936. There were certainly advocates of fascism within British political culture, but the continental shifts towards authoritarian regimes energized variants of British liberals, and pushed others towards the anti-fascist left. Julian Huxley was deeply interested in political and economic planning and a believer in state regulations and initiatives of various kinds, yet while he had many socialist and communist friends within his biology circle, he was well to the centre of their politics. On the one hand, he became president of a new trade union for workers in science (figure 8.6).[69] On the other, he was active in the Political and Economic Planning group – PEP – a think-tank for 'capitalist planning' organized in 1931 with ornithologist Max Nicholson and others.[70]

While his grandfather's political action in the world of biology concerned access, education, class and meritocracy, breaking down the hold of the Church and Oxbridge on knowledge and on positions, Julian and his biologist friends immersed themselves in the great embattled ideologies of the twentieth century. Some put themselves on the line for it.

Who was in this circle? Jack Haldane was a fellow of New College, Oxford, with Julian in the early 1920s. He was a socialist and by 1937 a supporter and later member of the British Communist Party. In 1957 he moved to India, heading a biometry unit in Kolkata, taking Indian citizenship in 1961. Experimental zoologist Lancelot Hogben had an equally politicized trajectory. He was a pacifist and conscientious objector in the First World War, became Chair of Social Biology at the London School of Economics and Political Science in 1930 and, unlike most of Julian's biological circle, was strongly opposed to eugenics. Joseph Needham, whom Julian was to entice into Unesco, was strongly involved in Christian socialism. Famed scholar of China, he was sympathetic to the Chinese Communist Party after 1949 but was never a member of the British Communist Party. Geneticist and embryologist Conrad Hal Waddington was sympathetic to Marxism also, certainly to central social and economic planning.[71] And Hermann Muller, Julian's early hire at Rice University, became so enchanted by communism that he moved to the Soviet Union. He brought the first stocks of *Drosophila* from Columbia University into Soviet laboratories in 1922.[72] Another of Julian and Muller's colleagues and friends, the geneticist Theodosius Dobzhansky, made a reverse professional choice. He was born and educated within the Russian Empire and worked in Soviet Russia on just those *Drosophila*, but soon left for the west, rightly perceiving better prospects for a Mendelian geneticist in the United States.

Alongside their laboratory and committee and institutional work,

Figure 8.6: Julian Huxley taking notes at the Association of Scientific Workers conference, 13 February 1943.

this set was committed to writing popular books on biology and society, on genetics in this world, and, for most of them, eugenics for future worlds. They admired Julian's *We Europeans*, which tackled Nazi race science head on, and were more than prepared to extend his public undoing of the politics of race and nationalism. They all popularized not just their science, but also its social and political implications and uses.

Over the years, Mendelian genetics brought them all together, one way or another. They were looking forward to meeting up at the Seventh International Genetical Congress held in Edinburgh in August 1939, during which war was declared between Britain and Germany for the second time that century. This group of political biologists had come intellectually armed to their professional meeting, delivering a joint anti-fascist manifesto. It bore Hermann Muller's authorship most strongly, but Huxley, Haldane, Waddington, Hogben and Dobzhansky were all signatories. What came to be known as the Geneticists' Manifesto was swiftly published in *Nature* on 16 September 1939 under the title 'Social Biology and Population Improvement'. Their joint mission of, and for, social biology was explicitly political in its aspirations. It required inequality to be addressed. They sought to flatten or even eliminate class privilege and expose the fallacies of racial prejudice. The Manifesto showed that ideas of 'race' were based on unscientific doctrines, the term printed in inverted commas after Huxley and Haddon's work.[73]

Other ideas in the Geneticists' Manifesto were less Huxleyan, but he nonetheless signed his name to them. Economic security should be guaranteed by major economic reconstruction. Julian's work with the Political and Economic Planning group inclined him some way towards state planning, but within a capitalist not a socialist framework. The manifesto also sought structural recognition that women were more central to childbearing and rearing than men, and so 'must be given special protection'. Yet this was not 'protection' of woman-as-mother, Victorian-style. This was protection of woman-as-human-being and as worker: 'to ensure that her reproductive duties do not interfere too greatly with her opportunities to participate in the life and work of the community at large'.[74] Julian and the others signed up to that idea, but hardly carried it through in their own partnerships and marriages.

Other aspects of the Geneticists' Manifesto they pursued more honestly. The legalization and scientific development of birth control was demanded, certainly Julian Huxley's ambition and mission, as we see in the next chapter. Birth control, among other things, would aid better conscious selection, as it was put, of future generations' physical and mental characters. This was eugenics.

The Geneticists' Manifesto reflected the political biology of Hermann Muller most closely. Working first at T. H. Morgan's laboratory at Columbia and then at Rice University, Muller was disillusioned personally and politically over the 1930s as he saw his lab, his science and his American society crumble with the Depression. Muller left the United States, first for Weimar Berlin in 1932, but shortly that turned out to be Nazi Berlin, and so he moved again, this time to the Soviet Union. By 1934 he was doing *Drosophila* work in Moscow, but unhappily so, since Soviet doctrine was beginning to favour, and later even to enforce, outdated and disproved genetics. And so Muller moved once more, this time to Edinburgh in September 1937. In 1940 he returned to the United States and was awarded a Nobel prize in 1946 for his work on X-rays and artificially induced gene mutation as shown in *Drosophila*.

The international geneticists' meeting was meant to have been held in Moscow in 1937, but genetics in the Soviet Union had come undone, and dramatically so, as Muller experienced first-hand. The Soviet state sought to align genetics with Marxist-Leninist ideology. The implications of neo-Mendelian laws of heredity and variation were at odds with the possibilities of environmental (or social) change. Prominent Soviet agronomist Ivan Michurin and subsequently the plant-breeding biologist Trofim Lysenko preferred the Lamarckian possibilities of environmentally acquired inheritance. This aligned far better with Marxist-Leninist doctrine, not least to the satisfaction of Josef Stalin.[75] And so, Mendelians were momentarily at war with a phenomenon called 'Michurinism', and a good deal more than better breeds of corn were at stake. Over time, dissenting geneticists (Mendelians) were shot or were exiled or disappeared.[76]

Mendelism and Michurinism were not equally paired, however, as Julian Huxley was quick to point out. The former was an established branch of science, the latter a doctrine. Indeed, basic Mendelism was so foundational, long-established, proven and reproven – and indeed so simple, Julian noted – that he had mastered its basics as a school-boy. Now even that elementary understanding was at risk. The controversy was larger than the specifics of genetics or Darwinism. For Julian Huxley – and for all his geneticist colleagues – the scientific method itself was at stake in what came to be called Lysenkoism. In this matter he was truly the grandson of T. H. Huxley, defender of the faith: 'Lysenko and his followers are not scientific in any proper sense of the word', he declared. '[T]hey do not adhere to recognized scientific method, or employ normal scientific precautions, or publish their results in a way which renders their scientific evaluation possible.' By declaring that Michurinism is true and Mendelism is untrue, the Soviets suppressed an entire branch of science: 'a denial of that freedom of the intellect which we fondly imagined had been laboriously won during the past three or four centuries', wrote Julian. Science had been grossly overridden by ideology. For T. H. Huxley that ideology had been the Church. For Julian Huxley, it was Soviet communism. In a phrase that could have been his brother's he called it 'the totalitarian regimentation of thought'.[77] Julian quickly wrote and published the book *Soviet Genetics and World Science*, and the original Chatto & Windus jacket challenged: 'Can the State dictate "Truth"?'

Julian's arguments against Lysenkoism held some weight precisely because it was well known how many of his biologist circle were socialists, and some members of the Communist Party. As he himself put it on the cusp of the Cold War in 1949, his critique of Soviet genetics was not crude 'red-baiting'.[78] He was, rather, on the record as being impressed with various Soviet achievements, having travelled there in 1931 and again in 1945. In the Huxley tradition, he turned the first trip into a well-titled travelogue, *A Scientist Among Soviets*. Though the trip was organized by the Soviet state, Julian found he could nonetheless walk and photograph more or less as he wished through many cities. More than that, he could film. His black and white sixteen-minute documentary, *A Month in Russia*, was released later in 1931, showing scenes in Leningrad and Moscow of crèches, football fields, airships, bathing huts and propaganda.[79]

By 1945 Julian found everything much changed: 'I had realized on our first visit that Russia was an empire: now it had become imperialist, with a ruthless coterie at its head, determined to stamp out all dissent.'[80] It all culminated at a dinner with a strongly guarded and remote Josef Stalin. But the most professionally pressing event for Julian was a lecture given by Lysenko, now decorated with the Order of Lenin, at the Academy of Agricultural Science. Julian listened, shocked. He had a private discussion with Lysenko after the lecture, and did not mince words: 'scientifically illiterate'. Worse, Lysenko was a fanatic, a 'Savonarola of science'.[81] By then, any hint of Mendel had been shut down resolutely by the Central Committee of the Soviet Communist Party resolving that Michurinism alone directed genetic research. The genetic knowledge that Julian Huxley and his friends had learned, then developed, then communicated to the world was officially 'bourgeois', and Julian returned the insult, condemning 'a party line in genetics'.[82]

The Soviet party line in genetics was an intellectual accompaniment to the emerging Cold War. With many others, Julian perceived the change in Soviet policy, writing in 1949 that the people of the USSR were being prepared 'for a long struggle, possibly involving war, with the capitalist world in general, and the USA in particular'. In the meantime, Jack Haldane resigned his Communist Party ticket in 1950, or let it lapse, according to Julian because of Lysenko and the Soviet party's insistence on undermining Mendelian genetics. By the mid-1950s, though, Lysenko was running out of favour, and when Julian wrote his *Memories* in 1970, he noted a rumour circulating among the world's geneticists that Lysenko had been taken to an asylum for the insane.[83]

THE FAMILY OF MAN

The Geneticists' Manifesto, sometimes called the Humanists' Manifesto and at other times the Eugenicists' Manifesto, argued for internationalism: 'some effective sort of federation of the whole world, based on the common interests of all its peoples'.[84] This was signature Julian Huxley, the kind of statement that recommended him for the top job with Unesco. He was surprised – 'flabbergasted' – at the offer of the position but accepted almost immediately. Bringing together science and culture across the world – could there have been a better job for a twentieth-century Huxley, schooled in the synthesis of knowledges? Unesco's broad mission stemmed from the League of Nations' International Institute of Intellectual Cooperation, set up in Paris in 1926. Julian claimed that it was he along with Joseph Needham who successfully added 'science' to the cultural and educational mission of the new enterprise.[85]

The preparatory commission for Unesco was based in London and moved to Paris in late summer 1946. Julian and Juliette were settled into its rooms and offices at the Hotel Majestic. It was a freezing first winter, and the rooms were difficult to heat. Food was still heavily rationed. But this was a richly symbolic location for the new enterprise. Much of the Paris peace negotiation had been held there in 1919, and in the Second World War it had been occupied by the German army: the Nazi commandant of Paris had been the immediate previous occupant of Julian's quarters.[86] From those haunted rooms, once grand, now shabby and chilly, Unesco was built, and Julian Huxley put his eminent team together.

The new United Nations and its agencies operated within a deeply divided global sphere, as the World War quickly turned into the Cold War. Unesco's responsibilities put Julian right in the path of international tensions, even as he upheld internationalism. In his two and half years of leadership, Julian and Juliette travelled almost continually, and to many sites of political and economic strife. They went to Haiti in the summer of 1947, a nation enduring chronic poverty with meagre plans for a way out. Julian tried to focus on education and literacy. They toured Perón's Argentina, Julian's diplomatic skills tested to the limit, and travelled across fifteen nations in South America that year, forming delegations in advance of Unesco's General Conference in Mexico City, 1947. Between banquets and meetings, the director-general couldn't stop being a zoologist and a Huxley. Charles Darwin's own South American journey was often on Julian's mind as he admired the armadillos, opossums and llamas, the monkeys with the prehensile tail acting as a fifth limb and the iguanas that had turned amphibian in their island isolation. Julian was shocked that some of the forests Darwin had described around Rio de Janeiro had been completely cleared. As part of his Unesco work, he was able and interested to trace connections between ecology, agriculture and economy. He went to the guano islands in Peru's north, crammed with birds, 'where the sea's fertility is transferred to the land'. He was drawn to a tropical species of gannet – the bird that had won him an Oscar – and noted quite rightly that 'wherever guano yields profit, sea-birds are protected'.[87] It all inspired his successful proposal for the conservation of nature to become a formal part of Unesco's work, and later his work founding the World Wildlife Fund.[88]

The next Unesco General Conference was held in Beirut in 1948. As with the Mexico City conference, Julian anticipated the Beirut meeting with a long tour across the region, to Jordan, Egypt, Tunisia and Turkey, as well as Lebanon, once again turning public sphere work into a personal travel book, *From an Antique Land*. In Beirut, Julian had to manage the local manifestation of the escalating Arab–Israeli conflict. Arab delegates objected to the presence of Israeli delegates, and while he defended Israel as what he called a de facto nation, as director-general he was bound to confirm that the Israelis' status was observational only. Privately, Julian supported Israel fully as a nation-state and yet felt deeply for the 'plight of the Muslim Palestinian refugees'.[89]

Over these post-war years, walls between Eastern and Western Europe, between Israel and Palestine, between India and Pakistan were erected violently. The Family of Man was falling apart. In its operations and in his travels, Unesco and its director-general had to adjudicate these high international tensions, and yet the very idea of Unesco emerged from an optimistic internationalism. Ideas of 'the commonwealth of man' and of world citizenship were indispensable to Julian Huxley, and if anything the Cold War fuelled the imperative for an idealized cosmopolitan one-ness.[90] Huxley saw before him a vision of efficient human-controlled and planned progress. He was hardly alone in viewing internationalism as a political, cultural and social remedy in a better and stronger form than the old League of Nations. There were advocates of internationalism of all stripes. H. G. Wells – dying in 1946 – had promoted liberal internationalism and a world state for much of his adult life.[91] For others, like the younger Jawaharlal Nehru, internationalism offered a step beyond imperial world orders, even as this new world necessarily incorporated post-colonial nation-states, like his own India, and political blocs like the Non-Aligned Movement emerging in the 1950s. For Julian Huxley it was the nation-state itself more than imperialism or colonialism that was retrograde. After the extremities of fascist nationalism of the inter-war years, Julian was resolutely anti-nationalist. Anti-fascism not anti-imperialism, in other words, was his route to internationalism. He noted gravely that his generation faced 'a task unique in history – of attempting to ensure that in the place of German and Japanese fascism new civilizations should arise'.[92] The possibility of world unification lay before his generation, Julian declared, 'for the first time in history'.[93] And he was one of its self-appointed thought leaders.

Charged with the leadership of Unesco, Julian focused his mind, his politics, his communication skills and his family's special expertise, as trustees of evolution, and put pen to paper. Unesco needed a 'world philosophy', one that would provide the right foundation for 'a single world culture'.[94] To think this through, he took a fortnight away from family and office distractions, at the Pembrokeshire coast. Having rehearsed manifesto writing with his geneticist friends just before the war, Julian wrote another at its end: *Unesco: Its Purpose and Philosophy*. Some form of humanism must underwrite the new venture, he announced, but this clearly could be neither religious nor anti-religious nor political dogma. Neither a world religion, nor Marxism, nor liberalism, nor even an ecumenical combination would do. In their place, science offered itself, 'by its nature opposed to dogmatic orthodoxies'. The apparent impartiality of T. H. Huxley's beloved scientific method lent itself to the twentieth-century international challenge. A little dogmatically, Julian declared that Unesco's foundational philosophy 'must' be evolutionary. A general theory of evolution 'not only shows us man's place in nature' – a nod to his grandfather – 'but allows us to distinguish desirable and undesirable trends'. With evolutionary progress as a touchstone, internationalism was much more useful and hopeful than nationalism, he argued. Knowledge needed to be pooled. Ideas that unified, not competed, needed to be placed together for all to assess and use. Yet science was not enough: Huxley's brief was also 'culture'. Unesco's philosophy cannot be simply materialistic, he understood, but must also 'embrace the spiritual and mental . . . a truly monistic, unitary philosophic axis'. And so, Julian's 'scientific humanism' was named, based on Darwin's theory of selection and extended to the exclusively human psycho-social sphere. *This* should be Unesco's philosophy and guide for action; a 'scientific world humanism, global in extent and evolutionary in background'.[95]

Julian's manifesto for Unesco is brilliant in many ways, but it was not judicious. His own dogma on evolution, on ways to limit population growth, on the subversion of religion, startled some and made others cautious.[96] Later its advocacy of the idea and even practice of eugenics came to be seen as its supreme problem, as we will see in the next chapter, but this was not criticized as its major shortcoming at the time. Julian misjudged the delicacies of writing a highly personal statement, and aspiring to its elevation within the new United Nations agency. But before he ever put pen to paper on the Pembrokeshire coast, careful work had already been directed to Unesco's principles. A constitution signed on 16 November 1945 was the work of many minds, guided most forcefully by poet and former United States Librarian of Congress Archibald MacLeish, who wrote a gripping preamble: 'since wars begin in the minds of men, it is in the minds of men that the defences of peace must be constructed'. It continued poetically:

> [T]he great and terrible war which has now ended was a war made possible by the denial of the democratic principles of the dignity, equality and mutual respect of men, and by the propagation, in their place, through ignorance and prejudice, of the doctrine of the inequality of men and races.[97]

This is a lost world when poets were statesmen and scientists were wordsmiths. What happened in the minds of men, how words and images could reshape those minds for good and ill, was as much the business of liberal progressives as it was of communist re-education camps.

Adjudicating the equality or inequality of 'men and races' was Huxley business, if anything was. The preamble to Unesco's constitution repudiated racism, and, as Julian noted in his own manifesto for Unesco, it repudiated any idea that there were or are inferior and superior races, nations or ethnic groups.[98] It all took the institutional form of a Unesco programme on 'the race question', leading to several formal statements on race that were released soon after Julian left Unesco, but with which he was nonetheless involved. In 1948 the UN Economic and Social Council instructed Unesco to collect the 'scientific facts' that would eliminate 'that which is commonly called race prejudice'. It was an enterprise which focused attention on the nature of expertise on human difference; biology on the one hand, culture on the other. The first statement on race, issued 18 July 1950, was pointedly the work of social and human scientists. This was a group of experts on 'man' schooled in cultural anthropology, as far removed from T. H. Huxley's methods as it was possible to be. And yet it was all revised after biologists, many of whom had presented the Geneticists' Manifesto in 1939, including Gunnar Dahlberg, Theodosius Dobzhansky, Hermann Muller, Joseph Needham and Julian S. Huxley, had their say on the Statement on Race.[99]

While the new orthodoxy was that human difference was not to be comprehended biologically, but culturally, in fact the first Unesco statement on race relied on the biological tradition. It restated what Thomas Henry Huxley had said long before: 'mankind is one . . . all men belong to the same species'. And it even cited Darwin from *The Descent of Man*, to authorize the idea of the unity of humankind.[100] And so, while Julian's anti-nationalism was one powerful impetus for his advocacy of internationalism, human unity had another intellectual backstory closer to home. He had tried to introduce both evolution and natural selection into Unesco's orbit, but it was in the later Statement on Race, so often seen to be a victory for cultural anthropology, not Darwinist biology, that the 'unity' offered by evolutionary theory became explicitly useful. This is counter-intuitive because a link between 'Darwinism' and 'biological racism' soon came to be orthodox critique within 'science studies' and popular culture alike. Yet for Julian Huxley, as for the architects of the Unesco statement, Darwin's own contribution and Huxley's Man Family *aligned* with One World, the Family of Man, and a post-war cosmopolitanism. Julian Huxley would constantly reiterate that both his grandfather and his grandfather's friend Charles Darwin together 'demonstrated the unity of all life' and provided 'rational and scientific explanation of that unity and that variety'.[101] The biological Man Family and the cultural Family of Man were as related and connected as Huxley and his grandson Julian.

The socio-cultural and obviously aspirational Family of Man reached its high point in the divided mid-twentieth century. In 1946 Julian Huxley and Jack Haldane worked with filmmaker Ivor Montagu on *Man – One Family*. It debunked master race theories and was translated into sixteen languages 'for the liberated peoples of Europe'. Julian was the film's narrator as well as scientific adviser, explaining, once again, the baselessness of Nazi and Japanese theories of racial superiority and inferiority. There are two ways forward for the world, Julian Huxley the narrator instructs: the path of prejudice or the path of democracy. The message was unity in diversity for the Family of Man, and in 1946 Ivor Montagu, a wealthy Jewish communist, was both keen and able to celebrate Soviet multiplicity as a model for the world. The film included footage of a Red Army parade of decorated heroes and heroines, 'more than 80 different peoples and nationalities', including Russians, Ukrainians, Armenians, Uzbeks, Jews, Lithuanians, Gypsies, Kurds, Poles, Bulgarians and Greeks, so Julian Huxley's voice-over announces. 'Those nations whose people stand together like brothers, cannot be conquered.' Soviet unity in diversity is the penultimate sequence, however, just one set of European liberators. The final sequence is of an American troop, 'all citizens of the United States'. They move forward symbolically towards a brave new world order. 'All of us together: Man – One Family', less African Americans, notably absent in this footage.[102] It was a sign that the cosmopolitan 'Family of Man' did not necessarily have a corollary in the principle of equality of all humans.

Very much in his grandfather's mould, Julian had earlier presumed his own wisdom over 'the Negro Problem' in 1920s North America. Fully aware of his grandfather's 'Emancipation' intervention, he tried to diagnose and even offer political and social solutions. His suite of articles in 1924 described with great would-be authority a difference between the black mind and the white mind, which he correlated to be as stark as between black and white bodies. Julian detailed the migrations out of the South, various race riots, 'Jim-Crow-ism' and lynchings. What are the solutions for this American problem? One possibility, he suggested, would be 'to continue on the present lines of theoretical equality: but the equality is so very theoretical, the racial prejudice so very strong and deeply-grounded'. True enough. But then, as so often, he undid himself again, adding that racial prejudice 'was not without a good deal of sound social and biological instinct'.[103]

And yet, all this was not enough to turn W. E. B. Du Bois away. The American sociologist and Pan-African activist had read Julian's *Africa View* with some interest and turned to him for support and possible contribution to an idea for an *Encyclopedia of the Negro*, or an *Encyclopedia Africana* to offset *Encyclopedia Britannica*. It was a large, ambitious and long-term project that never materialized, at least for that generation.[104] In late 1937, Julian declined Du Bois's request for involvement, explaining that he had no special knowledge of 'the American Negro', he had not been able to keep up with African matters, and in any case had hardly any time for writing.[105] It was disingenuous, and whatever his reason for refusing that particular invitation, Julian Huxley's investment and interest in Africa, Africans and the African diaspora intensified the older he got. He certainly felt himself an authority and a mediator. As a broadcaster he spoke regularly on the BBC African Service, in 1962, for example, discussing 'Wild Life in Africa' for the weekly *African Forum* with three Ugandan students then in London, one of them, Edward Rugumayo, a student of botany and ecology, later politician, environmentalist and ambassador (figure 8.7).

Du Bois's twentieth-century *Encyclopédie* was not arbitrarily named. It was a precise intervention first to add to knowledge in the great Enlightenment tradition, and second to counter those eighteenth-century views of Africans and the African diaspora. Just like the American Declaration of Independence, the Enlightenment *Encyclopédie* had managed to hold up 'rights' and 'racism' together. This incongruity was carried forward by the Huxleys and so many others, not least, perhaps especially, by those who imagined themselves to be progressive, liberal, and even, as in Julian's case, anti-racist.

Over the eighteenth, nineteenth and twentieth centuries, liberal and republican ideas of the rights of man and the rights of woman tracked alongside biologies of equality and difference, one step forwards, two steps back. Thomas Henry and Julian Huxley were so caught in this forward and back stepping that they were wrong-footing everything, tripping over themselves and bringing people down with them.

How were humans both the same as and different from each other? How were these similarities and differences to be valued? The Huxleys

Figure 8.7: Julian Huxley speaking on the weekly *African Forum*, 1962, with three Ugandans studying in London. Left to right, Paul Mussala, Edward Rugumayo, Julian Huxley, F. X. Katete.

thought these questions belonged as much to biology as to republican, liberal, communist or socialist political philosophies. It was with biology in mind that Thomas Henry Huxley carefully copied out the republican *Déclaration des droits de l'homme et du citoyen* and wondered long and hard about these revolutionary rights of man.[106] He insisted instead 'on the natural inequality of men', taking on Jean-Jacques Rousseau and setting his own nineteenth-century biology against eighteenth-century political philosophy and the republican principle that 'all men are born free and equal'.[107] Natural or bodily distinctions did not and should not matter, according to Rousseau, since the origin of inequality lay in social and moral conventions – the distinctions between men derived from the original claiming of private property (although women for Rousseau were as 'naturally' unequal to men as they were for Huxley, a major inconsistency in his argument about humans). Huxley declared, contra Rousseau, that men are not born equal: 'the doctrine that all men are, in any sense, or have been, at any time, free and equal, is an utterly baseless fiction'. Inequality *did* exist in a state of nature, for Huxley. He thought Rousseau's natural man was ill-conceived, reading the philosopher's fictional device too literally: 'it is a profound mistake to imagine that, in the nomadic condition, any more than in any other which has yet been observed, men are either "free" or "equal" in Rousseau's sense'.[108] T. H. Huxley argued, against Rousseau, that the social world should ameliorate such natural inequalities, even if they could not be totally undone. But this amelioration gesture was a mean one, with constant recourse to 'nature'. In 1890 he still agreed with his younger self, the author of 'Emancipation – Black and White' in the Civil War era. It may be incumbent on dominant men to remove social and political barriers, to 'emancipate', but natural roles – unequal ones – would nonetheless persist.

We might expect argument for 'the natural inequality of men' from the Victorian Huxley. But Julian was active in the United Nations public sphere precisely at the point when the principle of human equality was being transformed from national to international proclamation; universal human rights were declared in 1948. To his modest credit, Julian had some insight into the racism of his own 1920s self, and later applauded 'the conception that ethical principles apply to all humanity, irrespective of race, language, creed, or station'.[109] Yet this was completely different to claiming that all humans were equal, or even that all *men* were equal. Julian unflinchingly followed his grandfather's lead, refusing to concede that any singularity of humanness recommended the equality of all individuals, even at the level of aspirational principle.

> Human beings are not equal in respect of various desirable qualities. Some are strong, others weak; some healthy, others chronic invalids; some long-lived, others short-lived; some bright, others dull; some of high, others of low intelligence; some mathematically gifted, others very much the reverse; some kind and good, others cruel and selfish.[110]

Indeed, Julian seems to have been almost affronted by equality and radical democracy. He dismissed the popular idea of 'the Age of the Common Man', 'the Voice of the People', even 'majority rule'. All of these moved humanity towards mediocrity, he thought, not towards a more perfect future. For Unesco, he was even able to write a section in his manifesto on 'the principle of human equality and the biological fact of human inequality'. Thinking carefully and closely about *inequality* was a great objective for Unesco, he declared, and so 'to adjust the principle of democratic equality to the fact of biological inequality is a major task for the world, and one which will grow increasingly more urgent as we make progress towards realising equality of opportunity'.[111] Thus, just two years before the Universal Declaration of Human Rights, the director-general of a United Nations agency dismissed equality. 'Our new idea-system must jettison the democratic myth of equality. Human beings are not born equal in gifts or potentialities, and human progress stems largely from the very fact of their inequality.' He was even prepared to render this into a slogan for future humankind: '"Free but unequal" should be our motto.'[112] It was a family tradition so to argue.

9

Transhumans:
Sapiens of the Eugenic Future

In 1937, Julian Huxley was approached, business-like, by a stranger: 'Dear Sir, Would you consent to being the father of my wife's child, possibly by artificial insemination? I am writing this as a feeler and purely from the eugenic point of view.'[1] Although boldly personal, such an inquiry would hardly have startled Julian. As someone interested in the physiology of reproduction and in improvement, he might well have agreed; we do not know. Equally, as someone expert in genetics and inheritance, he might have declined, only too cognisant of the illness that ran in his family. It was a request that was entirely plausible in the 1930s. Artificial insemination had long been practised in animal breeding, and crude versions – by today's standards – were already under way among humans. This was a response to infertility in the first instance, but the rise of eugenics in the early twentieth century, plus a new awareness of mechanisms of inheritance, meant that some were actively seeking and selecting particular 'genes', as they understood them.[2] The Eugenics Society in London occasionally discussed artificial insemination of humans as part of its business in the 1930s.[3] By the time Julian Huxley was president, 1959–62, it was recurring, if not quite standard business for eugenicists of various kinds; hardly a shocking matter.

'Artificial' reproduction of humans unfolded apace in Julian's twentieth century, but it had an embryonic connection to his grandfather's nineteenth-century world, both fantastic and real, as the mysteries and wonders of reproduction across all orders of life revealed themselves. Charles Kingsley had early imagined 'babies' growing *in vitro* – in water, in glass bottles – one illustration of which caricatured T. H. Huxley alongside Richard Owen. The image became an icon of 'ectogenesis',

J. B. S. Haldane's later word, or 'exowombs', as some transhumanists would now say (figure 9.1).[4] *The Water Babies* enchanted nearly-five-year-old Julian in 1892. In one of the few conversations we have on record between the old biologist and the boy-naturalist, they discuss babies in bottles. 'Dear Grandpater,' Julian wrote in his juvenile hand. 'Have you seen a Waterbaby? Did you put it in a bottle? Did it wonder if it could get out? Could I see it some day? – Your loving Julian'. Grandpater took the opportunity to deliver an intergenerational lesson about what can be known, and what must remain unknown – a lesson in agnostics. 'I never could make sure about that Water Baby. I have seen Babies in water and Babies in bottles; but the Baby in the water was not in a bottle and the Baby in the bottle was not in water.' Publishing his father's letters, Leonard Huxley did not transcribe this one, but instead reproduced it in Grandpater's mock-juvenile script, printed not cursive for young Julian. It remains one of the very few legible letters by T. H. Huxley, whose hand is notoriously difficult.[5]

Precise biological knowledge about the reproduction of living organisms was monumentally complex and ill-understood, even in 1892, when Huxley wrote to his grandson. Yet reproduction was key to understanding continuity and change across generations, sending the past into the future. And through this mechanism the future might be planned, that is, selected. In one sense, breeding for improved futures was neither dramatic, nor outrageous, nor controversial. It was ordinary: 'artificial' selection of and for domestic breeds was the quotidian enterprise against which Darwin set 'natural' selection, the former pursued in his own garden, and cruder versions in thousands of yards, barns, cages and pastures across England. It is unsurprising at one level that the selective breeding of humans became so interesting and offered such possibilities to natural and social scientists, especially when the laws of inheritance and population genetics became clearer. All kinds of projects and plans for artificial selection were dreamed and occasionally implemented over the following generations, as they are in our own time. Some involved the prevention of conception and reproduction – selection out – while others involved the special promotion of conception – selection in. Throughout, mental ill-health was central to eugenics and its dark ambitions for healthier futures. The irony and difficulty therein hardly escaped the Huxleys.

Figure 9.1: Richard Owen and Thomas Henry Huxley examining a water-baby, illustration by Linley Sambourne in 1885 edition of Charles Kingsley, *The Water Babies*.

Julian thought and wrote often about 'quality' – of people, of life, of societies – but this was preceded by a quantitive breeding problem – overpopulation – and its solution, birth control. Like most Malthusians and eugenicists of his generation, Julian thought that the 'quantity' problem needed to be addressed before, or at least alongside, the eugenic 'quality' problem. Both would help deliver a better human future, and he was entirely sure of this before, after and in the light of Nazi eugenics. For him, the German programme was a dire abuse of eugenics, which itself should – Julian would say 'must' – be reclaimed for the best human future. This is why, in the immediate aftermath of that war, eugenics formed part of his manifesto for Unesco.

Over Thomas Henry's, then Leonard's and then Julian's joint lifespan, 1825–1975, England had gone through a so-called natural demographic transition from high mortality and high fertility to low mortality and low fertility. Some worried that fertility rates dropped too low, but the change was hailed by others as highly desirable, not just in health terms, but also economically speaking for individuals, families and the state, for the common weal. Might such a transition profitably take place elsewhere in the world and indeed across the whole world? This is what Victorian neo-Malthusians advocated. For them, as for Huxley himself in his late years, overpopulation meant poverty, and the control of fertility (by many different mechanisms) would be a reproductive and health solution to want and need, towards economic viability and sustainability at multiple scales. By Julian Huxley's mid-twentieth century, curbing population growth was urgent business.

Julian's particular concern was for a better environment, an ecologically more secure planet. More humans meant the need for more food, and that meant more land to be cleared and cultivated, more soil turned. His late-in-life concern was less for women reproducing eight, ten, thirteen, fifteen humans out of their worn-out bodies, and far more for the wild animals of the world, whose habitats were being put to the plough at an accelerating and unsustainable rate. A thousand other neo-Malthusians proclaimed limits to the earth in the mid-twentieth century.[6] Julian and Aldous Huxley were two of the more famous. In 1958 Aldous reassessed his *Brave New World* (1932) – *Brave New World*

Revisited – and named something far more terrible, 'the Age of Over-population'.[7] For the Huxley brothers, population growth was leading inexorably towards a catastrophic planetary future.

QUANTITY: POPULATION GROWTH AND HUMAN FUTURES

Around 1880 it was apparent that something was changing with regard to human mortality and fertility. In Huxley's England, mortality rates had been dropping for some time. When Thomas Robert Malthus wrote his principle of population around 1800, 329 children per thousand did not reach the age of five. The year Charles Darwin died (1882), and when Huxley started thinking about human population numbers, about 223 per 1,000 died. The great change was yet to come: a twentieth-century one. At the end of Julian Huxley's life, 1975, this had reduced to about 20.[8] This was a massive modern shift, and one corollary was that net population in Britain was increasing rapidly.

Huxley was one who early noted population growth and sounded the alarm. By 1888 he had decided that the Man Family was too fertile for its own good, drawing attention to the 36 million people then alive in the United Kingdom and to their excessive procreation – excessive, that is, to what could be provided by British soil for them to eat.[9] Yet as he was writing and worrying in the 1880s, fertility rates were actually dropping in Britain and continued to do so through most of the twentieth century. For reasons and in ways that are still not entirely clear, conception was being limited, likely a combination of later marriage, forms of control of reproductive sex and endemic sexually transmitted disease.[10] In 1800 the average British woman would have five children; by the mid-twentieth century the average was less than two. The Huxleys themselves followed this pattern across generations. Henrietta bore eight children, with Noel dying before the age of five. Julia Arnold Huxley bore five children at a time when the average number across Britain was 4.18, her firstborn dying at a few weeks.[11] Leonard Huxley and his second wife Rosalind Bruce bore another two long-lived children. Juliette Huxley bore two, right on average, one living to seventy-two, the other to ninety-three, increasingly normal longevity.

These issues were all-important and pressing for the Huxleys across generations not just substantively and politically, but intellectually, theoretically and scientifically. Patterns of fertility and mortality fell squarely into their domain as trustees of evolution. Malthus's principle of population had sparked both Darwin's and Wallace's theories of natural selection. All organisms give birth to more offspring than can, and do, survive. 'Infant' mortality – in plants, in reptiles, in both fast- and slow-breeding mammals alike – is a necessary part of evolution by natural selection. If all survived who were born, and went on to reproduce themselves, the limited space of the planet would have been overrun thousands of years earlier. The fittest survive in this necessary mismatch between fertility and mortality; that was the idea that Darwin and Wallace took forward, and that Julian Huxley and his biology friends synthesized with Mendelian inheritance. T. H. Huxley knew too that excessive fertility led to high mortality in the young of any and all species in the animal and vegetable worlds. With Darwin he saw this to be as true of humans, even as they both grieved their own beloved dead children. Darwin thought that those who could not avoid poverty should refrain from marriage.[12] This was 'Malthusian', not in the sense of advocating artificial birth control (Darwin and Huxley did not), but in the sense of delaying marriage and conception via an economic logic.

Malthus argued, and Darwin and Huxley agreed, that spatial and environmental limits – of food, of resources, of soil – naturally correct numbers of all living organisms back to a kind of balance, or really in cycles that balance over the longer term. But this balance was often achieved through undesirable mechanisms in human populations: death by disease, infanticide, war, famine. The Malthusian point was to insist that this misery was not in fact necessary, or at least not necessary for so many humans, so much of the time: there were ways to control fertility and mortality with far less suffering.

T. H. Huxley thought that humans could and should consciously manage their own fertility, and he thought this urgent, even in the 1880s. 'The political problem of problems is how to deal with over-population, and it faces us on all sides.'[13] The imbalance between food and people was continuing to increase. Great populations, especially of industrial centres, bring *'la misère'*.[14] Even as he insisted that the

Man Family was necessarily 'tied to the matter & the lower forms of Life', he well knew that humans were able to, and should, manage 'nature' in this regard and limit their own fertility. He folded this, like everything else, into a model of civilizational stages. For Huxley, in 'nature', and for the 'primitive savage', or for 'man as a mere member of the animal kingdom', fertility and mortality were unchecked and the struggle for existence was fought 'to the bitter end, like any other animal'. For 'ethical man' – social or civilized society – energies are devoted instead 'to the object of setting limits to the struggle'. One way of establishing limits to struggle was through limiting population growth. However, people across the world and 'even' civilized English people, he thought, had failed in this regard, heeding too well the commandment to 'increase and multiply'.[15]

Huxley was in all these ways Malthusian. Yet he was not an active 'neo-Malthusian', a late-Victorian descriptor for men and women who proceeded a step further, advocating not just an age-of-marriage means to limit fertility, but mechanical and chemical contraceptive means as well. Neo-Malthusians of Huxley's era were closely aligned to 'free-thought', the long English secular humanist tradition that gathered pace over the 1870s and 1880s. Huxley might well have engaged far more closely with this next generation of London- and Midlands-based secularism and atheism than he did. They certainly owned his 'agnosticism', as we see in the next chapter. But it was all too sharp and radical for him, even too unlawful: leaders in this field were not uncommonly gaoled. Freethought advocates of birth control Annie Besant and Charles Bradlaugh were tried in the 1880s for the republi-cation of an old medical self-help book, the strangely titled *Fruits of Philosophy*, which included an explanation of measures of birth con-trol.[16] Besant was a student of Huxley's, as it happened, part of his 'teacher's class' in South Kensington in 1882. He thought her 'a very well conducted lady like person; but I have never been able to get hold of the "Fruits of Philosophy", & do not know to what doctrine she has committed herself'.[17] He was vaguely interested in her commitments, unsettled by her recent arrest and appalled that she had been banned from botany classes at University College, but Huxley was never drawn into 'Malthusian' lobbying for contraceptive knowledge.

Julian Huxley *was* a birth control activist, however, and a committed one. He followed through with his grandfather's theoretical Malthusianism that sought a better future by limiting population growth across the world, as in Britain. But his twentieth-century version was far more forthcoming about new methods – early intra-uterine devices, rubber sheaths, chemicals. His was an *applied* neo-Malthusian agenda. As early as the 1920s, when he should have been concentrating on his Oxford teaching and research, he was already penning articles on birth control for all kinds of publishing outlets, articles that were political as much as scientific or medical. 'Birth Control and Hypocrisy' in *The Spectator*, for example, expressed outrage that the government threatened to withdraw funding from local health authorities if they gave information and practical advice on birth control. 'This is in democratic England in A.D. 1924,' he exclaimed, 'not under a theocracy ... nor yet a military autocracy.'[18] In return, there were all kinds of critiques levelled at birth control lobbyists, and beyond religious objections. A group called the National Council of Public Morals thought that birth control somehow threatened 'the continuation of the race', was a 'policy of annihilation'; that birth control was a crude hedonism, and one that heralded a 'class legislation', implying that the burden of unwillingly controlling fertility would be borne by working-class men and women.[19] By contrast, Julian considered new mechanical and chemical forms of conception control to be one of the great biological steps forward of their time, similar to the discovery of anaesthetics or asepsis. 'It points a practical way to a final satisfactory control by man of his own evolution, since the only other regulators of numbers which are not merely pious wishes are war, pestilence, famine, or overcrowding.' And then he channelled his grandfather and assassinated the National Council for Public Morals, continuing the long Huxley tradition of cutting canons and bishops down to size. Canon Lyttelton – his Eton teacher of Latin, Greek and Divinity, as it happened – sat on the committee of the National Council of Public Morals and declared entirely against the use of contraceptives, 'a frustration of God's design in Nature'. What a senseless piece of Rousseauism, Julian declared dismissively.[20] On this matter, and in another piece in *Nature*, he pitched Huxleyan progressivism against the 'medieval' church all over again,[21] less the Church of England, which was beginning to capitulate, and more the Roman Catholic Church, which held steadfast. In the 1920s, he regarded this as 'one of the big politico-religious questions of the near future'.[22]

The geopolitics of birth control escalated to its highest point in the Cold War, when population control policies were linked to US-led anti-communism and were also high on the agendas of leaders of newly independent nation-states, especially Jawaharlal Nehru. This was the era of 'the population bomb'. It is impossible to overstate the apocalyptic register and whole-Earth scale of the phenomenon in those years. Mature Julian's era watched regional and planetary population grow at a rate unimaginable even to his grandfather, who was already worried in 1888. In one of thousands of statements, in 1957 Julian put it in terms of world population doubling at an accelerating rate: 1 billion in the mid-eighteenth century, 2 billion by the mid-1920s, and at its 1950s rate he forecast the increasingly rapid doubling to 4 billion by the 1980s. His brother put the same idea differently. In a highly pessimistic *Brave New World Revisited*, Aldous wrote that 'on the first Christmas Day' there were about 250 million humans; this grew only slowly, so that when the Pilgrim Fathers landed at Plymouth Rock, there were perhaps 500 million.[23] But by the time he wrote *Brave New World* in 1931, there were almost 2 billion people on the planet. And only twenty-seven years later, as he revisited this brave new world, the number approached 3 billion. It was catastrophic for that generation. This whole arena of discussion and action was not just comparable to twenty-first-century earth-warnings of climate change and species-end, but was in fact its immediate precursor.

In so many ways it was Julian's great fortune to live in a catastrophic era. The planetary scope of human and natural threats called for signature Huxley scale and synthesis. He occupied an international communication role superbly, doing a great deal both to generate awareness of global catastrophe and direct high-level discussion about what might be done. By the 1950s, population control was central to his cosmopolitanism, his global political ecology and his planetary-oriented pacifism. And by the 1960s he was spokesman-of-choice about the world population crisis and for family planning. He lectured and wrote and filmed on the population problem in every possible medium, sharing the catastrophe with his brother, another A-list worrier. Juliette recounted Aldous's anxiety after a dinner in New York in

1947. Aldous was often silent – she thought him frightening some-times on this account – but that evening he was verbose. He 'talked about war dangers, biological warfare, national ferments, US, Russia . . . Said Einstein refuses to sit on any committee except concerned with food & production.' Aldous was entirely consumed with the problem of a world population that was increasing by 60,000 people a day, he told Juliette as they drove home together. This, combined with precar-ious harvests and insecure food production, was the greatest global crisis.[24] In California, Julian and Aldous together formed the core of a group called Population Limited, along with Hollywood director Fred Zinnemann, aiming to get a 'population' film produced.[25] At this point in their lives, and on this matter, the brothers were twinned; the same arguments, the same catastrophe, the same popular deployment of Huxley science and politics. While Julian was writing articles on over-population for *Playboy*, Aldous was writing the same for *Esquire*. While Julian was broadcasting on the BBC, Aldous was speaking to ABC on population and world resources. Both of them used the widely recognized formula of death control without birth control, that is, the easily comprehensible idea that the public health control of infectious disease had been implemented across the world with incredible success especially in the reduction of infant mortality, but without any accom-panying programmes for birth control. The 'balance is out' as Aldous put it in one ABC interview.[26]

It was the great international topic of the moment, from the White House to Whitehall, from the United Nations to the British Medical Association, whose plenary session in 1965 was on 'The Control of Global Population'. Julian Huxley, of course, was to speak first, setting out the data: the world's population had reached 3 billion, was increas-ing by 165,000 each day and would reach 6 billion by 1999. He was correct. And he repeated the straightforward Malthusian principle: 'The increase in food production was not keeping up . . . and unless the balance was restored by birth control then it would be restored by famine, disease, or war.'[27] Population policy *was* foreign policy in this era, and Julian Huxley spoke and wrote often enough in those circles. Yet 'population' on a large international scale also implied – more than that, it required – discussion about sex on an intimate and individual level.

It is difficult to appreciate the depth of the enduring objections to contraception in earlier generations. Discussion of birth control was seen by some to incite illicit sex, and Julian found himself often at the receiving end of apoplectic outrage from the beginning of his advocacy in the 1920s to his final statements in the mid-1970s. One 1920s opponent – hardly representative – was C. H. Sharpe, Anglican clergyman turned Catholic priest. Julian Huxley had recently spoken on a BBC programme on birth control, and the man of God was viscerally affronted and offended. Still galled by the grandfather's agnosticism, the Anglo-Catholic perceived a desecration and wrote: 'this Huxley business' was 'filth', 'this slimy thing', 'the morals of the poultry yard', 'wickedness'. God, he finished, is chastising the country, and little wonder. He considered birth control activism a 'lust-providing' movement.[28] It was the Church *versus* Huxley all over again. The same level and tenor of outrage followed Julian Huxley through the decades, sometimes prompted by completely different politics of offence and indignity. Some were simply appalled by public discussion of population and therefore of sex, and not just the older and more conservative.

In January 1965, *Playboy* published 'The Age of Overbreed', in which Julian set out – once again – his message of a catastrophic future for humankind and for the planet, unless population growth was contained. The line of argument was standard by then, even if the forum for publication was unusual. But that was the thing about *Playboy*: it was an unlikely mix of pornographic display of women with sometimes highbrow words by, and for, interested men.[29] Murray Fisher, *Playboy*'s associate editor, engaged closely with Julian and wanted some good copy on 'the future of man'. Ever since reading Olaf Stapledon (a contemporary of Julian's at Balliol), especially his 1930 science fiction *Last and First Men: A Story of the Near and Far Future*, the *Playboy* editor confessed to fascination with future changes to 'man'. Might Julian write another piece for *Playboy* speculating on evolutionary change to man's mind and body? And what was Julian Huxley's estimation for how long man, 'in any form we'd recognize', might last, either on Earth or in colonies in space?[30] It is a letter which shows how linked were the 1930s generation of science fiction writers to the 1960s catastrophists.

Julian was perhaps flattered, or tempted, or both. As always, he was dependent on money from writing. In any case, he obliged with 'The Age of Overbreed'. By then it was all familiar fare for ageing Julian, a reliable income. His 1965 rendition promoted reproductive rights of a sort, novel for *Playboy*. But this was an idea of women's rights to control of conception *sans* feminism. Contraception obviously enabled other freedoms too. The more sex and reproduction were disaggregated the better for the kind of entitled masculine libertarianism on which *Playboy* thrived, and this is why Julian was brought up short on receipt of an enraged letter. It was as long as it was blunt, as different in argument from Sharpe's 1920s Anglo-Catholic objection as it matched it in heights of apoplexy. Its sexual politics are harder to comprehend.

Julian's correspondent was a thirty-year-old Sussex man, self-described as 'a good-looking male and have been to bed with about 40 girls'. His letter must have embarrassed Huxley immensely.

> Before you write your next article for 'penthouse' etc Playboy or what have you, know the use of these magazines. All these magazines go down the public lavatories and are used for masturbation. Some go to Bed sitting rooms in the large cities but the articles are seldom looked at.

It got worse:

> Dear Sir Julian. Cannot I, a prime example of the debauched crying out from the public convenience with a copy of Penthouse on my knees, cannot I convince you that these books are evil, and destructive, and I see thousands of other men creeping away to eject their sperm over your articles, rather than read them ... Perhaps you have become senile, for you must be blind to the fact that for all their supposed liberating articles, the magazines are merely our general insightment [sic] to National Masturbation and degradation.[31]

This tortured Sussex man was trying to repress his sexual desires more than Julian himself ever would, or, we know, did.

> The most depressing thing about 20th Cent. Life is when people like you, lend your talents (sell them) to the mass titillation of the nation ... But I want, like nearly all men, to get away from my sexuality & desires ... You may think I am some kind of puritan ... in fact I am the

opposite … Sex is to be enjoyed, we all know this damn cliché, but offered to us all day long in magazines, films, TV, papers, advertising etc etc it becomes [boring?], disgusting, and destructive.

Julian was charged with prostituting himself – selling sex to Hugh Hefner. 'You are a pathetic tool in the hands of the clever and successful young Money Maker … Money through sex, the uncontrolled exploitation of men's weakness.' At the same time, in this chastising letter Huxley's unwelcome correspondent set out a version of feminist argument that was soon to become widely appreciated: 'Now wherever I go I see these magazines as the apex of the sexual <u>attack</u> on women.' He accused Julian Huxley of promoting and in fact living 'the sordid sinking of Western Urban man'.[32] On occasion, Juliette might have agreed. For his correspondent, wanton sexuality was itself the sign of 'national' degradation, what he called race deterioration, an old-fashioned term by the 1960s.

And so, as Julian worked his way through changing twentieth-century public and personal responses to sex, population and reproduction, one man in 1937 wanted his sperm for artificial insemination, as genetic security for the future, while another in 1965 would have bred out this morally unfit Huxley entirely, given the chance.

QUALITY: ARTIFICIAL SELECTION AND EUGENICS

T. H. Huxley was outspoken, late in his life, on the problem of population growth, but as a rule he was less so on the emerging ideas of human improvement. Occasionally and privately he explored some possibilities. He experimented with ideas occurring to him after having read Lyell's *Antiquity of Man*, writing to Charles Kingsley, his great sounding board:

> The advance of mankind has everywhere depended on the production of men of genius: and that production is a case of 'spontaneous variation' becoming hereditary, not by physical propagation, but by the help of language, letters and the printing press.

This was a strange mix of agricultural and horticultural commonplaces, and a vague gesture towards 'inherited' human culture – something that Darwin and his own grandson would both take forward, with more sophistication. 'Newton', Huxley gave as an example to his friend, 'was to all intents and purposes a "sport" of a dull agricultural stock, and his intellectual powers are to a certain extent propagated by the grafting of the "Principia," his brain-shoot onto us.'[33] Huxley built here from the familiar base of (human) artificial selection of plants and animal stock, linked to the new possibilities of evolutionary time, over which humans, like other species, had changed up to the present, and would yet change in the future.

Darwin's statistician-explorer half-cousin Francis Galton, by contrast, brought it all down to a three- or four-generation scale conceivable within a family tree. In *Hereditary Genius* (1869) Galton also had his eye on Darwin's new ideas on variation and domestic breeding of plants and animals. Might this be undertaken not just *by* humans, but *on* humans?[34] He came up with a new word: 'what is termed in Greek, *eugenes*, namely, good in stock, hereditarily endowed with noble qualities. This, and the allied words, *eugeneia*, etc., are equally applicable to men, brutes and plants.'[35] Eugenics was born.

Charles Darwin thought his cousin's idea of improving the human race a 'grand' one, meaning ambitious, but plainly utopian. He read Galton's ideas about heredity and the future with interest, though with some scepticism in regards to his cousin's knowledge of natural sciences and of the physiology of reproduction.[36] Darwin was well aware that Galton's work derived from his own observations of human selection for plant and animal breeding. It was all explicit.[37] Humans were already experienced and thoughtful selectors of, and for, other species, and so it is unsurprising that in *The Descent of Man* Darwin set down his thoughts about possible human futures if that experience was self-applied. First, Darwin reiterated Malthus's old political economy point: 'all ought to refrain from marriage who cannot avoid abject poverty for their children'. And he went further. Both sexes should refrain from marriage 'if they are in any marked degree inferior in body or mind'. We might well wonder what Huxley thought when he read that. But then Darwin caught himself: that kind of self-selection on a larger scale could have population-level implications if there was a pattern to who was doing the refraining, in class terms. '[A]s Mr Galton has remarked, if the prudent avoid marriage, whilst the reckless marry, the inferior members tend to supplant the better members of society.' As Darwin himself put it, selection when it comes to humans 'is a most intricate problem'.[38] That was the truth. A century or so later, when Julian Huxley delivered his second Galton Lecture to the Eugenics Society, he thought that the whole discussion of eugenics had been 'bedevilled by the false analogy between artificial and natural selection'. Artificial selection, manifest in agriculture or horticulture, aimed to produce a 'particular excellence', specialized breeds with less and less variation. But precisely because Darwin did not really entertain eugenics, Julian explained, he never pointed out the difficulties in applying that model to humans. Increasing specialization does not work for our species. 'Man owed much of his evolutionary success to his unique variability,' Julian Huxley explained, and human improvement must retain that diversity.[39]

Darwin foresaw the same problem. He was attracted to Galton's ideas about self-selection for improvement and yet knew that humans had reached their present evolutionary advantage through 'a struggle for existence consequent on his rapid multiplication'; the mortality that came with high fertility *was* the process of natural selection in humans as in every other species. If that was interfered with, 'man' might not advance or without struggle would 'sink into indolence, and the more gifted men would not be more successful in the battle of life than the less gifted'. The struggle of natural selection yielded the advantage, for humans as for all other species. Troubled still with that conclusion, Darwin intellectually contorted again, and reminded his readers that though this struggle is important, even more so are 'moral qualities' advanced not through natural selection but through the intergenerational social world, through social habits, through 'reasoning powers', even as human 'social instincts' were themselves conceivably acquired through natural selection. The one thing Darwin knew for sure was that this intricate problem of human artificial selection for better futures could only be sorted when 'the principles of breeding and inheritance are better understood'.[40] Darwin was concerned about his own consanguineous marriage. But for others, later, his family stock of 'outstanding capacities in philosophy, science, and art' became emblematic, especially over the 1930s, when eugenics was at its height. Director of the Cold Spring Harbor Eugenics Record Office and horse-breeder Harry Laughlin produced a famous family tree image for the 1932 Third International Congress of Eugenics (figure 9.2).

The Huxley family could hardly measure up, yet in 1932, when this poster was produced, there *was* a fledgling Huxley pedigree that was starting to fit this 'hereditary genius' idea. T. H. Huxley's sometime student Henry Fairfield Osborn gave the key address at that year's international eugenics meeting and, perhaps to Julian's consternation, signalled the philosophy, science and art that was passed from the great Victorian to Aldous Huxley and his just-published *Brave New World*.[41]

A Huxley pedigree is perhaps better represented in another famous eugenic tree: its secular 'tree of knowledge', designed and published in 1921 (figure 9.3). Like the biblical tree of knowledge – the irony seems to have escaped the designer – the Eugenic Tree of Knowledge would yield the fruits of sin and evil. For some humans deemed 'unfit' it already had, categorically. Yet this image, created for the International Eugenics Congress, is a brilliant design of intellectual genealogy. It might have been – but was not – commissioned by the Huxley family, representing the rich disciplinary roots and branches of their own minds, expertise and creative capacity.

We can read the Huxley generations back into in this epistemological family tree, the emergence of disciplines out of a polymath nineteenth century and into the more specialist twentieth-century human sciences and natural sciences. From T. H. Huxley came the 'bio' cluster: anatomy and physiology, biology and anthropometry – how ancient and present humans are literally to be measured. We see geology and history becoming anthropology. Julian's population genetics is there: 'genetics' and 'statistics'. The Huxley family is represented personally in the roots of genealogy, biography and psychiatry. This was all family intellectual business, the complex history of ideas that produced 'evolution' over the nineteenth and twentieth centuries, and eugenics as 'the self-direction of human evolution'. And while the Huxleys certainly can't be reduced to 'eugenics', this tree of knowledge explains how and why they were so invested and connected.

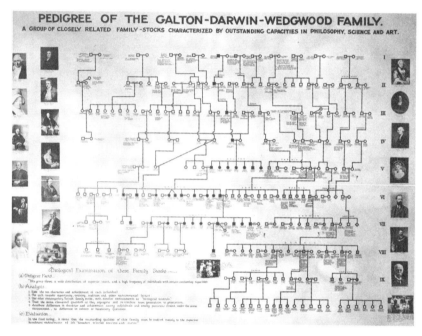

Figure 9.2: The Galton-Darwin-Wedgwood family tree of 'outstanding capacities in philosophy, science and art', 1932.

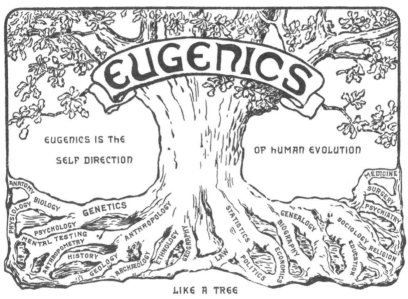

Figure 9.3: The Eugenic Tree of Knowledge, International Eugenics Congress, 1921.

Julian Huxley knew that phylogenetic trees and family trees were often misunderstood. He explained to his twentieth-century readers that 'with biological groups "common ancestry" does not imply descent from a single ancestral pair, as in human relationships; it means the gradual modification of a more or less sharply delimited group by the progressive substitution of some genes for others'.[42] And yet he also recognized the great popular resonance of genealogy. It is little wonder that the tree image was swiftly mobilized within twentieth-century eugenics, the whole point of which was to show and know populations in terms of breeding and intergenerational inheritance. The family tree became the graphic device of choice for the retrieval and visualization of eugenic data, both for research and for public pedagogy. Dysgenic family trees began to emerge as well, displaying predisposition to feeble-mindedness or alcoholism or insanity. In the US, the invented and pedagogical 'Kallikak' family and its apparent inheritance of 'feeble-mindedness' became iconic within early twentieth-century eugenics, but truth be told it might have been the Huxleys, with their maladies of thought.[43]

Of the three Huxley generations – Thomas, Leonard, Julian – Leonard was the least afflicted *by* mental illness, and yet expressed the purest eugenicist statement *about* mental illness, more so even than Julian, later president of the Eugenics Society. In 1926 Leonard stepped out from behind the eminent Victorians whose lives and letters earned him publishing income to make his own statement in print. Yet there was both irony and sadness to the topic of Leonard's book *Progress and the Unfit*. It had all become far more brutal and explicit than Galton's talk of the 'faulty' and the 'gifted', or the moderately phrased 'more and less suitable'. Now it was the physically and mentally 'fit' and the less lucky 'unfit', the latter to be rooted out and discarded: 'Their increase is like a fungoid growth on a sound tree, precursor of degradation and decay.' Leonard Huxley wondered at the poor selection of human raw material for future populations:

> we look on complacently while the feeble-minded multiply ... free to breed their like superabundantly, to bring down the general level of intelligence and character, and to be a life-long and growing burden on the rest of society.[44]

This Huxley caught the classic vocabulary of 1920s eugenics, then in full flight in London.

The year after Leonard Huxley published so crudely on the unfit, Julian went to Geneva to attend the extraordinary conference on world population that Margaret Sanger had put together to pressure the League of Nations on birth control. Julian Huxley was watching and listening closely, especially to a former employee of the Poor Law Authority, E. J. Lidbetter, who had been busy for some decades collating information on 'mental disease in successive generations' and publishing 'pedigrees' of insanity.[45] Julian was impressed, reporting his data that showed how 'mental and physical defect runs in families and how a quite small measure of negative eugenics could enormously reduce the burden of defective humanity which the race has to carry on its shoulders'.[46] What were these small measures? Mainly sterilization. Judiciously applied, this would minimize the burden, Julian was convinced, and this platform monopolized most of his eugenic energies over the 1930s.

There was one problem, however. The surgical procedure to sterilize men and women for this kind of preventive purpose was not strictly legal in Britain. A committee was appointed by the Minister of Health, aware that, as things stood, sterilization for health reasons was legal, but sterilization to 'prevent the propagation of unsound offspring' was not, when applied to those with a mental 'defect' or 'disorder'. The sterilization for this reason of those of sound mind was possibly admissible, but many authorities considered this also beyond the law. The committee's report recommended that, subject to safeguards, voluntary sterilization be legalized for people with a mental defect or disorder, for people with a grave physical disability that was known to be transmissible and for a person likely to transmit a mental disorder or defect.[47]

Julian sat on the Eugenics Society Committee for Legalizing Eugenic Sterilization, which had both lobbied for the (partisan) inquiry in the first place and then publicly responded to the so-called Brock Report, pressing its recommendations in favour of sterilization. He was joined on this committee by Lidbetter, Ronald Fisher and Lady Ellen Askwith, who had engaged in difficult newspaper flurries over the late 1920s regarding unfit parents, 'weak-minded mothers' and the state's responsibility to prevent 'people who are diseased, whether in mind or in body, from propagating children at all'. This, she had controversially claimed, was 'the truest mercy'.[48] The Committee clarified that no one and certainly not the Eugenics Society advocated *compulsory* sterilization, although it is apparent that the Committee contemplated a phased programme towards just that. Certainly by the 1960s Julian edged towards stronger methods of sterilization: comparing it with 'compulsory or semi-compulsory vaccination, inoculation and isolation' deployed commonly enough for other public health risks, he normalized the prospect.[49] Possibly thinking of his own family, Julian endorsed the Brock Committee's consideration not just of mental 'defectives', but also those who had suffered from the lesser mental 'disorder', those who had 'previously been insane' but had recovered and those who might simply be carriers of mental disorder. He noted that the prevention of parenthood either by segregation or by sterilization was the only method available by which 'the burden of defect in the hereditary constitution of this nation can be reduced'. This was not quite true, since the Nazi Party had already thought of other measures: by 'euthanasia', that is, by killing. But the German case was already seen by British eugenicists to be an entirely unacceptable outlier. Julian clarified that in Germany sterilization was compulsory not voluntary, and its whole 'crazy mentality' was entirely different to what was going on in his own country. Still, Julian praised the implementation of sterilization laws and policy in California, especially its implications for deinstitutionalization. Sterilization there, he claimed, was regarded by 'higher grade defectives' as 'an award for good conduct, and a ticket-of-leave, so to speak, to a freer life'.[50] In Julian's logic, which edged towards a kind of liberal eugenics, sterilization freed some people from the alternative, from carceral segregation.

There is no question that offsetting the costs and burdens and, as he would have it, suffering of mental disease in future generations was Julian Huxley's specific concern. He was in the process working hard to popularize the great ambitions of eugenics more widely and broadly.[51] What Darwin had called Galton's utopia was for this generation at the operational stage. For Julian, eugenics heralded an exciting possibility for an improved human future, now implemented sensibly in some polities (California) and misused and misapplied in others (Germany).[52]

It was through eugenics that the Huxleys and the Darwins continued to shadow one another. The dynasties were entwined and deeply integrated into the political biology of assessing future human population quantity, and with the emerging dreams – nightmares for many – of human improvement at a population level. The other Leonard – Charles Darwin's middle son – chaired the Eugenics Education Society between 1911 and 1928. On his retirement in 1928, Leonard Darwin was worried: 'In moments of despondency I have doubted whether anything was being accomplished for future generations by the Society, though I have always believed that great things were possible.'[53] In many ways similarly pessimistic was his nephew Charles Galton Darwin, the physicist who presided over the Eugenics Society in a different era, between 1953 and 1959, succeeded by Julian Huxley. This Darwin grandson was granted a 'Galton' name in honour of his godfather.

The original Galton left no direct issue, but he did leave an institutional one. The Eugenics Society changed its name to the Galton Institute in 1989, after hosting decades of Galton Lectures as part of its annual programme. These was also a declension of Darwins and Huxleys. The first was delivered in 1914 by another of Charles Darwin's middle sons, the botanist Francis Darwin. In 1917, Leonard Darwin spoke on the disabled sailor and soldier 'and the future of our race', and then again in 1929 on the growth of the eugenic movement. Julian's first Galton lecture was in 1936. He gave himself a large brief – 'Eugenics and Society' – and the company dined afterwards at the Waldorf Hotel, where Aldous raised a toast to his brother. The following year John Maynard Keynes (whose brother married a Darwin in 1917) spoke on some economic consequences of a declining population. And in 1939 Charles Galton Darwin offered his thoughts on 'Positive Eugenics Policy'.[54]

In 1962, Julian Huxley gave his second Galton Lecture, this time as retiring president. He had confided to Solly Zuckerman that he was 'bullied and cajoled into becoming President of the Eugenics Society', but having accepted tried to do more *scientific* work.[55] His lecture was delivered in an altogether different era to his first. In the intervening years, German eugenics had been violently implemented, then scrutinized and then tried under a new legal concept, 'genocide'. The twentieth-century history of political biology 'from eugenics to genocide' thereafter consolidated. Critics began to link Darwin (and by extension Huxley) and 'social Darwinism' to eugenics and then to Adolf Hitler and 'the mass murder of those deemed unfit'.[56] We can't say that the common-enough statement of a link between Darwin's *Descent of Man* and the worst manifestations of twentieth-century biological racism is false – it's not. And yet it is often a surface and sensationalist idea: as historically insufficient as it is politically usable, as easily (and this is important) for anti-evolutionists as for anti-racists.

We also need to recognize that this critique hardly escaped the historical actors themselves, often highly politicized biologists. Julian Huxley himself made it plain a long time ago. In his second Galton Lecture he named the naivety that presumes the superiority of one's own race or people. Such superiority was subscribed to by Darwin himself, he had no compunction in saying, especially in his comments on the Fuegians in *The Descent of Man*. Julian Huxley anticipated the common later critique by pointing to Galton's racist beliefs as well. 'This type of belief . . . was used to justify the Nazis' "Nordic" claims to world domination and their horrible campaign for the extermination of the "inferior, non-Aryan race" of Jews.' Julian Huxley brought this racism up to his own time, accusing first the Holy Writ, then the Dutch Reformed Church in South Africa and finally Verwoerd and his denial of full human rights to black South Africans.[57]

Notwithstanding this critique, Julian Huxley was positively excited by the new possibilities of, and for, humanity's artificial selection of itself for a better future. 'Man', as he still insisted on naming us all, had evolved like everything else, but we are a highly imperfect organism, he lectured in 1962. What were these inheritable imperfections, at this point? It was a long list, far more refined than 'mental defect and disorder', 1920s style. Julian included haemophilia, colour-blindness, intersexuality, 'some kinds of sexual deviation', much mental defect, 'dwarfism', 'Huntington's chorea'. He also included 'mongolism' in a draft of this lecture, but scored it through for reasons that remain unclear. Man, essentially is 'unfinished', he announced, and has nowhere near reached 'his' potential. And then, as so often, Julian pivoted from the quality issue back to the quantity issue: progress is threatened by the high rate of increase of world population. Genetic deterioration can and must be checked. For him, measures ranged from reducing man-made radiation, to reducing human over-multiplication, to evening out fertility and mortality rates across global regions and across classes. How could this be done in 1962? Certainly by economic and educational measures to 'level up society' he insisted, casting back to the Geneticists' Manifesto, but also by new possibilities in the control of human reproduction. Some of these were low-tech: vasectomy, now legal and easy enough, and new, cheaper mechanical and chemical contraceptives. The latter did not yet include the hormonal 'pill' that was about to take off. In fact, in 1962, Julian Huxley still put imagined oral or immunological contraception in future dreams for the implementation of eugenics, along with the 'preservation of sperm' and of 'living tissues by deep freezing'.[58] Future humans would be shaped by developing technologies, yet some were more proximate than this would-be futurist could divine.

TRANSHUMANISM:
FUTURE HUMAN POSSIBILITIES

People close to Julian Huxley were monumental science fiction writers, the genre that imagined distant, or more often near, futures; who conjured utopian and dystopian places-out-of-time. Charles Kingsley's fantasy genre for children was renewed for adults, and between Wells's *The War of the World* (1898) and Aldous Huxley's *Brave New World* (1932), Julian's circle held a monopoly on bestselling futurist fiction. Uncontrolled reproduction of humans and other organisms in limited space was, and is, a go-to device in science fiction, one that offers an in-built narrative tension and the possibility of imagining alternative origins of life. In Aldous Huxley's famous futurist dystopia, his brave new world, 'World State' humans are made in artificial wombs, managed by foetus technicians who work in the 'Central London Hatchery'. Once 'born', citizens of the World State are 'sleep conditioned' into a peaceful compliance and retained thus partly by the drug Soma. Abnormal biology, ghastly psychology and terrifying pharmacy: this was a Huxleyan brave new world written in the year of our lord 1931, but set AF (After Ford) 632.

Julian had a go at science fiction but often could neither finish nor publish his attempts. One was about a 'mythical substance which could predetermine a baby's sex'. He started this story in the summer of 1934, vacationing at the Swiss chalet that his grandfather's mountaineering friend John Tyndall had built, still filled with all the Victorian ephemera that the old man's widow could not throw away. But like ancient Mrs Tyndall and her poignant fifty-year effort to publish her husband's life and letters (she had unwittingly poisoned and killed him), Julian's draft was never published.[59] He had been momentarily successful, however, with what he called a 'biological fantasy'.[60] In 1926, 'The Tissue-Culture King' appeared first in *Cornhill* magazine, then in the sci-fi periodical *Amazing Stories*.

In 'The Tissue-Culture King', a white explorer – the first-person narrator – is captured by a tribe of African giants. He witnesses all kinds of monsters and mutant versions of familiar species: humans who are too tall or too short or too fat; conjoined toads; snakes with three heads (figure 9.4). He meets a London-trained endocrinologist who has also been captive for many years and learns that elaborate experiments in tissue culture, in experimental embryology, in endocrine treatment and in artificial parthenogenesis have taken place, experiments which had produced the 'menagerie of monstrosities' before him. The captive doctor was also culturing – mass-producing in explicitly Fordist terms – subcutaneous tissue from the all-powerful king, thus reproducing and multiplying him, dispersing his power, which remained magically resident in each piece of cultured tissue, across the country. The laboratory-factory was not a necropolis, Julian the narrator-figure offers, but a 'histopolis, if I may coin a word; not a cemetery, but a place of eternal growth'.[61] But the most novel and dangerous experiment was with minds, not bodies: with hypnosis and mass telepathy. The doctor of the story was seeking not a subconscious but a mass 'super-consciousness' through which ideas and commands and instructions could be silently and effectively communicated across the whole society, by whoever has the capacity to be master hypnotist. Danger unfolded as the doctor and narrator became themselves subject to this telepathy, all but unable to resist commands from the tissue-culture king who had watched and learned how to hypnotize. Endocrinology and weird reproduction on the one hand,

The TISSUE-CULTURE KING
By Julian Huxley
Reprinted from the Yale Review

"I thought I would see whether art could not improve upon nature, and set myself to recall exprimental embryology...I utilize the plasticity of the earliest stages to give double-headed and cyclopean monsters... But my specialities are three-headed snakes, and toads with an extra heaven-pointed head." Hanscombe called it "Home of the Living Fetishes."

Figure 9.4: Julian Huxley, 'The Tissue Culture King',
Amazing Stories (August 1927).

and psychology and altered consciousness on the other. These were the fantastic worlds of the 1920s, and Julian couldn't resist entering them both into this tale set in another dimension, in mythic out-of-time 'Africa'.

It's not a well-told short story, although Julian boasted that it was republished in *Great Science Fiction by Scientists* in 1962.[62] Aldous might have winced and suggested Julian keep his non-fiction day job. But the creation of mutations and modest monsters *was* Julian's day-job at that point. He was just then forcing the artificial metamorphosis of some living creatures and trying to extend the life-spans of others. He was administering thyroid hormones from various sources to axolotls, already hybrid creatures that transform weirdly, one day aquatic, the next terrestrial and air-breathing.[63] In Oxford over the 1920s, Julian's laboratory work was in experimental endocrine physiology. It was a kind of magic. Hormones regulated by these newly discovered 'glands' might make organisms larger or smaller, longer-lived or shorter-lived, more or less male or female. The press loved it more than his scientific colleagues, at that point doubting his laboratory staying-power. The *Evening Standard* ran a sequence of articles on the destiny of man, 'How the Race of the Future May Live', with Huxley headlines: 'Biological Control', 'Life's Span Doubled: Health 10 Times Better' and more.[64] 'Long-Sought Elixir of Life Is Found: Oxford Savant Experiments Successful. But Alas! With Flatworms', another reported.[65] What interested the Oxford savant especially in flatworm experiments was that they seemed not so much suspended in time, but under certain conditions became physiologically 'younger' rather than 'older'. All very well for flatworms and rats, but humans are tricky organisms, Huxley warned. 'We are so constructed that we cannot live upon our own tissues, nor can our temperature be altered.'[66] His hormone work drew a good deal of attention via another article on 'Glands, Growth and Race', drawing a personal plea.[67] Julian received a letter addressed to 'the Discoverer of the Thyroid Gland Action etc, Oxford Experimental Laboratories'. A Karachi man had read some of the press on his research and wondered about his own 'undeveloped or undergrown Male organ'. Might it be doubled in dimension? What would be the cost? And could Professor Huxley reply via the stamped self-addressed envelope, thereby assisting a fellow man in 'utter despair and despondence'.[68]

Hormones were the new magic substance of the 1920s, enchanting biologists and lay folk the world over. Hormones might stretch all kinds of seemingly fixed physiologies, even in humans. Time – or the age-markings of it in living organisms – might even be reversed. Some in the 1920s were experimenting with the 'rejuvenation' of human tissues. The Viennese endocrinologist Eugen Steinach was thinking about precisely this, and Julian followed his work closely, thoroughly intrigued.[69] Grim procedures of ligation and grafting made the onset of old age appear to be controllable. He was also a fan of the 'regeneration' work of Charles Manning Child, which challenged the idea that all somatic cells must die. Along with Alexis Carrel and his work on tissue culture (noted in his science fiction story), Julian was intrigued by the prospects of prolonging the life of any organism, by grafting, for example. 'We can save any particular part of an old plant from death by taking it for a cutting.' And likewise, 'it has been found possible to continue growing the cells of a single original piece of tissue not merely for weeks or months, but for over 7 years'. The cells of the tissue showed no sign of ageing: 'it would seem that tissues cultivated thus outside the body are probably immortal'.[70] Here were the real tissue-culture kings.

It all made Julian think conceptually about the relation between age and time. The year that Aldous published *Brave New World* Julian published *Problems in Relative Growth* (1932). Age was not a matter of the passage of external time, that is, it was not determined by the lapse of years. It was, more truly, a matter of physiology, which could be (and perhaps could be made to be) independent of external time. He was deeply interested in relative growth – time and tissue change, growth and decay – eventually studying so-called 'rate-genes', 'genes which determine the rate of a developmental process'.[71] The great extension of deep time that had awed his grandfather had become physiological time with curious reproductive possibilities. For this early twentieth-century generation, physiological research on embryos, on tissues and on reproduction turned on the possibilities of slowing, suspending or even reversing the effect of time on living tissues. 'Are we ever to know enough,' Julian pondered, 'to be in control of our own vital destinies . . . for greater growth, for greater vitality, for long life?'[72] In the meantime, Aldous's 1939 novel *After Many a Summer* cast doubts on long life: an eighteenth-century earl is still alive in the novel, aged by Aldous's hand to 201. All very well, but Aldous also had the character now resemble a 'foetal ape'.[73]

Vital destinies – imagining possible futures – entirely absorbed Julian Huxley. Like the medical doctors involved in rejuvenation, and entirely like proponents of human enhancement in our own time, he aimed for augmented individual capacity. From an early date, Julian's personal papers were organized in endless folders, notebooks, scraps of papers and lists on 'possibilities': 'possibilities of sex'; 'possibilities of biosociety'; 'possibilities of psycho societies'; 'possibilities of filter feeding'; 'biochemical possibilities'; 'ecological possibilities'. The 'Need for a Science of Human Possibilities' was his special plea at one point, along with the idea of university chairs of the future.[74] He wanted humanity to build its future 'by transcending its past'.[75]

Gauging the future from a high modern mix of catastrophe and optimism, Julian arrived at not just the idea but the very term 'transhumanism':

> The human species can, if it wishes, transcend itself – not just sporad-
> ically, an individual here in one way, an individual there in another way,
> but in its entirety, as humanity. We need a name for this new belief.
> Perhaps transhumanism will serve: man remaining man, but transcend-
> ing himself, by realizing new possibilities of and for his human nature.[76]

Julian recounted the birth and naming of 'transhumanism' much as Galton had recounted the word-birth of 'eugenics' in the previous century, and quite possibly with that account in mind. While 'eugen-ics' was never a controversial term for Julian, it was perhaps coming to be stale and old-fashioned by the 1950s. It signalled the futurism of his grandfather's generation. It belonged to the 1880s, when Darwin thought Galton's idea grand but utopian. In the shiny high modernity of the 1950s, on the cusp of the space age, 'transhumanism' signalled something more exciting for the generation which had spent its childhood consuming and composing science fiction possibilities.

Julian had been working towards transhumanism for decades. The final message of his major work, *Evolution: The Modern Synthesis*, was precisely about how future humans that might think, feel and act in ways exceeding and improving on current capacities. He drew on

Aldous's *Ends and Means* to explain the advancing possibilities of the human brain, of increased intelligence, of 'fuller control of emotional impulse'.[77] He noted Hermann Muller's fantastic idea of selecting a few highly endowed males to sire more of the next generation. And if only we could discover how to implement Haldane's idea in *Daedalus* and reproduce our species solely from selected germinal tissue cultures, then all kinds of new possibilities would emerge. Ultimately, Julian thought, the focus of transhumanism should be aimed at measures that would increase mean intelligence and extend the capacity of the organ that is the basis of human evolutionary dominance: the brain.[78]

Because the human brain (and only the *human* brain) can comprehend evolution by natural selection, humanity can install its own purpose into the process: 'The future of progressive evolution is the future of man.'[79] More modest and more realistic evolutionary biologists then, as now, never understood natural selection to be progressive in this way. It neither leads anywhere in particular, nor does it necessarily mean improvement for all or any species: all those dinosaurs that Richard Owen named, for example, were extinct. Julian Huxley, however, *did* hold a belief in evolutionary progress and was highly unusual for it.[80] For him, the future was therefore a grave responsibility of and for humans. And for Julian Huxley the future was a long one, notwithstanding the catastrophic register of the mid-twentieth century – atomic and demographic annihilation. Optimistically, Huxley signalled that as humans are a product of 2,000 million years of past biological evolution, so we can expect an equal or greater span ahead, accelerated by the pivot into the new stage of evolution.[81] He declined to prophesy, but this was a disingenuous posture. In fact he indulged in speculating on the future more and more. His new belief was his own invention:

> 'I believe in transhumanism': once there are enough people who can truly say that, the human species will be on the threshold of a new kind of existence, as different from ours as ours is from that of Pekin man. It will at last be consciously fulfilling its real destiny.[82]

This was 'evolution in action: the human phase'. On this earth, Julian speculated – 'and presumably in a few other isolated spots in the universe' – the single and comprehensive process of evolution had unfolded over three sectors and epochs. [83] In most of his publications from the 1950s he refined a kind of temporalized stratigraphy, or else a biological Genesis, creation in evolutionary time. The first layer/time of Earth history was the inorganic or sometimes 'cosmic' sector, the evolution of elements, stars, nebulae and planetary systems. The second layer/time was marked by the evolution of DNA and genes through which some material became self-varying and self-reproducing. He called this an organic or biological sector or phase in which the evolution of plant and animal organisms unfolded. A third layer or time shifted from the biological to what he called the psycho-social, up to his own era. Now – coincident with his own lifetime – was a further stage, in which accumulated knowledge of evolution itself could shape the future and consciously – selectively – so.

With the same kind of hubris with which Thomas Henry Huxley pronounced on 'man's place in nature', Julian Huxley took on much of the responsibility personally for this newly evolved capacity of humans to imagine and shape evolution itself. Since man is now trustee of evolution, this reserved an extra-insightful place for those humans who happened to be evolutionary biologists; that is to say, Julian himself. He also immodestly cast himself – or his like – at the centre of the great life-epic by insisting that his historical moment was special and unique. He happened to be living in a key transition point of vast evolutionary time.

THE FUTURE OF MAN AND OF WOMAN

In 1963 – 100 years after the publication of Huxley's *Man's Place in Nature* – Julian Huxley spoke at a London conference titled 'Man and his Future'. Many of the world's key scientists were there, across a range of natural and physical disciplines: Gregory Pincus spoke on the control of reproduction in mammals; Donald MacKay on humans and machines; A. S. Parkes on determining sex ratios in humans; Joshua Lederberg on biological futures; Hermann Muller on assisted reproductive technologies; Alex Comfort on longevity of human tissues. All these papers and futurist ideas were commented upon by

Paul Crick. It was a meeting of stellar minds, with a liberal sprinkling of Nobel Laureates, gathering in what they suspected to be a key moment in human biology. Many were in their prime, like Pincus, on the cusp of phenomenal worldwide success with his hormonal pill to control female fertility. The meeting was bookended by two elder statesmen of biology, both of whom brought vast timescales to the discussion. Julian Huxley opened the meeting and J. B. S. Haldane closed it.

These men – all men – discussed man's future the year that Gregory Pincus's pill was marketed, and to some extent this was the rationale for the meeting. Pincus himself thought it was time to take stock of the change that was imminent, and to gather with others whose applied research also seemed to be setting up an altogether different future. 'The world was unprepared socially, politically and ethically for the advent of nuclear power,' it was declared. 'Now, biological research is in a ferment, creating and promising methods of interference with "natural processes" which could destroy or could transform nearly every aspect of human life which we value.'[84] The meeting rightly sensed the early 1960s as a genetic and reproductive turning point.

Julian Huxley was well practised in leaping forwards in time and imagining exciting futures 10,000 years hence. He perceived a strange combination of medium-term decline – 'genetic erosion' – and imminent crisis (population bombs, H-bombs). But, as always, he saw possibilities as well. His was an opening conceptual talk about evolutionary humanism and the significance of their own era. It was a statement about ecology, 'the science of relations between organisms and the resource of their environment', and the new threat to their own habitat that humans had recently manufactured. He cited Rachel Carson's *Silent Spring* – the toxic silence where once there were sounds of bees and birds. He took the liberty of citing 'my brother Aldous', who commented after reading Carson's book that 'we are exterminating half the basis of English poetry'.[85]

Julian Huxley perhaps surprised the various chemists, geneticists and engineers in the audience with his exposition of 'three kinds of habitat': the planetary habitat, the social habitat – so far so good – and then the 'psychological habitat'. That's when he started to drift.

Coordinated research is needed for the future on 'all methods of attaining states of self-transcendent experience', he told them: yoga of various kinds, meditation and hypnotism, dreams 'and their possible control'. He even indicated the need to understand 'possession by an alien personality or spirit', as for shamans or in Haitian voodoo or with LSD. It seems, as an old man, that his brother (dead that year) and his son Francis were grafting their own imaginations onto his, like the telepathic tissue culture king. And then he returned to more familiar territory, overpopulation. This is what he thought endangered the future of humans most. 'What are we to do about it?' He dismissed ideas about transporting surplus populations to other planets; that was quantitive nonsense. He invoked in entirely general terms the need for birth control. '[A]ll advanced nations should devote an increasing amount of scientific and technological manpower to discovering and perfecting simple and acceptable methods of birth-control, and making their discoveries freely available to the rest of the world.'[86] Well might we imagine the looks between the younger men in his audience. Julian Huxley might have been in touch with LSD experimentation and modern hypnosis, but the old man was out of touch with the latest in reproductive physiology. There lay 1963's near-future bioscape.

When the meeting got down to business after Julian Huxley's conceptual horizons, they heard, or would have already known, 'that exceptionally effective methods of fertility control are at hand and that others are in the offing', as Gregory Pincus put it. He set out the work he had undertaken with Min Chueh Chang that regulated the ovum in the process of ovulation.[87] He had long been researching female mammalian reproduction – in rabbits, in rats – and success was just then manifesting as a sequence of trials of Enovid in female humans in various parts of the world.[88] The ethics of those trials, or of human trials in general, did not necessarily escape this group, even if they failed to conceptualize this in terms of sex and gender. One of their members, Marc Klein from the University of Strasbourg, brought his year in Auschwitz to the table and the sterilization experiments undertaken there by endocrinologist Carl Clauberg. 'Once you have seen with your own eyes where those problems can lead, you are very cautious,' he warned, 'even when you hear about the very beginnings of this type of experiment'.[89]

The launch of the contraceptive pill in the 1960s was one end point, or pivot point, of the fantastic hormonal imaginings of Julian Huxley's 1920s. Yet for all his efforts to prophesy human evolution, and to think 10,000 years hence, he was misreading the immediate future. He failed to perceive the kind of commercialized reproductive biology the pill was about to facilitate. He also missed the brave new world of neo-liberal bio-economies that was just around the corner, driven by highly commercialized IVF and organ commerce, for example. And finally, in 1963, he seemed unable or unwilling to discern a highly individualized reproductive constituency. His talk was all of species fulfilment and population-level dangers or possibilities. In seeking to understand 'the biological future of man', he was quite patently not paying enough attention to the future of women. *The Feminine Mystique* was also published in 1963,[90] and in so many ways it was Betty Friedan, with her psychology and her Marxism, not Julian Huxley, the scientific dreamer of possession, LSD and 'noospheres', who best caught the wave of the immediate reproductive and social future. Julian's problem was that he was still thinking about *man's* place in nature.

Generational change was catching up with him, and he was losing his capacity to assess the political present, to be in it. For all his inclination towards future thinking and his interest in progress, the present was overtaking him. But it wasn't just that Julian Huxley was getting old. It's that he was always more interested in male reproductive physiology than female, we might even say more interested in men than women. As a biological fantasist, Julian had long been drawn to human reproduction that bypassed women and even bypassed artificial wombs. It was reproduction through tissue culture that he had leaped to in the 1920s, not even the ectogenesis in *Brave New World*, which at least referenced an artificial womb. And when it came to new reproductive technologies forty years later, it was never the hormonal contraceptive pill for women and the regulation of the ovum in the uterus that captivated him, but a new kind of fertility selection that foregrounded men and their sperm. It is in this context that Julian saw a mechanism for the biological future of man: artificial insemination, using 'preferred' donor sperm that was frozen.

This wasn't a new proposition either for Julian personally or for his circle of biologists. His friend Muller had raised the preferred donor idea at least twenty-five years earlier and in the early 1960s was still pressing it forward strongly. Artificial insemination with semen derived from a donor, 'AID', was itself technically unremarkable, Muller noted, increasingly accepted and applied. In 1963 he estimated that this method created 5–10,000 American births per year. But this existing technological capacity was being wasted, and there was a better way: 'to utilize, where feasible, germinal material from donors of outstanding ability and vigour, persons whose genuine merits have been indicated in the trials of life'. Effective, systematic and widespread germinal choice of the donor was the next step.[91] Julian Huxley was excited.

Germinal choice followed standard artificial insemination, except that 'the germinal material here is to be chosen and applied primarily with a view to its eugenic potentialities. Preferably it should be selected from among banks of germ cells that have been subjected to long-term preservation.' For Muller, the choices were not to be imposed, but would be fully voluntary and 'democratic', so insisted the refugee from Stalinism. And with Julian Huxley's help he established a Foundation for Germinal Choice, thinking through the processes for implementation.[92] Muller wrote to Julian in 1964 with a draft plan. A couple would offer as much information as possible as to their request, and could narrow down a decision to three donors. But it was a Selection Committee, usefully termed, who would make the final decision about which sperm, which donor, would be used for the best population (not individual) future. And 'their choice would not be disclosed until after the donor's death'. How did Julian regard such a proposal, his old friend asked.[93] Muller and Huxley agreed that individuals would likely make the right choices in their liberal programme.

Julian had recently focused on germinal choice in his second Galton Lecture in 1962. He noted the need to collect, preserve and freeze sperm, a process that would retain its good genetic properties. There was a Cold War urgency to this venture, nuclear fallout threatening a universal genetic deterioration. Julian even pressed the need to build 'deep shelters' for the indefinite preservation and storage of this precious frozen sperm, shamelessly suggesting that 'shelters for sperm banks will give better genetic results than shelters for people, as well as being very much cheaper'. That was the kind of statement perhaps only tolerable to a Eugenics Society audience, to whom he experimentally offered some new terms: 'euselection' and 'EID', Eugenic Insemination by Donor.[94] Julian thought that proponents of EID might look to the history of the birth control movement; early pioneers were opposed, vilified, even gaoled, but rationality prevailed. He put all of this 'rational control of reproduction' in the same basket, all moving towards the promotion of wellbeing, as he would have it. This prevention of suffering and the improvement of future humankind was itself a moral imperative. To his fellow scientists at the Future of Man meeting he was insistent the time had come for artificial insemination or eugenic insemination to step up a gear. 'Such a policy will not be easy to execute,' he claimed. 'However, I confidently look forward to a time when eugenic improvement will become one of the major aims of mankind'.[95]

There they were, the Rice University colleagues grown old. Grebeloving Julian Huxley and *Drosophila*-loving Hermann Muller, fifty years later on the cusp of implementing a fantastic new biological program. The year the pill was released, they both pushed for Human Betterment via man more than woman.

When ageing T. H. Huxley indulged his little grandson about the strange idea of babies *in vitro*, babies floating inside glass, he did his own bit of fortune telling, his own divining. 'When you grow up I dare say you will be one of the great-deal seers, and see things more wonderful than the Water Babies where other folks can see nothing.'[96] Julian Huxley *was* a kind of seer. He continually looked to a future, and it was one created by his own family's past, in experimental physiology, in evolutionary thought, in imagined bio-political worlds. Julian Huxley's 1960s *was* the bizarre future of his grandfather's 1860s, one in which 'special' semen was frozen in bunkers and in which babies in bottles were experimentally conceived. The first human was born from *in vitro* fertilization in 1978, a threshold in human possibilities that neither Julian Huxley nor Hermann Muller quite lived to see crossed.

In turn, we live in Julian's future, no question. On the one hand, infant mortality and maternal mortality have dropped across the world, if differentially so. Fertility rates, too, have declined to an extent that would have staggered T. H. Huxley and would have made Julian Huxley think that the world had been wise, and listened to his lectures on birth control. Perhaps, in a way, we did, though this was down more to those at the experimental and clinical end of contraception research, and the household and family economic decisions made by millions of couples across the globe, than to the high-level diplomacy and catastrophic warnings of the thousand conferences and meetings at which Julian Huxley spoke. On the other hand, the drop in fertility has had an unexpected outcome, rather more like germinal choice than anyone would care to admit. The billion-dollar global industry in assisted reproductive technology would have amazed those at the Future of Man meeting, 1963, including the distinguished futurists who bookended it, Julian Huxley and J. B. S. Haldane. They would have been most astonished by the rapid normalization of assisted reproductive technologies across the global south and the global north.[97] Lest we feel analytically and politically superior, it is clear that political biologists like Huxley and Haldane would have been able to analyse, and *would have* analysed, the power dynamics and trends across the world. We in the present hold no monopoly on the capacity or drive or tools to understand biology as political, even if we might disagree with their presumptions and conclusions. If anything, the twenty-first century is more resigned, even in thrall, to the apparent imperative to realize individual reproductive desires and choices – even 'needs'. This is a neo-liberal fulfilment of Julian Huxley's future possibilities.

Much as we might wish away Julian's eugenics from the twenty-first century, the fact is that many of his generation's reproductive and 'transhuman' dreams are part of everyday lives, everyday decisions, 'fulfilment' that has been normalized, dreams that have been de-exceptionalized, for better and for worse. The extension of human life conceptualized as a transhuman possibility is becoming demograph-ically normal and is accelerating. The foetal diagnosis of not a few of Julian Huxley's 1920s list of the 'unfit', followed by legal abortion, became standard in many parts of the world decades ago. At the now low-tech end, semen is actively selected every day by individuals and couples, and like it or not these decisions are made straight out of pre-sumptions about race, class, intelligence, education, height or 'fitness'.

Less fit future humans are every day diagnosed and made unviable, *in utero*. This might be good, it might be bad, but it is certainly a fact. Tissue culturing has turned into germ line technologies, even human germ line engineering and 'editing'. A neo-liberal, choice-oriented eugenics has become more or less normalized, a continuation of a Darwin, Galton and Huxley world, much as we like to imagine that we live in a refusal of it.

PART IV

Spirits

10

Other Worlds:
The Re-enchantment of the Huxleys

SHE is gone: Death's unremitting, unremorseful hand hath taken her
  To his garths of desolation, silent home of spirits lone,
To a land where never voices from the living may awaken her
  She is gone.

From Leonard Huxley, 'First Loss', in *Anniversaries*
(London: John Murray, 1920), 50.

Thomas Henry and Julian Huxley were together a modern-day scientific Janus, the Roman god who could look backwards and forwards, as well as inside and outside, simultaneously. Janus was also the doorkeeper, the watcher over the threshold into the heavens. But that was a future place in which neither of them believed, at least which neither they nor anyone else could show to exist. One of the great debates about the modern world (for our purposes 1825–1975) concerns secularization. There is a thesis that we have become 'disenchanted', that we live in a world that once believed more fully, was governed more directly and was made more meaningful by faith, by belief in God, by the supernatural, by magic. The German sociologist Max Weber gave us the term disenchantment in 1918 – translated from *Entzauberung* – to describe the modern shift away from belief to the knowable, from magic to the provable, from the sacred to the profane.[1] He put this forward in the early twentieth century, taking stock of the scientific naturalism and rationalism of the nineteenth century, and of the anti-clerical Enlightenment that preceded it. This teleology of modernity is persuasive in many ways, in others easy enough to qualify. But how did these Huxleys fit into a disenchanting world, or at the least into the thesis of disenchantment? One answer is that they were not just affected by it, the family of agnostics were major players in the making of it.

Thomas Henry Huxley had an enormous impact on the possibilities of secularization, putting on the public agenda the question of what can be known and what can never be known about a Creator, about worlds beyond.[2] Julian Huxley compounded the impact of his grandfather's agnosticism, delivering the idea of evolutionary humanism, for him the great human stage after belief in god or gods: Janus or Jesus Christ. The Huxleys thus helped create the vocabulary and the story of secularization itself. And yet their own family story was manifestly not one from the sacred to the profane, and certainly not towards a secular worldview in which religion and mysticism were unnecessary. The reverse is a far truer trajectory. Thomas Henry Huxley lived in the religious Victorian period, countering it as the prophet of scientific naturalism. Twentieth-century Julian Huxley's humanism did not have to be insisted upon so stridently. Agnosticism was his starting point, largely because of the intellectual and public work his grandfather had done. Yet from that base, and in what *was* an increasingly secular twentieth-century Britain, Julian became interested in the domain of the sacred. His brother Aldous extended this quest not so much for religion as for other perceptions, other ways of being and feeling. And Julian's son Francis spent his entire anthropological life thinking about, and learning from, shamans and shamanism, mystics and mysticism.

This line of the Huxley dynasty, then, became more 'sacred' rather than less, even if other lines of the family doubled down on T. H. Huxley's naturalism, his physiology. Andrew Huxley received a Nobel prize for discoveries concerning the nerve cell membrane. Still, even there we might track an enduring connection with Thomas Henry's interest in nerves and sensation and Aldous Huxley's interest in perception and other doors into it.[3] The world, or at least the Huxleys' English-American world, might have been secularizing around them towards a normalization of agnostics and rationality, but over the nineteenth and twentieth centuries, the most famous line of the Huxley dynasty became increasingly enchanted by the sacred, by the beyond, by imagining the unknowable.

AGNOSTICS

In the late 1840s, the evangelical believer Henrietta Heathorn – not long out of her Moravian schooling – learned that her betrothed was going to test her in a dozen ways. Huxley was self-schooled in and by Midlands dissenters – that accorded well enough – but he was also reading the Enlightenment sceptics Hume and Kant and as early as his Sydney years was already actively wondering about the knowability of God. This certainly did not accord with her personal faith, and yet the Moravians were not orthodox. That, at least, was a starting point for life with the original agnostic. The Moravians believed that Christianity should be a religion of the heart, in which experience of faith trumped doctrine. 'When I feel grieved and lonely, as I do now, I turn to the past and review my treasures, until I grow calmer and hopeful, full of trust on our Great Father who has so long watched over us with such tender kindness, unworthy though I be so cared for.'[4] Thomas's doubts were troubling, not just for himself, but also for her, and it quickly became clear that she could not count on him for any spiritual matter whatsoever. 'I have so much need of leading unto holy things, am so dilatory luke-warm and dream-like myself that I fondly hoped he would have been the guide and instructor unto more perfect ways,'[5] she wrote as she waited for the *Rattlesnake* to return her prince. It was only the beginning of the trouble that Huxley's agnostics – as he later called it – was to bring. Henrietta's belief calmed her in the endless separations and the uncertainty about her engaged future. She noted in her 1849 diary that taking the sacrament 'generally makes me peacefully-minded',[6] but a lifetime thereafter with Thomas Henry Huxley shifted her from finding peace in the eucharist to dismissing it as pagan:

> The God whom man eats in the bread,
> Whose blood he drinks in wine
> Such pagan faith be far from me –
> I own a more divine.[7]

Her 'Agnostic Hymn' suggests the extent to which she came to align herself with Huxley over time. Julian certainly thought so, considering his grandmother, as an old lady, to have fully adopted the agnostic position.[8] She nonetheless attended church regularly throughout their London lives and took all her children with her, initially to Christ Church, Marylebone, where she listened to the sermons of the Christian Socialist Rev. Llewelyn Davies for decades, then the local St Johns Wood parish church, St Marks. All the family ceremonies of birth and death involved religious ritual. Leonard, for example, was baptized with Darwin and Hooker as godfathers. Huxley permitted this, just, but the men of science indicated they would ignore the religious strictures.[9] It must have been profoundly difficult for Huxley to accede to this christening, because it came so soon after the death of Noel, which by his own account hardened and clarified his antipathy to religious choreography and his philosophical near-scepticism about the knowability of God.

Huxley talked his doubts through not just with his scientific naturalist circle, especially John Tyndall, but also, and perhaps more importantly, with his friend Charles Kingsley, man of God. Kingsley offered deep sympathy, reflections and support for the all but inconsolable Huxley and Henrietta on their child's death. In September 1860, Huxley took up his pen, heavily, to thank him. What he wrote ended up being so much more than a courtesy acknowledgement of condolences. He confided that his son's sudden death made him wonder why he had 'stripped myself of the hopes and consolations of the mass of mankind'. If he had lived in previous centuries, even he would have thought his doubting to be the work of the devil. Instead, in this time of greatest hardship, and even with Kingsley's assurance of the immortality of his son's soul impressed upon him, 'still I will not lie'.[10] The scientific naturalist did not believe that a soul survived the physical body.[11] Huxley confided how shocked he was – 'inexpressibly shocked' – at his little son's burial when the minister read dutifully at the graveside: 'If the dead rise not again, let us eat and drink, for tomorrow we die.' He took great offence. 'Why, the very apes know better, and if you shoot their young, the poor brutes grieve their grief out and do not immediately seek distraction in a gorge.' As he assessed his own life at the death of his son's, he defied the notion that it was hope of immortality or future reward that had made his life what it was, for good or ill. No – there were three other touchstones: Thomas Carlyle's early nineteenth-century novel *Sartor Resartus*, which taught him that 'a deep sense of religion' was entirely compatible with 'absence of theology'; the scientific method, which gave him a 'resting place'; and love, which taught him 'the sanctity of human nature, and impressed me with a deep sense of responsibility'. These were his teachers and his guides, he explained to Kingsley, 'these agencies have worked upon me' and he had not for a split second been driven by a hope of immortality. 'I neither deny nor affirm the immortality of man. I see no reason for believing in it, but, on the other hand, I have no means of disproving it.'[12] Noel's death ended up affirming the significance of his own agnostics, and no wonder Leonard's christening was a phenomenal effort and concession to Henrietta. Leonard Huxley looked back to the tumultuous moment of Noel's death and his own more or less simultaneous birth and rightly perceived this confession of 1860 to be a kind of manifesto; this was his father's moment of clarity. His 'moral fire' was put into these words to Kingsley 'more completely than in almost any other writing'.[13]

It was a credo. '[T]he most sacred act of a man's life is to say and to feel, "I believe such and such to be true". All the greatest rewards and all the heaviest penalties of existence cling about that act.' Huxley's letter to Kingsley was filled with defiance and truth, a manifesto to himself, to his worthy friend, to the world, the measure of which was his grief, the cost of which was now-impossible consolation. Measured by the standard of his own credo – sacred he called it – 'what becomes the doctrine of immortality?' Blunt and with fierce honesty, he wrote to Kingsley that efforts to console do not help. What does help is his own belief in science:

> Science seems to me to teach in the highest and strongest manner the great truth which is embodied in the Christian conception of entire surrender to the will of God. Sit down before fact as a little child, be prepared to give up every preconceived notion, follow humbly wherever and to whatever abysses nature leads, or you shall learn nothing. I have only begun to learn content and peace of mind since I have resolved to all risks to do this.[14]

For Huxley, science fully and richly took the place of religion.

Agnosticism was thus born from a man's pain. It was christened in 1869 at a meeting of the Metaphysical Society, also conversation between men scientific and men religious, but now polite and genteel and learned, not emotional and private and anguished. Removed, calm and perhaps rested by his science, Huxley much later reflected with some rightful pride on how the new term resonated, yet he had to explain again and again that 'Agnosticism, in fact, is not a creed, but a method.'[15] Agnostics meant 'without knowledge', derived from the ancient Greek 'gnosis'. There was, and could be, no evidence of a Creator, judged through Huxley's demand that evidence be apparent to human senses. 'The proposition that a given thing has been created, whether true or false, is not capable of proof.'[16] For Huxley, it all rested on the method of only making claims from the possibility of evidence. The question was epistemological: what could be known and on what basis? And he turned this knowledge requirement to science itself, as much as religious belief. At the point when Darwin's *The Origin* was published, one might refuse to accept the Creation hypothesis as set out in Genesis, but Huxley was the first to indicate that alternative hypotheses had evidence-based problems too. 'In 1857,' he recalled, 'I had no answer ready and I do not think that anyone else had.'[17] With Darwin's work, a provable (or falsifiable) hypothesis was suddenly available. Leonard Huxley later put it this way: just as there was no evidence for Creation as set out in the book of Genesis, so, before Darwin, there was insufficient evidence on the transmutation of species. Now there was.[18]

Agnosticism was thus not of the same order as atheism. Indeed it originally bore very little resemblance at all. There *were* a-theists in Huxley's Victorian world, sometimes crossing over with 'freethinkers', whose opposition to the Church was of a different measure. He distanced himself from them as a rule. Huxley always engaged more with the theologically minded than with radical secularists and was unimpressed when his term was borrowed. One journal, originally titled *Secular Review*, was in 1888 changed to *The Agnostic Journal*. '[T]he term is not patented,' Huxley wrote curtly to the atheistic editor, 'but I confess I prefer to think that its essence lies in the determination – common to all agnostics – not to go beyond evidence.'[19] Huxley battled them almost as much as he battled the Church, especially resisting the co-option of 'agnostics' by those he viewed as militant rationalists. The editor of another journal, *The Freethinker*, proclaimed a 'relentless war against superstition in general, and against Christian superstition in particular'.[20] Yet this kind of aggressive and plain hostility to religion was never Huxley's position, especially not in his retired life.[21] He reserved 'the greatest possible antipathy to all the atheistic and infidel school' and yet was commonly called an atheist, then as now.[22] Not just his own self, but also Darwin's theory were drawn into a long, mired, public debate about creationism and evolutionism that extends to our own time. There is, now, the possibility of declaring one's own credo in evolutionary attire: 'atheist shirts'. One T-shirt prints the common misrepresentation of Huxley's sequence of primates, the frontispiece to *Evidence as to Man's Place in Nature* (figure 10.1). Huxley's agnosticism lurks here, confused with unbelief. He would have been the first to take issue.

The origin of life, genesis, was as much Huxley's business as that of all the theologians around him, including those who offered spurious theses on human origins. For the Adamites, whom Huxley dismissed, Adam was made out of earth somewhere in Asia, about 6,000 years ago. Eve was modelled from one of his ribs, and their progeny, reduced to eight persons after the universal deluge, then reproduced all humankind. 'Five-sixths of the public are taught this Adamitic Monogenism, as if it were an established truth, and believe it.'[23] It was easy enough for him to scorn such literalism: some of his Christian friends did too, even within his own family. But the larger point was to dismiss any claim that life had a supernatural origin, a Creator, because we can never know this. Critiquing both orthodox and heterodox Christian Creation and origin stories was not the limit of his thought on beginnings, on genesis. Huxley looked to classical philosophers too, and the axiom that 'life may, and does, proceed from that which has no life'. Rather, it was now known that life did not and never came from dead matter, but always from other life, from 'living germs'.[24] Huxley provocatively called this 'Biogenesis', a process dependent on oxygen, as eighteenth-century physiologists had discovered. He argued and was convinced that 'no such thing as Abiogenesis [that living matter may be produced by not living matter] ever has taken place in the past, or ever will take place in the future'.[25]

Some believers were desperate to make evolution by natural selection and supernatural acts of Creation compatible. One correspondent

Figure 10.1: 'The Atheist T-Shirt', a misrepresentation of T. H. Huxley's comparative primate skeletons.

tried to persuade him that that was possible since '"natural evolution" is only a method, and not a cause'.[26] For the rest of his life, letters, arguments, pleadings and threats of damnation were delivered almost daily to Huxley's desk in London and later to his study in Eastbourne. He was uninterested in most of them. But he *was* deeply concerned with calm and learned theological debate. Huxley's clearest rejection of the truth of the Christian Bible in no way meant that he was not interested in it, still less that he was uneducated in the complexities of early, high and late Victorian theological and doctrinal debate within the Church of England, or between Protestant and Catholic denominations, or between Judaism, Islam and Christianity.[27] For this agnostic, theologies and cosmologies were almost as fascinating as primate skulls and the cells of the Hydra.

As a younger man on the *Rattlesnake* Huxley had encountered radically different cosmologies. He had had long talks with the Scottish 'castaway' Barbara Crawford Thompson, who had been shipwrecked and lived with Torres Strait Islanders for about five years before her rescue in 1849. He inquired after the Islanders' religious belief, and she informed him that they had no understanding of a Supreme Being, and indeed laughed at her for that idea. But she also explained their belief in what Huxley 'translated' as 'all sorts of varieties of ghosts':

> When the heavy squall of these latitudes gathers and the clouds topple over one another in huge fantastic masses they say the 'marki' (ghosts) are looking out for turtle and they profess that one comes down and fetches a supply for the rest. So again they believe that porpoises and sharks are a sort of enchanted beings and they never injure them.[28]

At the other end of his life, Huxley's religious writing extended to Buddhism and Hinduism. In his 'Evolution and Ethics' Romanes Lecture, he was looking for ideas of 'evolution' in other systems of thought. He suggested that theories of the universe 'in which the conception of evolution plays a leading part' were evident across the world, from early Indian to early Aegean philosophers, from Stoicism to Hinduism. He set out Buddhism in particular, as 'a system which knows no God in the western sense'. For many Indian philosophers, karma 'passed from life to life and linked them in the chain of transmigrations ... The substance of the cosmos was "Brahma".'[29] Although Julian Huxley, and certainly his brother Aldous, later engaged directly with leading Indian philosophers,[30] for their grandfather the conversation was mediated by British orientalists. Huxley relied in particular on Pali scholar Rhys Davids's work on Buddhism and its relation to early Hindu thought.[31] At the same time as recognizing other cosmologies, the ultimate point for Huxley was still evolutionary unity. Buddhism, Hinduism and Stoicism were important because they aligned with the one master narrative, evolution.

FAMILY HETERODOXIES

The chronic discussion of creationism v. evolutionism is often reduced and caricatured, and when stakeholders look for an origin story, it is found in the so-called 'Oxford debate' between Bishop Wilberforce and Huxley in 1860, which was really just a discussion in the wake of someone else's paper. We lack any direct record of what was actually said between them, and in fact this birth of 'creationism v. evolutionism' is reconstructed on hearsay and in the end was at least as much about professional and personal agendas as about science and Christianity.[32] There was another debate about belief and agnosticism that was more measured, less mythologized, and that penetrated inside the Huxley family as well as emanating from it. The two great intellectual forces, Thomas Henry Huxley and Mary Augusta Ward, both set out their views on heterodoxy, orthodoxy and agnosticism over the 1880s. They enjoyed each other's company. Her favourite word for him was 'racy'. He liked that she both praised and scolded him intellectually: 'I do not know whether I like the praise or the scolding better. They, like pastry, need to be done with a light hand.'[33]

Believer Mary Ward was not orthodox. Far from it. Her formative years were in an Oxford that was still taking stock of the so-called 'Tractarians' of the 1840s, conservative High Church leaders who resisted modernization, who created Anglo-Catholicism, with one of their number, Cardinal Newman, converting to the Roman Catholic Church. By the 1860s, when teenaged Mary and younger Julia Arnold grew up in Oxford, the Broad Church liberal response was in the ascendant, associated with Balliol College through soon-to-be-master Benjamin Jowett. In 1860 Jowett was a contributor to one of the most controversial books of that decade, initially outstripping sales of *The Origin of Species*. The explosive book that defied its title, *Essays and Reviews*, set out new liberal and some radical theologies, including suggested alignments with Darwin's theory and with evidence from geology, including the work of so-called higher critics on 'simply' reading the Bible as a historical text, including the argument that miracles are not possible, and even that belief in New Testament miracles is atheistic.[34] Oxford of the 1860s was defined by the implications of these statements, across colleges and across political sympathies. Mary Ward was drawn in. More than that, she was entirely immersed, partly through Uncle Matthew Arnold's Balliol-oriented sympathy with Higher Criticism, partly by her historical work in the Bodleian Library on early Spanish Church history (she was writing for Henry Wace's *Dictionary of Christian Biography*) and partly by her particular sensitivity to Anglican and Roman Catholic allegiances. Her father, Tom Arnold, couldn't make up his mind.

There was deep analysis of the 'evidence' of the Bible at that point, a question of method as much as doctrine, and Mary Ward brought her reading of documents to the problem. She was strongly persuaded by Higher Criticism, sometimes called historical criticism, a contextual interpretation of the Bible as a historical compendium. Higher Criticism was originally a German intervention, and Huxley and Ward shared that tradition of rationalist approaches, from Kant to Fichte to Hegel. Much later, in 1906, young Julian Huxley was trying to understand this tradition of biblical scholarship that flowed through both sides of his family, and asked his mother Julia to explain:

> It means that you apply to the Bible exactly the same kind & degree of criticism as you would apply to any other collection of historical documents. I.e. you do not say, as people always did say in old days, & as many people say now 'The Bible is the direct word of God – communicated by men directly inspired by God's spirit – [therefore] it is wicked to question the truth of a single word of it.' But you say: 'The Bible is a collection of documents. It is obvious they cannot all be inspired, as they give different

accounts of the same events ... we must not accept statements on any less good authority than that on which we accept any other historical statement.

And then she added what her sister Mary had long believed: 'This tends naturally to eliminate the miraculous in the vulgar sense.'[35] Here, the Arnold sisters were following uncle Matthew Arnold's religious criticism. In *God and the Bible* (1875), Matthew Arnold claimed that while the story of the crucifixion and salvation by Jesus Christ is powerful and deep, 'the story is not true; it never really happened'.[36] Unsurprisingly, it was this whole question of evidence, interpretation and fact, then crossing both science *and* religion, that brought T. H. Huxley and Matthew Arnold together well before Leonard Huxley and Julia Arnold met in the Oxford of the 1880s.

In his retirement, Thomas Henry Huxley became a late-in-life Higher Critic of an agnostic kind. He delved deeply into biblical history, into how genealogy and history worked and didn't work in Genesis, into the history of the Jews and into the early history of Christianity. He read Old Testament scholars, he read the Proceedings of the Society of Biblical Archaeology, and from the Hebrew Bible he worked out his own 'Genealogy of the Horites and Genealogy of the Edomites'. His notes on the Old Testament were voluminous, far exceeding his notes on the crayfish, the jellyfish or even the gorilla. The generations and the chronologies of Genesis, up to the deluge, didn't match and made no sense. It was fascinating as an account of a dream or an allegory, he thought, but inconceivable as historical reality, let alone universal historical reality. Good agnostic, though, he stated that just because he could observe – show – the account of the deluge to be untrue didn't mean that other statements in the Old Testament were also untrue.[37] The prospect seemed to grip him with excitement, one that might keep him going for another lifetime. He worked in his late years as a kind of cultural anthropologist of the Old Testament, of totem clans, of exogamy and female descent, of ancestor worship and animal worship, of creeping beasts and abominations, of forbidden foods and blood feuds.[38]

By the later nineteenth century, Mary Ward was probably the most theologically and philosophically instructed and astute thinker in

Huxley's extended family. They enjoyed comparing different hetero-doxies, hers from a position of belief, his from a position of agnostics. They discussed together Rousseau's *La Profession de foi du vicaire savoyard* (1765; *The Profession of Faith of a Savoyard Vicar*) and the derivative intellectual line, whom Huxley called sentimental deists. They discussed Charles Voysey, who was cast out in 1871, having offended ecclesiastical law and the articles of the Church of England. Mary Ward considered his highly unorthodox Christian Theism to be 'a reasonable faith'. (His son, the Arts & Crafts architect and designer built Julia and Leonard's house in Godalming, what became and remains the centre of Prior's Field School.) She was attracted to the Unitarian tradition of thought from Joseph Priestley to her nineteenth-century contemporary the transcendentalist James Martineau. They discussed American Unitarian William Ellery Channing: 'I was a great reader of Channing in my boyhood,' Huxley told her, 'and was much taken in by his theosophic confectionery.' He had fun with Mary Ward, bringing her favoured Unitarian down. 'At present I have as much (intellectual) antipathy to him as St. John had to the Nicolaitans.'[39] The theological thinkers to whom Ward was most drawn – 'our English idealists', Green, Caird and Martineau – were those with whom Huxley stood ready to argue, as much as he did the High Church. Learned as he was, Mary Ward suggested that he acquire more knowledge, even go back to basics with Scottish theologian John Caird's *Introduction to the Philosophy of Religion* (1880). 'I will get it and study it,' he assured her. 'But as a rule, "Philosophies of Religion" in my experience, turn out to be only "Religions of Philosophers"– quite another business, as you will admit.' It was a fair comment, and fair as well to say that Thomas Henry Huxley himself was a late-in-life philosopher of religion. They both were. He admitted to Mary Ward occasionally empathizing with, and even experiencing, something that she would call Christian feeling.

> Beneath the cooled logical upper strata of my microcosm there is a fused mass of prophetism and mysticism, and the Lord knows what might happen to me, in case a moral earthquake cracked the superincumbent deposit, and permitted an eruption of the demonic element below.

It was an altogether different tone and register to his honesty with Charles Kingsley all those years before: that *was* the moment of a moral earthquake. But Huxley's mission or belief in science was steady. The message was the same, and he was constant to his own credo. 'One must stick to one's trade,' he wrote to Mary Ward. 'It is my business to the best of my ability to fight for scientific clearness – that is what the world lacks. Feeling, Christian or otherwise, is superabundant.'[40] We know how the Huxleys were intermittently overcome with superabundant feeling.

In 1889, Huxley was deep inside another public battle on Christianity, or really on agnosticism, after the death of another child, Mady. This time it was between himself and Henry Wace of St Paul's Cathedral and principal of King's College, London. Wace read a paper 'On Agnosticism' at the Manchester Church Congress, deriding the idea. Huxley responded in *The Nineteenth Century*, and it was all unleashed, with rejoinder and replies and more rejoinders, another article by Wace on 'Christianity and Agnosticism' followed by Huxley's 'Agnosticism and Christianity', titled to irritate. There were so many words on Christianity and agnosticism that year that it all had to be collected into a book plainly subtitled 'a controversy'. The final essay was by Mrs Humphry Ward, 'The New Reformation', a fictionalized dialogue between a High Church Oxford man and a Higher Critic.[41] Mary Ward was definitely more an ally than an opponent of Huxley in this round and she herself was sometimes identified as agnostic, at the very least her sympathy was apparent.[42] One evening in 1889 there was a chance meeting in her Russell Square drawing room between Huxley and Wace. She was entertaining Huxley – whom she called 'Pater' – and engaged in lively talk, 'he in his raciest and most amusing form'. Dr Wace was announced unexpectedly and 'Huxley gave me a merry look.' Huxley and Ward were on the same side. They might have shared an amused antipathy to Wace and they might together have loved both reason and German reasoning, but even with regard to her own heterodox Christianity she assessed him as 'no more kind to my *sacra* as to other people's'.[43]

Mary Ward's written words had an enormous impact in her own lifetime, as much as anything Huxley had to write or say. Her heterodox Christian views were most influentially channelled through her character Robert Elsmere, who held the title role in her most famous novel (1888). It was described as shaking the orthodox world to its foundations. Half a million copies were immediately sold in the US, in England 5,000 copies each fortnight rushed off the shelves, and over time hundreds of thousands were circulated 'in the sixpenny and sevenpenny editions'. Mary Ward and her various publishers literally lost count.[44] Elsmere is a troubled Anglican vicar, who eventually leaves his country parish to establish his own highly irregular Christian social work in London. His long path to that end is strewn with one of everything, doctrinally speaking, quite like the Metaphysical Society. Ward put in place a puritanical Quaker, a Higher Critic, a sceptic of Christian mythology and a ritualistic priest. Dogmatic, she has the latter reduce Robert Elsmere. 'Trample on yourself,' the priest lectures him. 'Pray down the demon, fast, scourge. Kill the body that the soul may live. What are we, miserable worms that we should defy the Most High ... Submit – submit – humble yourself, my brother.'[45] Mary Ward objected to this authority as strongly as Huxley ever did. And through him Robert Elsmere is meant to appear principled, if tortured. But others read Elsmere, with as much objection, as 'agnostic'.[46] Through one character she restated then-current arguments that scriptural miracles were neither substantiated nor able to be substantiated, a position that her poet uncle Matthew Arnold had also notoriously taken. Huxley, too, took this proposition to the Metaphysical Society in 1876: Jesus's death on the cross and subsequent resurrection were impossible to verify: he likely did not die but was subject to a 'somatic death' not a molecular death, from which it was possible to naturally revive. There was no supernatural miracle here.[47]

Little wonder that in the novel Robert Elsmere struggles to the core of his soul, keeping his (Anglican) heresies from his devout wife. The problem, but not the secrecy, was part of the Huxley household and the Darwin household too, difficult for Thomas and Henrietta Huxley, Charles and Emma Darwin. Yet, unlike the men of science, Robert Elsmere ultimately gives up his Anglican livelihood for his new faith. His eventual credo is a 'human Christ', on whom no miracles were worked, but still (like Ward) a faith and trust in God survives. And like Ward, Robert Elsmere eventually founds in London a kind of secularized Christian social work to displace orthodox Anglicanism.

She pursued hers through the Unitarian-oriented University Hall Settlement and then the Passmore Edwards Settlement, later called the Mary Ward Settlement. The London 'settlement' trend of the 1880s and '90s was practical philanthropy, which enabled young professional men to reside in dedicated housing in exchange for their education of local working-class men and women. Settlements were urban planning uplift.

There was enough in the novel for it to border on, but fall just this side of, heresy. Thomas Henry Huxley read it in March 1888, soon after she sent him a copy, and he followed the reviews closely.[48] The great prime-ministerial believer W. E. Gladstone reviewed it for the *Nineteenth Century* in May 1888, just then uncharacteristically out of office, with time on his hands, and he and Mary Ward entered into a close discussion on theological matters.[49] In the US, Gladstone's review was republished in a symposium dedicated to analysis of 'Robert Elsmere's Mental Struggles'.[50] Indeed Gladstone's *Robert Elsmere and the Battle of Belief* was published there as a book.[51]

Mary Ward first sketched out the controversial *Robert Elsmere* in notes sent to her sister, Julia. It was inspired not just by the controversies of their Oxford youth, but also by their father, Tom Arnold, Julian and Aldous's other grandfather. He was a doubter too, morally, religiously and emotionally restless, a 'wanderer in religion'.[52] Tom Arnold led a strange life, torn by different convictions: first Anglican then a sceptic then a Roman Catholic. He moved around the world and back again, not unlike Henrietta Huxley's father, but for entirely different reasons, and within very different circles. He emigrated first to New Zealand in 1848 and then to Tasmania, where he was inspector of schools, marrying (in June 1850) Julia Sorell, whose first and family names were inherited by successive Huxley generations, most directly by Julian Sorell Huxley, who by preference published with this full name. Troubled Tom Arnold also found Catholicism in Tasmania, converting in 1856. Forced out of the deeply sectarian colonies, he returned to England with Mary (born in Hobart) and then to Ireland, where he held a chair in English literature at the new Catholic University in Dublin. He then quit and moved to Birmingham, where he reverted to the Church of England, and ran a tutoring establishment in Oxford. In 1882 it was back to Dublin, finally, as Chair of

English Language and Literature at University College Dublin, where he returned also to the Church of Rome. *Passages in a Wandering Life* (1900) was quite appropriately the title of his memoir.[53] This theological and global wanderer's mental and spiritual struggles were every bit as tortured as Robert Elsmere's.

The sisters Mary and Julia Arnold shared their faith with each other, especially when it mattered most. When Julia was dying in 1908, Mary Ward sent an honest and raw letter to Julian and Trev.

> I was with her last week for a little while. She asked me to read to her from St Augustine's Confession, and I did, from a little book she had herself marked, and it was evident that the wonderful expression of faith ... was just the voice of her own heart.

Mary consoled them with knowledge of their mother's faith: 'through all pain & weakness, God is with her, & the hope of immortality'.[54] The extended Huxley family were all confident in Julia's belief, even if they were not about their own. Julia wrote to her sister in early November, '7.15 in the morning' she noted in her letter since at this stage every hour was precious.

> I want you not to be too unhappy about me darling. Of course it is terrible to leave you all & the world which I love ... instead of feeling as I did at first that I was caught in some ghastly trap – I know that the trap is God's arms which can hold me up thro every thing – about the next life – or rather the next stage of this – because after all it is only one life right on thro' eternity. I have no fears.

Mary would have recognized her next sentence instantly: 'I don't see much beyond "my soothed encouraged soul – in the eternal Father's smile" – to realise Him.' It was from Matthew Arnold's poem 'A Farewell'. Julia Arnold finished with what her Huxley father-in-law could never say, feel or believe. It was counter to every cell in Thomas the Doubter's body, but she knew her sister understood: 'one can only trust and submit'.[55] Julia Huxley née Arnold had passed to the other side.

SPIRITUALISM AND OTHER WORLDS

One kind of belief, even a faith, very much alive over the second half of the nineteenth century and into the twentieth was spiritualism. Those who had passed, like Julia Huxley, were conceptualized by spiritualists as 'alive' but in other spheres, other layers, accessible through a living human in this world, the medium. People sought communication with the dead through spirit mediums who invoked trances. Some made a good living touring their skill in intercession with nether worlds and other beings.

T. H. Huxley was drawn into spiritualism when it was at its high point, but not as a believer. He started to go to séances himself as early as 1854, just a few years after spiritualism was named and became well known. At Huxley's first séance, the medium was an American woman, and he was called in as 'scientific witness and doubter general'. He was to think of a dead friend, and she spelled out his name on an alphabet, as he passed his pencil along the letters. When the letters of the name were reached, the spirit would 'rap' a sound. Huxley explained much later in the *Pall Mall Gazette* how it all worked. He liked this American woman, however, and was happy enough not to expose her: 'Fraud is often genius out of place, and I confess that I have never been able to get over a certain sneaking admiration for Mrs. X.' But he disliked others, took a righteous exception to their fraudulence and considered their exposure his particular responsibility.[56] He went to another séance in January 1874, incognito as 'Mr Henry', taking detailed notes, including a sketch of the table and how everyone was arranged.[57] It was with one of Darwin's sons, George, whose uncle had succumbed to spiritualism. George Darwin looked forward to the fun and to uncovering 'the imposture' and wanted to show the account to Francis Galton (another agnostic of sorts). He did show it to his father Charles Darwin, who delighted at the elaborate entrapment, and the part played by his friend Huxley.[58]

Sometimes Huxley declared his agnosticism on this matter too, disingenuously parading an apparently open mind, intending to participate in séances experimentally and to adjudicate observationally. Huxley's agnosticism helped and hindered. It left him open to possibilities, but closed if there was no way of knowing. In the 1860s he wrote: 'I am quite ready to admit that there may be some place, "other side of nowhere," *par exemple*, where 2 + 2 = 5, and all bodies naturally repel one another instead of gravitating together.'[59] But he needed this to be provable. What he *could* prove was fraudulence, and he delighted in doing so. In truth, Huxley knew, and all who knew him knew, that he was going in for the kill. It was all, for him, a lesson in the powers of accurate observation, and the imperative to 'interpret observations with due caution'. The living room séance was a more challenging field for the application of proper observational powers than 'the calm seclusion of a laboratory or the solitude of a tropical forest'. Nonetheless, that's what was needed, and that's what he possessed in spades, he claimed immodestly.[60]

As a much older man, Huxley engaged in a more interesting and interested manner; we might say more genuinely agnostic. For some that meant a guarantee of open-mindedness, and for this reason strangers wrote to Huxley, not infrequently, detailing ghostly visits and asking for his assessment of the evidence. A Sheffield surgeon, for instance, recounted his wife's encounter in Yorkshire with a girl she had known in London, but who had died four days before, unbeknown to his wife. 'The child was dressed as usual ... stood by the door for a moment & then disappeared. My wife was so surprised that she at once left the room & communicated the occurrence to others.' The correspondent was at pains to set out for Huxley that he and his wife 'have both been agnostics for many years'. His wife was 'well educated, an agnostic in religion, calm & deliberate by nature'.[61]

Medical doctor Elizabeth Blackwell also wrote to him, in July 1892. Her interest was in ongoing communication across death and the existence of consciousness 'after this life'.[62] From the middle of the century in American and British contexts, spiritualism was strongly connected to women's reformist politics, often abolitionist with regard to slavery in the US, and often suffragist.[63] The Blackwell sisters belonged to just this non-conformist Anglo-American political type. Two of their number, Elizabeth and Emily, were pioneering women doctors in the US and in the UK. One of their number, Anna, was a well-known spiritualist.[64] Elizabeth, the doctor, was seeking Huxley's assessment of an experiment she had designed with a number of elderly friends, all facing death. She took pains to explain that this group was sensible, well educated and inquiring. Huxley always sought truth, she flattered, and asked him to respond to the whole question in that manner. His reputation as an experimentalist recommended him: 'Could you lay down any satisfactory tests; or suggest some plan of collection of facts which would prepare the way for acquiring certainty in this matter?' Blackwell was seeking an adjudication of her methodology. She appealed to his reputation as an open-minded truth seeker, and to the well-known active scepticism that underwrote his agnosticism: 'If it be possible for conscious individual existence to continue after dissolution of the body, it is certainly a great fact to know – with the same sort of assurance that we know that the sun will rise tomorrow.'[65] Blackwell was trying to prove what Huxley thought neither falsifiable nor unfalsifiable. Her appeal worked and Huxley responded with interest and with some propositions. They were contemporaries, and Elizabeth Blackwell asked him: should she die before him, would he carefully note the time of her death?[66] As it turned out, Blackwell lived until 1910, fifteen years beyond Huxley, and the opportunity for this experiment went with him.

There was more of a crossover between spiritualists and evolutionists than we might expect. Some considered the possibility that spiritualism itself was an extension, or a final phase, of human evolution. Alfred Russel Wallace, whose brilliant mind and dogged fieldwork had uncovered the idea of natural selection in the first place, alongside Darwin, soon after turned spiritualist. He wondered aloud about the connection between evolution and spirits, eventually believing that humans have non-material or supernatural forces working on and within and about them. In 1873 Wallace wrote to *The Times* that 'every branch of philosophy must suffer till they [spiritual forces] are honestly and seriously investigated, and dealt with as constituting an essential portion of the phenomena of human nature'. He set out his mounting evidence for spiritualism. Sounds and movements 'not traceable to any known or conceivable physical cause' were nonetheless to him 'undoubtedly genuine phenomena'.[67] To T. H. Huxley's horror, he wrote *The Scientific Aspect of the Supernatural* (1866) and later *Miracles and Modern Spiritualism* (1875). Huxley mocked him: 'I am neither shocked nor disposed to issue a Commission of Lunacy against you. It may be all true, for anything I know to the contrary, but really I cannot get up any interest in the subject.' It got worse: 'I never cared for gossip in my life, and disembodied gossip, such as these worthy ghosts supply their friends with, is not more interesting to me than any other.'[68] This was harsh, especially given Huxley's deep appreciation of Wallace's contribution. Huxley well knew how important his work in South-east Asia was, discovering new species and mapping them bio-geographically. Indeed it was Huxley himself who installed the spiritualist into and onto biogeographical maps, the now-famous dis-continuity between Asia and Australasia that Huxley named 'Wallace's Line' in 1868, just a few years after Wallace's first séance in 1865.[69]

Taking a position against something was Huxley's favourite pas-time, and an indelible part of his character. He thus chalked up 'anti-spiritualist' alongside his many other oppositional identities. Outspoken and clear on what he disliked, he served as a touchstone and anti-touchstone for many. Gerald Massey, for example, was a much-published early English spiritualist, who happened to have been born at Tring. He often quoted Huxley on agnosticism, on belief, on evidence. But this spiritualist was also trying to *connect* evolution and spirit worlds. Like Wallace, he thought that the theory of evolution was incomplete *without* an assessment of spiritualism.[70] The spiritual-ists were claiming evolutionary theory. Worse as far as Huxley was concerned, in Wallace's case an evolutionist was claiming spiritualism. And it was all becoming organized and public, to Huxley's sometime amusement, sometime horror. The Ghost Club was founded in Lon-don in 1862, and the Society for Psychical Research was founded in London in 1882. Huxley had to respond. It was a perfect problem for the agnostic, as theatrical for him as it was for the mediums.

When Julian Huxley joined forces with H. G. Wells, they were also on occasion merciless in exposing the naive and the fraudulent. They mocked Rebecca West's séances, and Conan Doyle's communication from a spirit or 'control' who was apparently thousands of years old, reserving 'platitudinous' and 'uninteresting' for Doyle's accounts.[71] Their point, though, unlike Thomas Henry Huxley's, was not simply to negate and expose, but to present an alternative explanation. There certainly were conjurers but there were also those who genuinely communicated, not from spirits, but from the 'outpourings of the operator's own subconscious mind'. It was not outer worlds but inner worlds whence this communication originated and manifested. They added another layer, 'the underworld' of the unconscious.[72]

In *The Science of Life*, spiritualism, hypnosis, trances and religious exaltation were all treated in the surprising end to the entire work, 'Borderland Science'. The authors sifted through what we know 'beyond life', and the possible survival of 'the personality' after death. Thomas Henry Huxley would have scorned the very idea of science's borderlands – either an inquiry is, or it is not, scientific. Certainly he would have laughed at the photographs of 'ghosts', and the 'flashlight photograph of a materialized head formed of teleplasm that has exuded from the medium known as "Eva C"'[73] (figure 10.2). But there is more of a crossover than first appears between T. H. Huxley's Victorian work on sense and feeling, perception and hypnosis, and his grandson's 1920s work on 'Behaviour, Feeling, and Thought'. Wells and Julian Huxley set out current knowledge of reflex, sensation and consciousness in animals, including the phenomenon of animal hypnosis, of dogs, frogs, snakes, adding a bizarre colour plate of hypnotized lobsters – no wonder the books sold so well (figure 10.3). And then they moved on to the hypnosis of humans. Hypnosis, they explained, used to rely on the idea that an 'animal magnetism' passed between hypnotist and patient, but no longer. Hypnosis and the human mind had since been revolutionized by Freud. The patient passes into a state 'in which, as in dreamless sleep, there is no remembering'. It is not sleep, however, as the patient is still *en rapport* with the hypnotist and will obey, with disassociated suggestibility. From there, not just hysteria but also 'exaltation' could be explained. The exalted may believe themselves to be 'possessed' or, alternatively, 'inspired'. Thus the religious mystic finds an inspiring force in his 'Divinity'. But equally, the exalted naturalist finds that he is 'at one' with Nature. And thus 'Kubla Khan' came to Coleridge in a dream, and Shelley's 'Ode to the West Wind' was inspired by 'some spirit or being'. The link between such exalted states of being, the unconscious and the physiological body was close, hence the stigmatization of St Francis of Assisi or of the poets' near contemporary Louise Lateau, the Belgian girl whose stigmata bled every Friday. Like Thomas Henry Huxley, the Roman Catholic Church pronounced some of these men and women with was what first drew the people's attention to him." It would. Such feats confer distinction on the most obscure. Afterwards people remarked "his glorious character." At Ur he used to write "through pieces of leafy stuff"—disdaining, it would seem, the clay

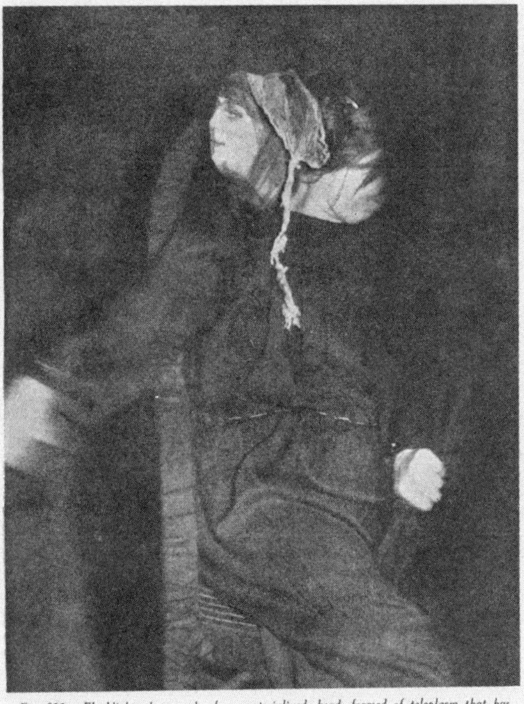

Fig. 952. Flashlight photograph of a materialized head, formed of teleplasm that has exuded from the medium known as "Eva C." (Courtesy of Baron von Schrenck-Notzing, from his "Phenomena of Materialisation." Kegan Paul, Trench, Trubner & Co., Ltd.)

tablet commonly used in that part of the world for cuneiform writing. There is a bookful of communications of this type from and about Phineas published by Sir Conan Doyle. A large part of this mass is vaguely platitudinous and uninteresting.

This Phineas matter is an extreme instance of the type of material produced by a medium in the clairvoyant state. Generally the "control" represents

itself as coming from less remote sources than Pseudonymia or Ur. More often than not, it claims to be the ghost of some recently deceased person known to some of the investigators. Sir Oliver Lodge has been the means of publication of various books about lost sons, including his own son Raymond. The Raymond "communications" are made up into a book together with Raymond's very vivid letters home from the Western front, where he was killed in 1915. The contrast between the interest and conviction of these letters and his essentially futile communications through the medium is very great. Certain recognitions and identifications are achieved, and beyond that the alleged Raymond is either weakly incredible or platitudinous. And it would seem that his messages have now faded out. Earlier communications professing to come from the late Frederic William Myers and the late W. Richard Hodgson, through such celebrated mediums as Mrs. Piper and Mrs. Verrall, have proved equally lacking in novelty or profundity and equally have they faded out. The alleged posthumous minds of these acute persons appear as if flattened and faded. And they desist. They cannot keep it up—important though the business must be to them. But occasionally in the minor matters used for identification, the grasp of the medium upon small intimate points has been remarkable. Such a grasp is no evidence whatever for this mythology about controls and spirits which seems to be necessary for the operations of most (but not all) mediums; in many instances it has been of a type to sustain a provisional belief in telepathy, and a case may be made out for further systematic investigation by competent persons of possible elements of an unknown nature still awaiting recognition in the clairvoyant state.

Figure 10.2: Spiritualism in *The Science of Life*. Flashlight photograph of a materialized head, formed of teleplasm, that has exuded from the medium known as 'Eva C.'.

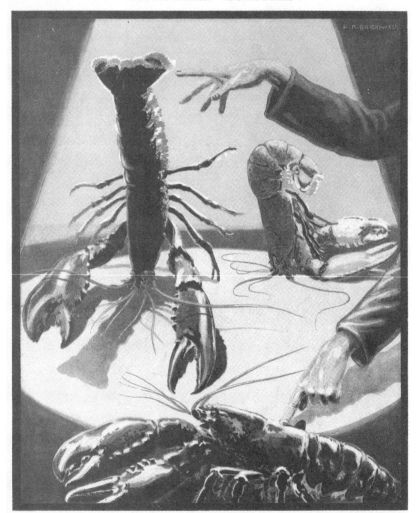

A number of different kinds of animals can be "hypnotized." If lobsters are firmly stroked along their backs, from behind forwards, they pass into a cataleptic state; their muscles become stiff, and they will stay motionless in any attitude in which they are put. The phenomenon of "shamming dead," seen for instance in many spiders, is apparently related to this.

Figure 10.3: Lobster hypnosis and 'borderland science' in *The Science of Life*.

stigmata 'impostures'. But unlike Huxley, the Church pronounced others to be true. The authors of *The Science of Life* had a different approach. They explained it all via the unconscious.[74]

'Automatism' – automatic writing by mediums – could and should also be explained this way. They diminished automatic writing as 'only phenomena of the split mind', in which the person's dissociated, repressed or buried consciousness is accessed. There is no ground on which to claim that automatic writing is the work of extraneous 'spirits', which they put in inverted commas. The medium too should be understood 'simply' as relaxing into a trance-like state, like a self-hypnosis. Yet, while some mediums simulate this fraudulently, others speak genuinely from that trance, from their authentic unconscious, which they may equally authentically perceive to be an external spirit. This was legitimately part of what they beautifully put as 'the co-ordinated realities we call "life" in which our own consciousness floats'. They posited that the stuff of the mind is 'co-extensive with the material stuff out of which life has elaborated itself'. That is, the phenomenon of mind has evolved.[75]

They set out the telepathic experiments of Professor Gilbert Murray and his daughter, and of Sir Oliver Lodge and Professor Charles Richet, their credibility underscored by their honorifics. The latter had put forward the hypothesis of a sixth sense, but Julian noted dubiously that a sense without any sensory organ was a difficult prospect. And in fact, like Thomas Henry Huxley, he explained away these experiments methodologically. But Julian was always more open than his grandfather. We should both keep our heads when hearing of telepathic stories, *and* remain tolerant of the idea's possibility. Once 'half-evidence and pseudo-evidence' is sifted out, there still remains something to explore. He suggested an 'attitude of critical indecision'. There *might* be such a phenomenon as a sixth sense, and if so we should be able to find some evolutionary benefit, not just in humans but also in the animal world.[76]

In the end, for Julian Huxley, the question was not about the reality of ectoplasm, or telekinesis, or clairvoyance, or table rapping, but about the mind that produced these apparent phenomena. Inquiry into such phenomena is needed, even if the spiritualist hypothesis itself is declined. Like mesmerism one hundred years earlier, Julian claimed, there was some charlatanism, some self-deception, but also a residuum that turned into 'the facts of Hypnotism'.[77] Similarly, Julian discussed 'auras' with Havelock Ellis, and not with the intention of discounting the phenomenon. 'I quite agree that eventually we must search for, and I hope find, a normal basis for abnormal phenomena.'[78] Sensible individuals and organizations of the right-minded could assist in the disaggregation of good evidence from pseudo-evidence, neither implacably sceptical nor credulously faithful. Julian Huxley was one of them, becoming a member of the Society for Psychical Research in 1928. At the very least, even if other-worldliness faded away, all kinds of new knowledge about suggestibility, hallucinations, perceptions, memories and new sensations would be forthcoming.

Publishing his *Memories* in 1970, Julian seemed a little embarrassed by his earlier fascinations.[79] And yet he retained a more expansive view than his grandfather of evolution itself that held a place for the interior world of the mind. Occasionally, he even wondered in public whether evolutionary progress that was now directed by humans might open up 'a future life in a supernatural world'.[80] Time and again, Julian noted how knowledge, dimensions and even time itself had expanded and on occasion in kind, not just in degree.[81] Even at his most humanist, this Huxley was not disenchanted. He yearned for comprehension of other possibilities, even other worlds, in a way that his grandfather did not. Often it was evolution itself that offered him this profound wonder.

JULIAN HUXLEY'S HUMANISM

Julian Huxley inherited at least two beliefs, one from each side of his family. One grandfather was a fervent believer in verifiable truth. The other grandfather fervently believed in God but did so in a state of constant doctrinal torture. Julian noted that his own parents, Leonard and Julia, were more moderate; religious education was minimal in their Surrey home. Julia was irreverent towards, and dismissive of, clerical authority, in a manner that must have delighted her father-in-law. The Bishop of Winchester talked drivel, she wrote to Julian then at Eton.

I shd like to shut 19/20 at least of the bishops up in a House of Correction – and have them taught & lectured to by sensible people – men & women – for a year at the end of wh. time they might come out if they wd promise to reform![82]

Unsurprisingly, then, the Huxley children rarely attended any church, although the boys were required to at Eton. Julian remembered chapel there as a kind of cultural ecstasy, although it had nothing to do with Christian belief: 'theologically considered', the whole Christian schema seemed to him incomprehensible.[83] His mother's school, by contrast, was non-sectarian, and Julia Arnold Huxley herself taught not just English literature, but divinity, and delivered her own 'Sunday sermons'.[84]

Julian was clear that his own mental struggles were as much inherited from the Arnold as the Huxley line. He was upfront about interior moral disquiet that crossed over into illness, tortured by 'passionate fervours' over belief, over faith, he confessed to the world. What to believe was unresolved and created an internal conflict that he described, publicly and intimately, as 'profoundly miserable': 'paralysing my energies by threatening to tear my mental being in half'.[85] Julian need not have felt quite so tortured. He was surrounded by more doubters, if not sceptics, than his grandfather had ever been. Uncle John Collier was one. He had just written *The Religion of an Artist* (1926), in which he looked forward to a time when ethics would take the place of religion, when the benefits of religion could be attained in other ways, more reasonable and less strife-ridden.[86] Julian discussed religion with Collier directly and often. His uncle-artist was unashamed: 'I should say that most adherents of any religion are much more concerned with its teaching than with any purely spiritual experience,' he wrote to Julian.[87] Herbert George Wells was another. Their *Science of Life* was conceptualized as a 'vaster Genesis' in its own introduction, authored by a trinity: 'three in one, constitutes the author of this work'.[88] Wells could happily make such provocative announcements, having set out his agnostics a decade earlier, in *God the Invisible King* (1917).[89]

In 1927 Julian was writing a science book by day and a belief book by night. He was commissioned to contribute to the series 'What I Believe', personal religious statements made 'with absolute candour by twentieth-century men and women'. The suffragist and preacher Maude Royden offered *I Believe in God*. The Oxford Anglo-Catholic and then Roman Catholic priest and broadcaster Ronald Knox explained *The Belief of Catholics*.[90] Julian Huxley offered a strange belief to the mix, *Religion without Revelation*. This book was dedicated to the memory of his mother but in substance presented a synthesis between the two familial lines. 'I hold, then, that all our life long we are oscillating between conviction and caution, faith and agnosticism, belief and suspension of belief . . . each has its right occasions.' Even as he confessed how little calm such synthesis brought him, he held that 'beliefs are essential tools of the human mind'. 'I believe', explained Thomas Henry Huxley's then most famous grandson, 'not that there *is* nothing, for that I do not know, but that we quite assuredly at present *know* nothing beyond this world and natural experience.'[91] He recapitulated the family credo, part of the Huxley liturgy, but at least he was ecumenical about his agnosticism: 'A personal God, be he Jehovah, or Allah, or Apollo, or Amen-Ra, or without name but simply God, I *know* nothing of.' And anticipating the atheist T-shirt, he argued that belief in future lives is a remnant of primitive or savage or superstitious histories.[92]

Julian recalled being in Colorado Springs during his Rice years, casually picking up some essays by the Victorian liberal Lord Morley, and being inspired by his call: 'The next great task of science will be to create a religion for humanity.'[93] This was the beginning of his own crusade, a calling to flesh out his own religion, evolutionary humanism. Refining and repeating the humanist credo was his life's work, the great synthesizer bringing secular wonder back to nature via the extraordinary new human understanding of evolution itself. He returned once more to 'what I believe' in a BBC broadcast in 1957: 'I do believe in the possibilities of man,' his profoundly humanist faith. 'Progress indeed has been miraculous', he broadcast, 'in the proper sense of that word – something to wonder.'[94]

Julian Huxley absolutely needed evolution to be progressive. He needed aspirational direction, for himself, for humankind, for the poets. He needed it because he was witnessing the disenchantment of the world, heralded not least by state atheism. When Julian and Juliette visited Moscow in 1931, the magnificent cathedral was being used as an anti-religion museum. Julian was not impressed.[95] He constantly searched for a world that allowed for wonder, for a beyond. A late-in-life letter confessed to a friend that while he was a humanist, like his grandfather, that was insufficient. 'I would say that something essentially <u>religious</u> should enter into one's beliefs, though never based on supernatural dogma.'[96] Materialism was never enough.

In 1946, Julian Huxley found his own mode of religiosity through the Jesuit priest Teilhard de Chardin, whom he met in Paris in his Unesco directorship. He considered their paths to be parallel, that they were on the same quest toward a cosmic unification. Teilhard the man of God was his intellectual and spiritual inside-out twin, and when he died in 1955, it was Julian Huxley who wrote the preface to the English edition of *The Phenomenon of Man*, a book that the Vatican had forbidden to be published.[97] Julian described how Teilhard became a 'prophet' and 'almost a saint' in France, his work a 'symbol of the reconciliation between scientific humanism and Catholic orthodoxy'. He recounted that the priest never entirely credited the mechanism of natural selection, but that they nonetheless found real alignment and accord in the idea of a total and progressive natural and human history on a universal scale.[98] Julian was besotted with Teilhard's capacity to think about an expansive whole not as static, but as a process. He thought Teilhard's 'neo-humanism' was similar to his own evolutionary humanism, with the implication that two genius minds had separately and simultaneously offered humanity a vision, like a twentieth-century Darwin and Wallace.

Here was a vision that on the one hand satisfied and served Julian's agnostic science and on the other offered a cosmology so expansive and profound as to be a kind of unification of all phenomena, a cosmic holism. It was an attempt to bring knowledge of the universe into one order; as Julian put it, 'the universe in its entirety must be regarded as one gigantic process, a process of becoming, of attaining new levels of existence and organisation, which can properly be called a genesis or an evolution'. Teilhard himself used the word 'cosmogenesis' rather than cosmology, to denote a continual process of becoming. And like Julian, Teilhard looked to an almost post-human future. He spoke of 'ultra-hominisation', a little like Julian's 'transhumanism', to capture the process 'in which man will have so far transcended himself as to demand some new appellation'. Like Nietzsche, Teilhard and Julian Huxley thought 'man' was unfinished.[99]

Teilhard was interested in ways of thinking about collective energy in his cosmology, not just physical energy, but a combined 'psychic energy', and Julian thought a new term like 'neurenergy' or 'psychergy' might assist. Teilhard saw wholes and understood a convergence not from a genesis beginning, but from the 'Alpha' (of elementary material particles) to an 'Omega' an attainment point of convergence, a point of total unification, including of the 'noosphere', the sphere of human thought. Yet for an evolutionist, such a cosmology must also be continual, and so Julian was dissatisfied with Teilhard's end point, Omega. Might that better be comprehended as a point 'beyond which the human imagination cannot at present pierce'? Currently incomprehensible states of being and practices – like extra-sensory perception, a sixth sense and 'the infant science of parapsychology' – might be small windows into such states.[100] For Teilhard, Omega was a kind of divine, 'Christogenesis', he sometimes called it, at other times the Logos or 'the sacred Word'. Julian confessed to not quite following, but the Church of Rome certainly did, and was not pleased.

The young Julian had earlier been drawn to philosopher Henri Bergson, and his idea that the 'absolute' might be grasped through intuition. Mature Julian found in the Jesuit a far more expansive synthesis of the material and the physical, the world of the mind, *and* the world of the spirit. It was a synthesis beyond the admissible remit of any agnostic Huxley, and indeed it made Julian's own expansive work look almost thin, but Père Teilhard's ideas certainly let him speculate profoundly. 'The religiously-minded', Julian wrote, 'can no longer turn their backs upon the natural world, or seek escape from its imperfections in a super-natural world.' Yet at the same time, speaking of and to himself (and perhaps his grandfather), 'nor can the materialistically-minded deny importance to spiritual existence and religious feeling'.[101] It was as if Julian allowed or willed the dead Teilhard to inhabit his Huxley self, serving as a kind of medium who gave him permission to talk about the world of the spirit afresh, who allowed him to incorporate a spirit into a universal unity.

Aldous Huxley too was always looking for something beyond and above the everyday, as Julian himself noted: 'some people accused him of mysticism in the pejorative sense but this was not true – he too was a Humanist, wanting maximum fulfilment for mind and body alike'.[102] Aldous certainly wanted it all, offering his own vision in his final novel, *Island* (1962). Near the end of his life, he was a longstanding practitioner of Vedanta, led by Swami Prabhavananda. He was a meditator, an advocate of *ahimsa*, a principle and practice and virtue of non-violence to all living things. Aldous was one of many who 'looked east' for re-enchantment, and seriously so.[103] His island utopia, 'Pala', was packed with all his hopes and dreams, and the mechanisms to make them manifest: mysticism, science, religion, psychedelia, Eastern and Western knowledges together. On Pala, science helps verify, not defraud, spirit worlds – his grandfather's mean approach – and Aldous had his characters instead imagine 'worlds of still unrealized potentialities'.[104] 'Human Potentialities' was Aldous's concluding statement for Julian's book *The Humanist Frame* (1961). He detailed 'potentialities' achievable by a meditative perceptual awareness that will

> help you to go beyond discursive reasoning in terms of symbols and come to what the Buddhists call 'the wisdom of the other shore', to the unitive knowledge, obscure but self-evident, wordless and therefore profound, of the oneness in diversity.

What he called 'true yoga' enlightenment was another version, perhaps, of Père Teilhard's Omega.[105] But it also might be understood as another version of the Monist idea 'eon', which young Thomas Henry Huxley had grappled with in 1840, which brought everything into One.

> Let us suppose that an Eon is something with no quality but that of existence ... that in the highest degree, is God ... This eon in its state is matter in the abstract. Matter then is Eon in the simplest form in which it possesses qualities appreciable by the senses. Out of this matter by the superimposition of fresh qualities was made all things that are.[106]

For this family of synthesizers, any idea this side of the supernatural that folded everything into One was worth considering. It was Monism to which Ernst Haeckel had turned, himself trying to connect science and religion. T. H. Huxley took notes on his friend's late-in-life contemplation, *Monism as Connecting Religion and Science: The Confession of Faith of a Man of Science*, a book that ended up on Julian's shelves.[107] It helped Julian's personal mission to press evolution into cosmic terms, and cosmology into evolutionary terms, enabling his own visions to be upscaled to a kind of sublime.

THE NEO-ROMANTICISM OF HUXLEY HIGH MODERNITY

Thomas Henry Huxley's scientific naturalism was a reaction to metaphysical and supernatural explanations in the religious domain, and to German 'nature philosophy' of the early nineteenth century, far too mystical for the agnostic.[108] Thomas Henry Huxley might have had an anti-Romantic dream of progressive disenchantment, yet his own grandson's vision of nature-through-science was far more akin to that *romantische Naturphilosophie* than opposed to it.

Julian Huxley's evolutionary humanism was a kind of transcendental naturalism, but with a teleology, a direction, even a purpose that he himself invented and then insisted upon. Nature enabled a Romantic wonder, and science as nature's explanator could bask in that wonder too. After all his grandfather's insistence on natural not supernatural explanation, by Julian's hand and mind evolutionary humanism itself became a neo-Romantic vision.[109]

In truth, Thomas Henry Huxley's disinclination to the supernatural – his disenchantment – never had a chance with this line of twentieth-century Huxleys, neither scientific Julian nor literary Aldous. From the beginning, both were continually enchanted not just by Romantic nature but also by her trustees, the poets.

In the Prior's Field summer of 1908, before their mother passed to another world, Aldous was sick again, this time in bed with mumps. He wrote to his older brother, sending him pressed violets, already passing on magical wisdom on language and meaning: 'I wonder if you know the language of the flowers? If you give peach-blossom it means "I am your captive". White Hyacinth means "unobtrusive loneliness".'[110] As a much older man, approaching death, Aldous Huxley tried to explain that a wordless and profound transcendence and oneness could lead to perceiving 'a World in a Grain of Sand / And a Heaven in a Wild Flower'.[111] The 'wordless' state was to be found in William Blake's poetry: 'Hold Infinity in the palm of your hand / And Eternity in an hour'.[112] For Julian's part, when he was seeking 'sacramental transcendence of the boundaries of the self', he also found this best caught by Blake: 'the soul expands her wing'.[113] *Auguries of Innocence* was a touchstone for both of them.[114] Julian and Aldous both returned to Blake, time and again, trying to capture transcendence through him, for themselves. And Aldous, we know, actively sought ways to experience this state, through words, through meditation, and finally through his death-bed injection, 'a sacrament'.[115]

Julian found a beauty, a wonder and a magic – his term – in lyric poetry, in literature, in art. Yet far more profound was what lay in 'Nature and natural beauty'. Here he found not just rest and refreshment, but also something 'truly mystical' and 'sacramental', and continually searched for those who had best put this state of being into words (figure 10.4). He saw in himself Richard Jefferies, the English nature writer, a contemporary of his grandfather's, who had developed a 'pantheism' out of nature. And he often likened his occasional sacramental feeling to that caught by Wordsworth. His point, however, was not simply to indicate a kind of ecstasy from nature, but that that ecstasy could be cut off 'when most needed'. Julian more or less described himself as a 'nature-mystic', who like Wordsworth first experiences and then loses the divine, the loss creating a 'terrible blackness', the 'shut-in solitariness appalling the soul', that showed that 'not only heaven but hell is within us'.[116] Wordsworth lamented 'the failing of the radiant experience'.

> The things which I have seen I now can see no more.

> Whither is fled the visionary gleam?
> Where is it now, the glory and the dream?

Not just aesthetic feeling, Julian described, but 'a special feeling', an awe, an 'atmosphere of holiness', even though he knew that 'in any and every case the atmosphere of holiness has come where it is owing to human mind having put it there'.[117] And yet Julian did not subscribe to Matthew Arnold's vision of religion as 'morality tinged with

Figure 10.4: Julian Huxley reading at Prior's Field, 1906. The school motto is from Wordsworth, *The Excursion*: 'We live by Admiration, Hope, and Love'.

emotion'.[118] That was too dismissive, not least because 'emotion' did not catch the profundity of secular 'revelations' that he had himself experienced.

Julian sometimes wrote of his various thought-experiences as visions, and himself as an occasional visionary. Late one night in the First World War, he was training into the early hours, Aldershot the unlikely site for what he called 'meditation'; relentless, monotonous and quiet. Suddenly, he recounted, an understanding came to him that *both* religion and science were good, both 'two solid honesties', and yet when the two came into contact, the former moved so slowly that it dragged the latter. He could see, or 'divine', science and religion reconciled, both part of an evolutionary trajectory. He disclosed another 'vision', a 'seeing with the mind, a seeing of a great slice of this Earth and its beauties, all compressed into an almost instantaneous experience'. That was written in 1927, and we can see how and why Teilhard de Chardin struck him so powerfully twenty years later. Such revelations as he himself had experienced did not 'come from' a supernatural force, but from within the mind, like a dream, a 'vital experience on a new level, above that of the ordinary self . . . a single organized moment of spiritual perception'.[119]

In the counter-culture London of the 1960s, a guest in Francis Huxley's house picked up his father's *Religion without Revelation* from the shelves. Roberta Russell, lover of R. D. Laing, found therein nothing less than 'the answer to my long-buried question about the existence of God and His mystifying behavior'. Julian Huxley *was* a kind of Moses, his books the tablets for a humanist world. The key was that Julian looked first to human culture, human need, and therein saw the invention of God. Humans made God, in other words, not the other way around. Roberta Russell found in this a great clarity:

> He said that the term divine did not originally imply the existence of gods; on the contrary, he said, gods were constructed to interpret man's experiences of this quality. The divine is what compels man's awe. I was awed. Where had Julian Huxley found the strength and perspective to rise above the accepted religious systems of his times with such clarity?[120]

Roberta Russell imagined her own 1960s generation to be the original counter-cultural critics. Yet as Francis Huxley well knew, his own and his father's entire familial heritage specialized in 'rising above' spurious truth claims, even as Julian reached for a new kind of wonder.

What was God for Julian Huxley? Certainly not metaphysical, and thus far he and his grandfather were one. Human belief and religious culture were part of evolution. What man has worshipped over time are the various powers operating in and through 'outer nature and his own life'. These powers have been 'personified as supernatural beings', whether they be river-gods or the classical Pantheon or God of the monotheistic religions. Now all these divinities should be stripped of the personality which humans have projected onto them, Julian insisted. Indeed God – as a human invention, if a complex one – might well be understood through the methods of natural science. Julian proposed 'to classify and compare these Gods and these ideas; to analyse them in terms of all available kinds of knowledge'. What was needed was detailed study not on what humankind *ought* to have worshipped, 'but what it is which man *has* actually worshipped'.[121]

Julian could have been describing his son's later twentieth-century work *The Way of the Sacred* (1974), a rich comparative anthropology of life and death, bodies and spirits, souls and beginnings; a universal history of the Sacred. Francis found his own wonder, not in Romantic poetry or in nature, but in rites and symbols, beliefs and tabus 'that men have held in awe and wonder through the ages'. It retained a nod to his father's work on ritualization, converted into the deepest human mind-culture by his son.[122] If Julian had described human culture progressing from 'heaven' to 'earth', and thus from 'religion' to 'science', Francis journeyed 'back' in would-be evolutionary terms to just this world of animist and religious explanation and experience, folded into mid-twentieth-century psychoanalysis. *The Way of the Sacred* is a beguiling long essay, both the words and the images defying its coffee-table production.

In the family tradition, Francis Huxley was having yet another go at explaining everything, bodies and souls, here and beyond, interiors and externals, life and death. And yet, perhaps refreshingly, this Huxley's analysis proceeded in and through all manner of knowledge-making *except* science, *except* evolutionary theory. It worked through the symbols and stories of the monotheistic books; through Buddhism; through

Christian allegories; through psychoanalysis. And Francis perceived the sacred through attention to the material: stone, wooden, paper, oil, stucco representations and images, from as many times and places as there are stars in the heavens. His great-grandfather's scientific method is reduced to the smallest player in this accumulation of human knowledge, less than that. Far towards the end of his wonderful book, Francis added in passing that science has 'done much to dispel' the sacred. More interestingly, he brought psychoanalysis to the 'doubt' by which Thomas Henry Huxley lived. Descartes' *Cogito ergo sum* has a counterpart, *Dubito ergo sum* – I doubt, therefore I am. 'The axiom underlines all science, which can verify its theories only by attempting to falsify them.' This is what really interested Francis Huxley. And so, as Freud said, science has destroyed 'man's house of symbolism', and with it, Francis added, 'any idea of a world-soul'. And yet, the scientific method is more like religion than it thinks. 'For both, nature is but a grandiose appearance behind which can be glimpsed the workings of a deeper reality.' But therein lie other appearances, endlessly. It is chimerical, and Francis characteristically turned simultaneously to both Shakespeare and Zen Buddhism to explain just that.[123] And along with his father and his uncle, William Blake was foundational, the 'messianic prophet', the poet born in an age that 'frowned upon Enthusiasm – literally, possession or inspiration by a god'.[124]

Francis Huxley began his book with the claim that 'the man who goes beyond appearances is a searcher after truth'. But there lie ever more appearances, an unknowable sublime. He was announcing himself, contra T. H. Huxley, as a believer in the impossibility of finite knowing, opening his book with William Blake and closing it with the Tamil metaphysician Coomaraswamy, on the mind of the Buddha. His great-grandfather's search for knowledge came nowhere near – didn't even approach – a sufficient or satisfactory method. He saw the problem in an engraving of an astronomer's search for truth, the first image of his book (he might well have remembered it from his father's *Science of Life*) (figure 10.5).[125] Our search for the sacred is like the astronomer depicted on all fours 'thrusting his head through the adamantine vault of heaven in order to see backstage into the secret machinery of the universe'. And yet, all that he sees are other appearances that confirm a mechanical universe. But God, as William Blake

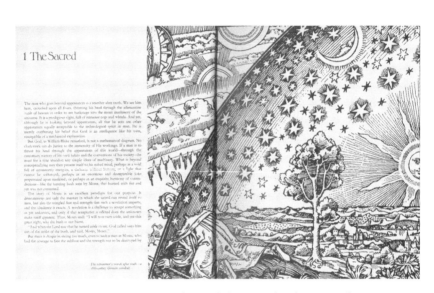

Figure 10.5: Francis Huxley and the Sacred. 'The man who goes beyond appearances is a searcher after truth'.

remarked via his medium Francis Huxley, 'is not a mathematical diagram'.

> What is beyond conceptualizing may then present itself to his naked mind, perhaps as a void full of unmannerly energies, a darkness without bottom, or a light that cannot be withstood; perhaps as an enormous and disreputable joke perpetrated upon mankind; or perhaps as an exquisite harmony of contradictions – like the burning bush seen by Moses, that burned with fire and yet was not consumed.[126]

Francis may yet be uncovered as the greatest Huxley wordsmith of all.

Julian and his son both knew – believed – that human culture had projected a thousand complex personalities, rituals, stories, symbols, tabus onto multiple natural forces. This was Francis Huxley's domain and he stayed there; these 'cultures' should never be stripped back, and in any case could never be, even though we continually try to see through the vault of heaven to the backstage of the universe. Julian thought, rather, that the 'personalities' ascribed to nature's powers *should* all be stripped back and revealed for what they are, products of a human mind overlying a material reality, products that he therefore comprehended with less sophistication and insight than his son. And yet even for Julian Huxley, this did not necessarily entail stepping away from wonder. Rather, the humanist tried to tell us how we humans 'may touch the absolute [and] sacramentally transcend ourselves'. He was a poetic and magical believer in his own species, whom he regarded with awe.[127]

> Here is a mass of a few kilograms, of substance that is indivisibly one (both its matter and spirit), by nature and by origin, with the rest of the universe, which can weigh the sun and measure light's speed . . . which can transform sexual desire into the love of a Dante for his Beatrice; which can not only be raised to ineffable heights at the sight of natural beauty or find 'thoughts too deep for tears' [Wordsworth again] in a common flower, but can create new countries and even heavens of its own, through music, poetry, and art.[128]

Julian Huxley's last work of art – a poem – was inspired by a cathedral. It came to him in the night after hearing Bach's B Minor Mass at Ely Cathedral in early 1974. Layers of sonic sensory experience triggered memories of his boyhood ecstasies in the chapel at Eton. Neo-Romantic to the end, it was a poem of enchantment in the 'ship of God', music that sings of 'God and Man' and of the angels. 'I do sometimes try the equivalent of prayer', he confessed to Juliette as an ageing man, 'to feel at one with the general current of life and its direction; and in quite another way through poetry and music.'[129] This was religion without revelation.[130]

Francis Huxley returned the dynasty, in a most unlikely turn, fully to the cosmopolitan sacred. *The Way of the Sacred* (1974) is a long way from *Evidence as to Man's Place in Nature* (1863), if equally universal in aspiration and claim (figure 10.6). It is an unexpected coda to the Huxleys' enduring engagement with religion and rationality, with culture and nature. 'Francis was himself a high animist,' explains one who knew him.[131] And yet Thomas Henry Huxley had already granted space for his descendants' yearning, or rather nature did: 'in relation to the human mind Nature is boundless; and though nowhere inaccessible, she is everywhere unfathomable'.[132]

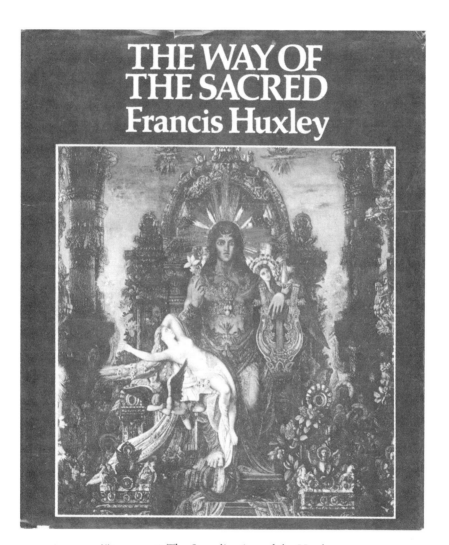

Figure 10.6: The Sacralization of the Huxleys.

Epilogue: Birth, Death, Afterlives

> And if there be no meeting past the grave,
> If all is darkness, silence, yet 'tis rest.
> Be not afraid, ye waiting hearts that weep,
> For God still giveth his beloved sleep,
> And if an endless sleep he wills, – so best.

> From Henrietta A. Huxley, 'Browning's Funeral', Epitaph,
> by his request, on the tombstone of Thomas Henry Huxley.[1]

The great crested grebe was Julian Huxley's familiar. In high summer 1973 he was delighted that the birds had returned to Hampstead after nearly fifty years' absence, and sixty years after his breakthrough paper on their courtship. By now Julian was only taking slow and short walks on the Heath, but still enjoying doing so daily.[2] Perhaps the grebes had returned to wish him well in his old age. Perhaps they were wishing him farewell. When he died in February 1975, Juliette and their sons, Anthony and Francis, designed grebes into and onto his memorial (figure 11.1). The scientific community – those from his own generation and those whom he had taught – recognized him with their own rituals, modest but meaningful. A special section of *Nature* summarized his contributions. An issue of the *Journal of the Zoological Society* was dedicated to him, and later the Eugenics Society celebrated the centenary of his birth with *Evolutionary Studies*.[3]

That death is part of life is a mawkish condolence, but for many Huxley mourners – evolutionary biologists and population geneticists – the fact was and is foundational. In the end, the whole of the Huxleys' intellectual work rested on the implications of birth and death, for jellyfish

MEMORIAL

TO

SIR JULIAN HUXLEY FRS

(1887–1975)

Figure 11.1: Memorial meeting sheet for Julian Sorell Huxley, designed by
Juliette, Anthony and Francis Huxley.

and axolotls, for Elsa the lioness and Guy the gorilla, for lobsters and courting grebes and for science-and-culture-producing Huxleys.

Understanding the implications of individual births and deaths, in and for the science of life, changed radically between 1825 and 1975. 'The individual animal is one beat of the pendulum of life,' young Thomas Henry Huxley declared one Friday evening in spring 1852. '[B]irth and death are the two points of rest, and the vital force is like the velocity of the pendulum, a constantly varying quantity between these two zero points.'[4] Just back from his South Sea journey, Huxley was proving himself a vital life force, already instructing his elders, the learned scientists of the Royal Institution.

At about the same point in his own career, young Julian also caught the philosophy of birth and death. Before the grebes, before courtship, before ritual, he published on 'The Meaning of Death' (1911) in his father's *Cornhill* magazine. It was about biological individuality. When a man dies, does *all of him* die, asked Julian? It is not necessarily so. If that person has reproduced, the special reproductive cells that turn into the next human have escaped the death of the original individual: 'by division into two, [he] has built up the body of his son, and, included in this body, new reproductive cells for future generations'. This 'germ-plasm', he called it after Weismann, is 'potentially immortal', separate from each generational 'body or Soma' that was the container for every individual person, or what he called 'Morphé'.[5]

For Julian, evolution itself was consoling: 'Our lives do not end with death; they stream on, not merely in direct offspring, but more importantly perhaps in the influence they have had on the rest of life.'[6] Some in Thomas Henry Huxley's era argued that just this human capacity to conceptualize an afterlife – of any kind – was what most distinguished humans from animals. Julian would agree that the capacity to conjure and then articulate a beyond, especially through a word-device called 'a metaphor', was uniquely human. He always preferred the 'stream' metaphor to the more common 'tree of life'. Either way, it was clear that a 'pendulum of life' swinging between the neutral points of birth and death was now a wholly insufficient metaphor: mechanical and antique. All births and deaths were necessarily connected in 'the continuity of life', a 'single flow', Julian offered, that advances 'along the plane of Time through many channels'.[7]

How did the Huxleys live on in this stream flowing along the plane of time? Out of respect for their agnostics we can't possibly imagine them in a heaven we can't know, but their earthly afterlives are clear enough.

They live on in 'culture', in hundreds of paintings, portraits and statues, even if most visitors to the Natural History Museum never get to Thomas Henry Huxley, arrested instead by Guy or Charles Darwin. A more modest bust of Julian Huxley was commissioned and cast for the Unesco building in Paris. Some of John Collier's portraiture is part of more intimate Huxley and Darwin worlds. T. H. Huxley and Charles Darwin live on as an etched family pair in the corridor of the home of Angela Darwin, the great-grandaughter of Thomas Henry Huxley, who married the great-grandson of Charles Darwin in 1964.

These science communicators' afterlives lie in their millions of well-crafted words. Neo-Romantic evolutionist Julian considered speech, books, writing and printing to be the means by which humans perpetuate themselves 'in spite of death'. This was socio-cultural evolution across generations: 'A row of black marks on a page can move a man to tears, though the bones of him that wrote it are long ago crumbled to dust.'[8] Julian's personal library is a little lifeless now, ignored by most, treasured by a few. In a thousand ways, this *Intimate History of Evolution* is an effort to interpret that library's meaning, an intellectual ecosystem of the modern natural and human sciences. Once a proud backdrop for his *Life* portraits, Julian's books are now sequestered in Rice University Archives, guarded diligently, but little known. Many are his grandfather's, bestowed on the Inheritor as the next trustee of evolution. Charles Lyell's *Principles of Geology or the Modern Changes of the Earth and Its Inhabitants* sits there, unopened for decades, perhaps a century, two volumes given to T. H. Huxley and affectionately inscribed 'from the author, 9 Nov. 1866', the most meaningful gift-giving ritual between Victorian scientists. Ernst Haeckel's books are also on those shelves, the German originals of course. And from an earlier era altogether, there is T. H. Huxley's copy of Buffon's canonical *Natural History*, the volumes from which Huxley drew images and information on eighteenth-century versions of 'man-like apes'. What a superb natural history collection, telling not just the story of ideas about humans and other animals, the science of life, but of how that came to be part of popular culture too. The sombre leather of Buffon's canonical work clashes with the screaming 1960s colour of George Schaller's *The Year of the Gorilla*, inspired by Julian's ethology. There is Julian's much-loved brother clinging to life through a scrappy paperback copy of *Ape and Essence*, read well and many times over. A copy of *Science and Synthesis*, proceedings of a Unesco-organized event on Albert Einstein and Teilhard de Chardin to which Julian offered his own synthesis, sits alongside a dozen other books by and about the Jesuit, that phenomenal man. Julian's library collects studies of religion, of belief, of visions, of occult, of extra-sensory perception and *The Doors of Perception*. Books on new genetics by T. H. Morgan nestle up to grandfather Huxley's books on old breeding by Francis Galton. Charles Bell's natural theology grafts onto Thomas Henry Huxley's provocatively titled *Lay Sermons*. It is a library that shows how geology and anthropology, biology and the-ology, ecology and sociology were linked all along, the natural and the human sciences arriving in the twentieth century together, con-joined twins throughout the modern period. They were ageing, not born with the 'sociobiology' of the 1970s. The next great commun-icator of evolution in humans and animals, E. O. Wilson, created his *Sociobiology: The New Synthesis* out of *Evolution: The Modern Syn-thesis* and published it in 1975, the year Julian Huxley died.

The Huxleys live on through their own family memories as well, many private, some public. They breathed life into each other. Julian even conjured his grandfather into his own wartime present, writing an imaginary interview between them for the BBC in 1942. It was a Huxley resurrection, a séance of sorts, to which everyone was invited.

> JULIAN (*rather crossly*): The fellow who runs these interviews has told me to come here and exchange a few words with my grandfather, Thomas Henry Huxley, who died in 1895 at the beginning of his seventy-first year. That's all very well, but how can even the B.B.C. put one in touch with a world of departed spirits – in the existence of which my grandfather no more believed than I do, though he was scrupulously undogmatic in all merely speculative judgments of that kind? I remember him very vividly, as a child does.

414

THOMAS HENRY: And I remember you, young Julian.
JULIAN: But what *are* you?
THOMAS HENRY: A projection of your private fancy.[9]

In 1987, another Huxley delivered another Galton Lecture. Andrew Huxley – son of Leonard and his second wife Rosalind Bruce – spoke on his half-brother Sir Julian. It was the centenary of Julian's birth, of the night that Leonard Huxley wrote 'The Inheritor'. Quite rightly this lecture included a Huxley genealogy, and by that point, and by any measure, not least Professor Sir Andrew's Nobel Laureate, it was a dynasty 'outstanding in philosophy, science, and the arts', rivalling the Darwins, the Wedgwoods and the Galtons. Andrew Huxley offered 'a family view' and warmly recommended the widow's view, Juliette having just published her autobiography.[10] But Juliette herself was more privately wondering if Julian was still alive in some ghostly way. 'Is he at peace now?' she worried to May Sarton, 'or still reaching out with flaming sword to reach some other sphere? It is a disturbing thought.'[11]

As we might expect, the Huxleys lived on in less conventional ways too, even in a manner that gives some credence to Juliette's concern about ghost-presences. When Julian died, someone took a photograph of him in his death-bed, not dying, but already gone. Quite possibly this was Francis Huxley's impulse, the anthropologically interested animist who had just published his own deep thought on life, death and the sacred. The photograph of Julian Huxley's corpse is shocking at first, but on reflection contemplation of the body of the individual after death raises core Huxley business, if anything does. And in any case, the photograph is a vestige of a common enough practice from Thomas Henry and Henrietta Huxley's time. In their Victorian era, the post-mortem photograph was a formalized order of portraiture, stylized daguerrotypes in which the dead were sometimes posed as if they were sleeping, sometimes as if they were still alive, a grim keepsake. As a rule, the Victorian images were far better than this one, a washed-out polaroid 1970s snapshot, poor quality, poorly composed, poorly developed. Victorian photographers might have wondered why such a disrespectful image was permitted, but in the end it is a post-mortem photograph more true to life and death than most Victorians' domestic resurrections.

Francis perhaps considered his father's body continually spirited. In fact Julian had early thought through the theoretical possibilities of immortality, both philosophically and physiologically. 'No actual impossibility stands in the way of the individual's continuance,' Julian the futurist and optimist declared in 1911. But the fact is that 'Life' or 'Nature', 'gifted with reproductive powers', found it more efficient and easier to choose the death of individual organisms, over and over, rather than the inefficiencies of sustained ageing.[12]

The Huxleys live on in their poetry as well, and much of that was about birth and death, the great essentials of evolution as well as its intimate history. It is not clear which child inspired Henrietta's 'Thoughts on a Birth', but in it she caught generations and regeneration, 'By Nature's law forced to exist'.

> Childhood, youth, maturity,
> Marriage, then paternity;
> Age, and then to all 'Good-bye.'
> The round goes on without cessation,
> Generation to generation.
>
> Wherefore and why? 'Tis just we must,
> We never know the reason why,
> By Nature's law forced to exist,
> Till Death of Life incorporate
> Conducts us to the barrier gate,
> And leave us, just because he must,
> Ashes and dust.[13]

Henrietta seemed resigned to nature, its fertility and mortality. Perhaps she was just resigned to life with (and without) Thomas Henry Huxley. By contrast, he had written a far more energized and intense poem on death, inspired by Tennyson's funeral in 1892, to bring Westminster Abbey to sentient life. 'Bring me my dead!' commands the ancient ship of God:

> BRING me my dead!
> To me that have grown
> Stone laid upon stone,
> As the stormy brood

> Of English blood
> Has waxed and spread
> And filled the world,
> With sails unfurled;
> With men that may not lie;
> With thoughts that cannot die.

'All must fall asleep', he finished.[14] In the end, sleeping is how Thomas Henry and Henrietta Huxley preferred to think about death, a kind of nothing-rest, despite everything perhaps a zero-point. Julian did too. He concluded his early biological philosophy on the meaning of death with sleep; imagining death 'welcomed perhaps with the same welcome that the tired body gives to sleep'.[15]

Thomas Henry Huxley had buried his dear sleeping friend Charles Darwin in the Abbey a decade earlier, but the Church of England funeral came to haunt the agnostic. After Darwin's death, Huxley was sometimes asked to verify the rumour that on his death-bed Darwin had renounced 'Evolution', and 'sought safety in the blood of the Saviour', even that he 'abjectly whined for a minister', as one journalist disrespectfully put it to him. Is there any truth in this? False and without foundation, Thomas Henry Huxley said dismissively.[16] His own burial was private. It was meant to be small, but 200 friends and colleagues turned up uninvited but welcomed. He was buried with little Noel in East Finchley Cemetery. The epitaph from her poem seems to raise 'He' and 'His' to upper case, whether by her own hand or the stonemason's we don't know. Henrietta joined her prince and her first-born later.

There is another cluster of dead Huxleys lying scattered in secular Surrey ground at the Arts & Crafts Compton Cemetery near Godalming, first Julia, then poor Trevenen. Leonard and Rosalind, Aldous and Maria, Julian and Juliette all lie there, as does their first son, Anthony (figures 11.2 and 11.3). They are visited still by the teachers and girls of Prior's Field School. It is a ritual that Julian would have both appreciated and analysed, an annual commemoration on the day of Julia's birth.

Figures 11.2 and 11.3: The Huxley Family Grave at Watts Cemetery, Compton, Surrey, near Prior's Field School.

In Loving Memory
Noel Trevenen Huxley
1889 - 1914
Leonard Huxley
1860 - 1933
Aldous Leonard Huxley
1894 - 1963
And Maria his wife
1898 - 1955
Julian Sorell Huxley
1887 · 1975
Marie Juliette his wife
née Baillot, 1896 · 1994

Acknowledgements

This book began as the Wiles Lectures, Queen's University, Belfast, in 2018. My thanks to Peter Gray and the Wiles Trust for hosting the lecture series, and to the professorial guests, friends and colleagues who commented so fully on early versions of chapters herein: Pratik Chakrabarti, Saul Dubow, Ruth Harris, Philippa Levine, Lucy Rhymer, Todd Weir and Queen's professors Peter Bowler and David Livingstone. I'm grateful to friends in the history of science world who have suggested and corrected: Janet Browne, Sarah Franklin, David Haig, Ian Hesketh, Emily Kern, Erika Milam, Evelleen Richards, Helen Tilley; and other historians, anthropologists and philosophers who have willingly reviewed my ideas on the Huxleys and their many contexts: Adam Bobbette, Frances Clarke, Bridget Griffen-Foley, Jarrod Hore, Craig Munro, David Napier, Anne O'Brien, Naomi Parkinson, Markos Valaris. For research and editing assistance, I'm grateful to Michelle Bootcov, James Keating, Marie McKenzie and David Watson. Chris Brunton, Oscar Brunton, Nerida Little, Ed Little and Chloe Ferres have contributed in multiple ways, for which thanks.

James Pullen, of the Wylie Agency, affirmed and inspired early ideas, and post-meeting notes scribbled between London and Cambridge have remained the steady leg of this writer's compass. Simon Winder at Penguin has read, reread, exclaimed and questioned. My thanks for such close attention, and to Eva Hodgkin for skilled assistance. I recall talking with Karen Darling of the University of Chicago Press about a T. H. and Julian Huxley book at a History of Science conference, perhaps back in 2010. Her keen response then, and interest all these years later, is warmly appreciated. T. H. Huxley wouldn't understand why he had to wait his turn, and behind the Reverend Malthus!

The Australian Research Council has generously supported research (DP190100498). At Rice University, whose Woodson Research Center, Fondren Library, holds the Huxley Papers as well as Julian Huxley's personal library, I'm grateful for the close attention and permissions to publish extended by Amanda Focke, Head of Special Collections, and Gabrielle Parker for assistance scanning and checking. At Imperial College London, which holds the vast Thomas Henry Huxley Collection, Anne Barrett has generously enabled access, permissions and ways to negotiate the complex papers. My thanks also to Alex Fisher and Lucy Shepherd. At Imperial College I met Angela Darwin, expert on, and member of, the Huxley and Darwin families, learned especially about her great-grandfather Thomas Henry and great-grandmother Henrietta. I appreciate the directions, corrections and lunches so willingly offered, as well as permission to reproduce the etchings of Collier's portraits of Huxley and Darwin in her home. Victoria Huxley and Susan Ray, Julian and Juliette's granddaughters, have generously granted permission for reproduction of images, published poems and quotations from Julian and Juliette Huxley's private papers, and from all Huxley publications. Thanks also to Adele Getty, Michael Williams and especially David Napier, who shared memories of Francis, who was for me an unexpected gift in the Huxley story. Justine Mann and Bridget Gillies of the University of East Anglia Archives made the useful Solly Zuckerman papers available, for which thanks. The principal and archivists of Prior's Field School, Godalming, Surrey, have shown great interest, especially Joanne Halford and Polly Murray. David Attenborough kindly offered memories of Julian Huxley and the early world of natural history broadcasting and took time to correct some 'small inaccuracies'. I acknowledge permissions to reproduce images kindly granted by Angela Darwin, Victoria Huxley and Susan Ray, Peter Suschitzky, Prior's Field School, Robert Forster, Andrew Smart of A. C. Cooper Ltd, the National Portrait Gallery, London, the Science Photo Library, the Galton Institute, the

University of Chicago Library, the Truman State University Library, the State Library of New South Wales, the British Psychological Association, the Imperial War Museum, the Cold Spring Harbor Laboratory and the University of East Anglia Archives.

AB, 'Alegria', summer, 2022

Notes

ABBREVIATIONS

BL	British Library
CWP	Charlotte Wolff Papers, Wellcome Library, London
DCP	Darwin Correspondence Project
JSH	Julian Sorell Huxley
JSH Papers	Julian Sorell Huxley Papers, 1899–1980, MS50, Woodson Research Center, Fondren Library, Rice University
J&JH Papers	Julian and Juliette Huxley Papers, 1899–1988, MS512, Woodson Research Center, Fondren Library, Rice University
JH Papers	Juliette Huxley Papers, 1895–1994, MS474, Woodson Research Center, Fondren Library, Rice University
ICL	Imperial College London
L&L	Leonard Huxley, *Life and Letters of Thomas Henry Huxley* (2 vols., London: Macmillan, 1900)
MAWL	Mary Augusta Ward Letters, British Library RP8974 [Mrs Humphry Ward Letters]
RU	Rice University, Houston, Texas
SZ	Solly Zuckerman
THH	Thomas Henry Huxley
THHC	Thomas Henry Huxley Collection
UEA	University of East Anglia

INTRODUCTION

1. 'Let us not forget that we men are the trustees of evolution, and that to refuse to face this problem is to betray the trust put into our hands by the powers of the universe.' Julian Sorell Huxley [hereafter JSH], *The Stream of Life* (London: Watts & Co., 1926), 56.

2. JSH, *Memories*, vol. I (London: George, Allen & Unwin, 1970) [hereafter *Memories* I], 272.

3. In 1858 Huxley boldly displaced Richard Owen's thesis on the archetypes of the skull in his Croonian Lecture. Thomas Henry Huxley [hereafter THH], 'On the Theory of the Vertebrate Skull', *Proceedings of the Royal Society*, 9 (1859): 381–457. THH, *Evidence as to Man's Place in Nature* (London: Williams & Norgate, 1863).

4. *The Times*, 23 January 1867, 7.

5. R. S. Deese, *We Are Amphibians: Julian and Aldous Huxley on the Future of Our Species* (Berkeley: University of California Press, 2014).

6. C. S. Nicholls, *Elspeth Huxley: A Biography* (London: HarperCollins, 2003), 270.

7. May Sarton, draft introduction to her letters to Juliette Huxley, and May Sarton to Juliette Huxley, 29 October 1978, in Susan Sherman (ed.), *Dear Juliette: Letters of May Sarton to Juliette Huxley* (New York: Norton, 1999), 344, 306.

8. Walt Whitman quoted in JSH, *Religion without Revelation* (London: Ernest Benn, 1927), 120.

9. Aldous Huxley, *On the Margin: Notes and Essays* (London: Chatto & Windus, 1923), 19.

10. THH, 'Are Men Born Free and Equal?,' letters to *Daily Telegraph*, 27, 29, 30 January 1890; THH, 'The Natural Inequality of Man', *Nineteenth Century*, January 1890, republished in THH, *Method and Results: Collected Essays* (London: Macmillan, 1893), 290–335.

11. Ivor Montagu, *Man – One Family* (Ealing Studios, UK, 1946).

12. JSH, 'The Courtship Habits of the Great Crested Grebe: With an Addition to the Theory of Sexual Selection', *Proceedings of the Zoological Society of London*, 35 (1914): 491–562.

13. Charles Darwin, *The Expression of the Emotions in Man and Animals* (London: John Murray, 1872).

14. Joanna Bourke, *What It Means to Be Human* (London: Virago, 2011), 11.

15. JSH to Cyril Bibby, 21 July 1964, Julian Sorell Huxley Papers, 1899–1980, MS50, Woodson Research Center, Fondren Library, Rice University [hereafter JSH Papers, RU, MS50], Box 37: 1.

16. JSH, 'Conclusion', in THH and JSH, *Evolution and Ethics* (London: Pilot Press, 1947), 181.

17. JSH, 'The Future of Man – Evolutionary Aspects', in Gordon Wolstenholme (ed.), *The Future of Man* (London: J&A Churchill, 1963), 1.

18. JSH, 'The Humanist Frame', in JSH (ed.), *The Humanist Frame* (London: George Allen & Unwin, 1961), 17.

19. Lecture draft, 'The Evolutionary Process and Man's Place in It', for BBC series *Humanity and Evolution*, JSH Papers, RU, MS50, Box 69: 5.

20. Genesis, 1:28. Old Testament, English Revised Version, first published 1885.

21. JSH, 'Evolutionary Ethics', in THH and JSH, *Evolution and Ethics*, 120.

22. THH, 'Prolegomena' (1894), in THH and JSH, *Evolution and Ethics*, 36.

CHAPTER 1: GENERATIONS

1. The manuscript of this poem 'The Inheritor, On JSH, b. 1887' is in Juliette Huxley Papers, Woodson Research Center, Fondren Library, MS474, Rice University [hereafter JH Papers, RU, MS474], Box 10: 7.

2. Ronald W. Clark, *The Huxleys* (New York: McGraw-Hill, 1968); Harold Cyril Bibby, *Scientist Extraordinary: The Life and Scientific Work of Thomas Henry Huxley, 1825–1895* (New York: St Martin's Press, 1972); J. Vernon Jensen, *Thomas Henry Huxley: Communicating for Science* (Newark: University of Delaware Press, 1991); Adrian Desmond, *Huxley: From Devil's Disciple to Evolution's High Priest* (Harmondsworth: Penguin, 1998); Paul White, *Thomas Huxley: Making the 'Man of Science'* (Cambridge: Cambridge University Press, 2003).

3. THH, 'Autobiography', in THH, *Method and Results*, 4.

4. Leonard Huxley (ed.), *Life and Letters of Thomas Henry Huxley*, vol. I [hereafter *L&L*, I] (London: Macmillan, 1900), 5.

5. Desmond, *Huxley*, ch. 1.

6. Charles Bray, *The Education of the Body: An Address to the Working Classes* (Coventry: H. Merridew, 1836); Adrian Desmond and Angela Darwin, 'T. H. Huxley's Turbulent Apprenticeship Years', *Archives of Natural History*, 47, no. 1 (2021): 215–26.

7. Martin Huxley Cooke, *The Evolution of Nettie Huxley, 1825–1914* (Chichester: Phillimore, 2008), 5–6.

8. [Anon], 'Visit to the Moravian Establishments on the Continent', *Chambers' Edinburgh Journal*, 11 (21 May 1842): 142–3; Marianne Doerfel, 'British Pupils in a German Boarding School: Neuwied/Rhine 1820–1913', *British Journal of Educational Studies*, 34, no. 1 (1986): 79–96.

9. From Henrietta A. Huxley, 'Cinderella', in *Poems of Henrietta A. Huxley with Three of Thomas Henry Huxley* (London: Duckworth, 1913), 13–14.

10. Henrietta Heathorn to THH, 27, 28 January and 4 February 1852, Thomas Henry Huxley Collection, Imperial College London [hereafter THHC, ICL], Series 22, Letter 184.

11. Henrietta Heathorn to THH, 20 March 1850, THHC, ICL, Series 22, Letter 78.

12. THH to Henrietta Heathorn, 9 February 1851, THHC, ICL, Series 22, Letter 138.

13. Henrietta Heathorn to THH, 4 December 1853, THHC, ICL, Series 22, Letter 255.

14. THH, 'Autobiography', 15.

15. Henrietta Huxley, n.d., Reminiscences of early married life, THHC, ICL, Series 3, Box 31: 91.

16. Henrietta Huxley, 'Now and Then, 1860', *Poems* (1913), 36.

17. Herbert Spencer to John Tyndall (1858 or 1859), Royal Institution Archives, Tyndall Collection, vol. 3, 1028. See Mark Francis, *Herbert Spencer and the Invention of Modern Life* (London: Routledge, 2014).

18. Cooke, *Nettie Huxley*, 86.

19. Henrietta Huxley, 'Leonard', *Poems* (privately printed, 1899), 48. British Library [hereafter BL], RB.23.a.27520.

20. Benjamin Jowett to THH, 19 July 1877 and 7 May 1878, THHC, ICL, Series 1J, 7:13 and 7:15.

21. Matthew Arnold to THH, 13 February 1873, THHC, ICL, Series 1A, 10.155.

22. Matthew Arnold to THH, 8 December 1875, 17 October 1880, THHC, ICL, Series 1A, 10.159, 10.163.

23. Leonard Huxley (ed.), *Life and Letters of Thomas Henry Huxley*, vol. II [hereafter *L&L*, II] (London: Macmillan, 1900), 435.

24. Leonard Huxley to JSH and Trev Huxley, 6 May 1896, JH Papers, RU, MS474, Box 10: 1.

25. Aldous Huxley and JSH to Julia Huxley, 10 January 1898, JH Papers, RU, MS474, Box 9: 13.

26. Cooke, *Nettie Huxley*, 28, 17; Henrietta Heathorn to THH, 28 October 1851, THHC, ICL, Series 22, Letter 171; THH, 'Nature: Aphorisms by Goethe', *Nature*, 1, no. 1 (4 November 1869), 9–11.

27. Julia Huxley to Mary Ward, n.d. 1908, Mary Augusta Ward Letters [hereafter MAWL], BL, RP 8974.

28. Julia Huxley to Mary Ward, 9 June 1899, MAWL, BL, RP 8974.

29. Julia Huxley to JSH, 27 March 1904, JH Papers, RU, MS474, Box 9: 13.

30. Julia Huxley to JSH, 13 June 1902, JH Papers, RU, MS474, Box 9: 13.

31. Trev Huxley to Aldous Huxley, 9 October 1903, JH Papers, RU, MS474, Box 11: 1.

32. Julia Huxley to Mary Ward, 16 December 1898 and 9 June 1899, MAWL, BL, RP 8974.

33. Julia Huxley to Mary Ward, 9 June 1899, MAWL, BL, RP 8974.

34. Julia Huxley to JSH, 3 May 1902, JH Papers, RU, MS474, Box 9: 13.

35. Julia Huxley to Mary Ward, 10 October 1908, MAWL, BL, RP 8974.

36. Julia Huxley to Mary Ward, 2 February 1906; Julia Huxley to Mary Ward, 9 September 1908, MAWL, BL, RP 8974.

37. Leonard Huxley to JSH, 18 November 1908, JH Papers, RU, MS474, Box 10: 1.

38. Julia Huxley to Mary Ward, 4 November 1908, MAWL, BL, RP 8974.

39. E. G. Conklin to JSH, 10 December 1910, and JSH to E. G. Conklin, 15 January 1911, President Edgar Odell Lovett's Papers, Woodson Research Center, Fondren Library, Rice University, TX, UA 014, Box 3: 8.

40. Mary Ward to Julian Huxley, 26 September 1916, MAWL, BL, RP 8974.

41. JSH, *Religion without Revelation*, 130.

42. JSH, *Memories* I, 110–13.

43. Juliette Huxley, *Leaves of the Tulip Tree* (London: John Murray, 1986).

44. H. G. Wells to JSH, 20 July 1932, JH Papers, RU, MS474, Box 4: 12.

45. Mary Ward to JSH, 27 January 1919, JH Papers, RU, MS474, Box 10: 19.

46. Henrietta A. Huxley (ed.), *Aphorisms and Reflections: From the Works of T. H. Huxley* (London: Macmillan, 1907).

47. JSH, *Memories* I, 197.

48. [THH] 'The Origin of Species', *Westminster Review*, 17 (n.s.) (1860): 541–70; JSH, 'The Age of Overbreed', *Playboy*, 12, no. 1 (January 1965): 103–4, 106, 177, 179–81.

49. JSH and Douglas Cleverdon, *Julian Huxley on T. H. Huxley: A New Judgment* (London: Watts & Co., 1945), 25; Aldous Huxley, *T. H. Huxley as a Man of Letters* (London: Macmillan, 1932).

50. David Riede, *Allegories of One's Own Mind: Melancholy in Victorian Poetry* (Columbus, OH: Ohio State University Press, 2005), 12.

51. Henrietta Huxley, n.d., Reminiscences written by Henrietta Huxley in her old age, THHC, ICL, Series 3, 31: 87–8.

52. THH, *Lay Sermons, Addresses and Reviews* (New York: D. Appleton, 1871).

53. THH, *Lectures on the Elements of Comparative Anatomy* (London: John Churchill and Sons), 1864.

54. Julia Huxley to Mary Ward, 9 June 1899, MAWL, BL, RP 8974.

55. Leonard Huxley, *Thomas Henry Huxley: A Character Sketch* (London: Watts & Co., 1920).

56. THH, *T. H. Huxley's Diary of the Voyage of H.M.S. Rattlesnake*, ed. JSH (London: Chatto & Windus, 1935), 3 [Hereafter THH, *Rattlesnake Diary*].

57. Report by 'MDH', 12 December 1904, Julian and Juliette Huxley Papers, 1899–1988, MS512, Woodson Research Center, Fondren Library, Rice University [hereafter J&JH Papers, RU, MS512], Box 37: 3.

58. JSH to Kathleen Fordham, 30 June 1917, J&JH Papers, RU, MS512, Box 37: 11.

59. JSH, 'The Tissue Culture King', *Amazing Stories*, 2, no. 5 (August 1927): 451–9.

60. JSH, 'Man's Place in the Universe', n.d. [1930s], RU, MS50 Box 63: 7; draft of 'Man's Place in the University', RU, MS50 Box 64: 8.

61. THH, 'From Shanklin' (1887), in Henrietta A. Huxley, *Poems* (1913), 1.

62. Henrietta A. Huxley, 'The Southern Cross', in *Poems* (1913), 7.

63. Henrietta Huxley, 22 November 1855, quoted in THH, *Rattlesnake Diary*, 314; and Cooke, *Nettie Huxley*, 5. Cooke indicates that the original has not been found (135).

64. Mrs Humphry Ward, *A Writer's Recollections*, vol. II (New York and London: Harper & Brothers, 1918), 122.

65. Leonard Huxley, 'Poets to Be', in *Anniversaries*, 2–3.

66. Henrietta Huxley to JSH, 20 March 1910, JH Papers, RU, MS474, Box 9: 11.

67. JSH, *Holyrood: The Newdigate Poem, 1908* (Oxford: Blackwell, 1908).

68. Henrietta Huxley to JSH, 24 June 1908; Henrietta Huxley to JSH, 20 March 1910, JH Papers, RU, MS474, Box 9: 11.

69. Julia Huxley to JSH, 24 November 1904, JH Papers, RU, MS474, Box 9: 13.

70. Julia Huxley to JSH, 7 March 1903, JH Papers, RU, MS474, Box 9: 13.

71. Julia Huxley to Mary Ward, 28 August 1905, MAWL, BL, RP 8974.

72. Julian Huxley to Mary Augusta Ward, 22 April 1910, MAWL, BL, RP 8974.

73. Mary Augusta Ward to JSH, 30 April 1910, JH Papers, RU, MS474, Box 10: 19.

74. Mary Augusta Ward to JSH, 30 March, no year; Mary Augusta Ward to JSH, 26 September 1916, MAWL, BL, RP 8974.

75. 'It will easily be seen how much I owe to M. Bergson', JSH, *The Individual in the Animal Kingdom* (Cambridge: Cambridge University Press, 1912), vii; Emily Herring, 'Great Is Darwin and Bergson His Poet': Julian Huxley's Other Evolutionary Synthesis', *Annals of Science*, 75, no. 1 (2018): 40–54.

76. Jeanette L. Gilder, 'Mrs Ward Describes the Genesis of "Robert Elsmere"', *Chicago Tribune*, 25 December 1909, 7.

77. JSH to Henrietta Huxley, 30 July 1911, JH Papers, RU, MS474, Box 9: 11.

78. Henrietta Heathorn to THH, 8 June 1851, THHC, ICL, Series 22, Letter 150; Cooke, *Nettie Huxley*, 38.

79. Marian Collier, *John Collier*, oil on canvas, c. 1882–3, National Portrait Gallery 6811.

80. John Collier to JSH, 19 May 1926, JH Papers, RU, MS474, Box 9: 2.

81. John Collier to JSH, 15 January 1930, JH Papers, RU, MS474, Box 9: 2.

82. Ethel Collier to JSH, 25 July 1930, JH Papers, RU, MS474, Box 9: 2.

CHAPTER 2: TRUSTEES OF EVOLUTION

1. Iain McCalman, *Darwin's Armada: Four Voyages and the Battle for the Theory of Evolution* (New York: Norton, 2009).

2. Henry Fairfield Osborn, 'Memorial Tribute to Professor Thomas H. Huxley', *Transactions of the New York Academy of Sciences*, 15 (1895): 40–50. See also Osborn, 'A Student's Reminiscences of Huxley', *Biological Lectures Delivered at the Marine Biological Laboratory of Wood's Holl* (Boston: Ginn & Company, 1896), 29–42.

3. Charles Darwin to THH, 25 December 1859, in Francis Darwin (ed.), *The Life and Letters of Charles Darwin*, vol. II (London: John Murray, 1887), 251.

4. Janet Browne, *Charles Darwin: The Power of Place* (Princeton: Princeton University Press, 2004), 104–6; Peter J. Bowler, *Monkey Trials and Gorilla Sermons: Evolution and Christianity from Darwin to Intelligent Design* (Cambridge, MA: Harvard University Press, 2007), 79–133. For sexual selection see Evelleen Richards, *Darwin and the Making of Sexual Selection* (Chicago: University of Chicago Press, 2017).

5. Mario Di Gregorio, *T. H. Huxley's Place in Natural Science* (New Haven: Yale University Press, 1984), xvi; Michael Bartholomew, 'Huxley's Defence of Darwin', *Annals of Science*, 32, no. 6 (1975): 525–35. For Huxley's 'vigilant scepticism', see William Irvine, *Thomas Henry Huxley* (London: Longman's Green, 1960), 812.

6. Peter J. Bowler, *Evolution: The History of an Idea*, 3rd edn (Berkeley: University of California Press, 2003), ch. 3.

7. THH, *Rattlesnake Diary*.

8. Peter J. Bowler, *The Eclipse of Darwinism* (Baltimore and London: Johns Hopkins University Press, 1983); Mark A. Largent, 'The

So-Called Eclipse of Darwinism', in Joe Cain and Michael Ruse (eds.), *Descended from Darwin* (Philadelphia: American Philosophical Society, 2009), 3–21; C. Kenneth Waters and Albert Van Helden (eds.), *Julian Huxley: Biologist and Statesman of Science* (Houston: Rice University Press, 1992).

9. For Darwin's early education, family learning and the voyage of the *Beagle*, see Janet Browne, *Charles Darwin: Voyaging* (New York: Knopf, 1995).

10. Erasmus Darwin, *Zoonomia: or, the Laws of Organic Life* (London: J. Johnson, 1794), I.

11. Adrian Desmond, *The Politics of Evolution: Morphology, Medicine and Reform in Radical London* (Chicago: University of Chicago Press, 1989).

12. Charles Darwin, 'Autobiography', in Francis Darwin (ed.), *Life and Letters*, I, 38.

13. Darwin, 'Autobiography', 83.

14. THH, 'The Rede Lecture: The Origin of the Existing Forms of Animal Life: Construction or Evolution', *Nature*, 28, no. 712 (21 June 1883): 187–9.

15. THH, Notes on Paley Natural Theology, THHC, ICL, Series 19, vol. 82: 125 ff.

16. Charles Bell, *The Hand; Its Mechanism and Vital Endowments as Evincing Design* (London: William Pickering, 1833), 15. See Jonathan Topham, 'Science and Popular Education in the 1830s: The Role of the Bridgewater Treatises', *British Journal for the History of Science*, 25, no. 4 (1992): 397–430.

17. Martin J. S. Rudwick, *Bursting the Limits of Time: The Reconstruction of Geohistory in the Age of Revolution* (Chicago: University of Chicago Press, 2007).

18. In fact, as James Secord shows, Lyell did not present a history of Earth's development, but his work was cast as that by William Whewell, who described his own 'catastrophism' vis-à-vis Lyell's 'uniformitarianism'. James A. Secord, Introduction to Charles Lyell, *Principles of Geology* (Harmondsworth: Penguin, 1997), xix.

19. Secord, Introduction, *Principles of Geology*, xxxvii.

20. Robert King to THH, 8 July 1863, THHC, ICL, Series 1k, 19.155.

21. Desmond, *The Politics of Evolution*, 5; Peter Bowler, 'The Changing Meaning of "Evolution"', *Journal of the History of Ideas*, 36, no. 1 (1975), 100. Julian Huxley, also a historian of the term, attributed the modern use of 'evolution' to Herbert Spencer. JSH, 'Darwin and the

Idea of Evolution', in JSH et al., *A Book that Shook the World: Essays on Charles Darwin's Origin of Species* (Pittsburgh: University of Pittsburgh Press, 1958), 2. Huxley himself later noted that the important thing about 'evolution' was that it was not used in the original editions of *On the Origin of Species*. THH, 'Evolution in Biology (1878)', in THH, *Darwiniana: Collected Essays* (London: Macmillan, 1893), 187–226.

22. *L&L*, I (London: Macmillan, 1900), 288–90; note in JSH Papers, RU, MS50, Box 121: 1; Robert J. Richards, *The Tragic Sense of Life: Ernst Haeckel and the Struggle over Evolutionary Thought* (Chicago: University of Chicago Press, 2008); Nick Hopwood, *Haeckel's Embryos: Images, Evolution, and Fraud* (Chicago: University of Chicago Press, 2015).

23. THH (trans.), 'Fragments Relating to Philosophical Zoology, Selected from the Words of K. E. von Baer', in THH and Arthur Henfrey (eds.), *Scientific Memoirs* (London: Taylor and Francis, 1853), 176–238.

24. [Herbert Spencer], 'The Development Hypothesis', *The Leader*, 20 March 1852. Reprinted in his *Essays: Scientific, Political, and Speculative*, vol. I (London: Williams and Norgate, 1891), 1–7.

25. THH, 'On the Reception of the "Origin of Species"', in Francis Darwin (ed.), *Life and Letters of Charles Darwin*, II, 188.

26. THH, 'On the Reception of the "Origin of Species"', 187.

27. THH, *Rattlesnake Diary*, 161.

28. Gowan Dawson, *Show Me the Bone: Reconstructing Prehistoric Monsters in Nineteenth-Century Britain and America* (Chicago: University of Chicago Press, 2016).

29. Richard Owen, *On the Nature of Limbs* (London: John van Voorst, 1849), 85–6.

30. Evelleen Richards, 'A Question of Property Rights: Richard Owen's Evolutionism Reassessed', *British Journal for the History of Science*, 20, no. 2 (1987): 129–71; Nicolaas A. Rupke, *Richard Owen: Victorian Naturalist* (New Haven: Yale University Press, 1994).

31. Rupke, *Richard Owen*, 93–4.

32. For Huxley's early use of 'archetype' see THH, 'On the Morphology of the Cephalous Mollusca, as Illustrated by the Anatomy of Certain Heteropoda and Pteropoda', *Philosophical Transactions of the Royal Society of London*, 143 (1853): 50.

33. Di Gregorio, *T. H. Huxley's Place in Natural Science*, xvii.

34. THH, 'Six Lectures on the General Laws of Life and on the Leading Points in the Structure, Production and Forms of Living Beings',

printed syllabus, London Institution, 1855, THHC, ICL, Series 20, 122.7–8.

35. Charles Darwin to THH, 23 April 1853, Darwin Correspondence Project, University of Cambridge [hereafter DCP], Letter 1480.

36. For Owen's and Huxley's models, see Adrian Desmond, *Archetypes and Ancestors: Palaeontology in Victorian London, 1850–1875* (Chicago: University of Chicago Press, 1982).

37. [THH], 'Review VII. Vestiges of the Natural History of Creation. Tenth Edition. London 1853', *British and Foreign Medico-Chirurgical Review*, 13, no. 26 (1854): 425–39.

38. James A. Secord, *Victorian Sensation: The Extraordinary Publication, Reception, and Secret Authorship of Vestiges of the Natural History of Creation* (Chicago: University of Chicago Press, 2000).

39. Joel S. Schwartz, 'Darwin, Wallace, and Huxley, and "Vestiges of the Natural History of Creation"', *Journal of the History of Biology*, 23, no. 1 (1990): 127–53.

40. Richards, 'A Question of Property Rights'.

41. [THH], 'Review of Vestiges of Creation', 439.

42. Charles Darwin to THH, 2 September 1854, DCP, Letter 1587.

43. THH, 'On the Reception of the "Origin of Species"', 188–9.

44. Charles Darwin to J. D. Hooker, 21 May 1856, DCP, Letter 1876; Adrian Desmond, *Huxley: From Devil's Disciple to Evolution's High Priest* (Harmondsworth: Penguin, 1998), 226.

45. Charles Darwin to THH, 17 July 1851, DCP, Letter 1442.

46. Charles Darwin to THH, 11 April 1853, DCP, Letter 1514.

47. Charles Darwin to THH, 23 April 1853, DCP, Letter 1480.

48. Charles Darwin to THH, 2 April 1856, DCP, Letter 1847.

49. THH to Charles Darwin, [before 3 October 1857], DCP, Letter 2144.

50. THH to Charles Lyell, 25 June 1859, in *L&L*, I, 173–4.

51. THH to Charles Darwin, 23 November 1859, DCP, Letter 2544.

52. Charles Darwin, *On the Origin of Species by Means of Natural Selection, or the Preservation of Favoured Races in the Struggle for Life* (London: John Murray, 1859), 60–61.

53. THH to Charles Kingsley, 30 April 1863, THHC, ICL, Series 1k, 19. 212–13.

54. THH, *Evidence as to Man's Place in Nature*, 108.

55. Di Gregorio, *T. H. Huxley's Place in Natural Science*, xviii.

56. THH, 'Past and Present', *Nature*, 51 (November 1894), 1–3.

57. Ian Hesketh, *Of Apes and Ancestors: Evolution, Christianity, and the Oxford Debate* (Toronto: University of Toronto Press, 2009).

58. THH, *Man's Place in Nature*, 107.

59. THH, *Rattlesnake* Diary, 31 December 1856 in *L&L*, I, 151.

60. JSH, *Religion without Revelation*, 103.

61. THH to Charles Darwin, 2 July 1863, DCP, Letter 4228.

62. Ruth Barton, *The X Club: Power and Authority in Victorian Science* (Chicago: University of Chicago Press, 2018).

63. John Lubbock to Charles Darwin, 23 August 1862, DCP, Letter 3698.

64. Charles Darwin to THH, 18 September 1860, DCP, Letter 2920B.

65. Henrietta Huxley to Charles Darwin, 1 January 1865, DCP, Letter 4733. In fact, Darwin was correct.

66. Charles Darwin to THH before 25 February 1863, DCP, Letter 3896.

67. THH to Charles Darwin, 25 February 1863, DCP, Letter 4010.

68. Francis Darwin to THH, 20 April 1882, THHC, ICL, Series 1d, 13.10.

69. William Erasmus Darwin and G. H. Darwin, 28 April 1882, THHC, ICL, Series 1d, 13.106–7.

70. 'Unveiling the Statue of the Late Charles Darwin in the Natural History Museum, South Kensington', *The Graphic*, 20 June 1885, 621–2.

71. See letters between Francis Darwin and THH over 1886 and 1887, THHC, ICL, Series 1d, 13.

72. THH, 'On the Reception of the "Origin of Species"', 179–81.

73. JSH, *Evolution: The Modern Synthesis* (London: George Allen & Unwin, 1942), 20.

74. Charles Darwin, 'Provisional Hypothesis of Pangenesis', in *The Variation of Animals and Plants under Domestication*, vol. II (London: John Murray, 1868), 357–404.

75. William Bateson, *Materials for the Study of Variation Treated with Especial Regard to Discontinuity in the Origin of Species* (London and New York: Macmillan, 1894).

76. Hugo de Vries, *Species and Varieties: Their Origin by Mutation* (Chicago: Open Court, 1905); Sherrie L. Lyons, 'T. H. Huxley's Saltationism: History in Darwin's Shadow', *Journal of the History of Biology*, 28, no. 3 (1995): 463–94.

77. JSH, *Evolution: The Modern Synthesis*, 20.

78. JSH to Lovett, 9 August 1912, Rice Institute President Edgar Odell Lovett's Papers, Woodson Research Center, Fondren Library, Rice University, UA 014, Box 3: 8.

79. Thomas Hunt Morgan, *A Critique of the Theory of Evolution* (Princeton: Princeton University Press, 1916).

80. Thomas Hunt Morgan et al., *The Mechanism of Mendelian Heredity* (New York: Henry Holt, 1915); Ernst Mayr, *The Growth of Biological Thought* (Cambridge, MA: Belknap Press of Harvard University Press, 1982).

81. JSH, *Evolution: The Modern Synthesis*, 21.

82. E. B. Ford, 'Some Recollections Pertaining to the Evolutionary Synthesis', in Ernst Mayr and William B. Provine (eds.), *The Evolutionary Synthesis*, 2nd edn (Cambridge, MA: Harvard University Press, 1998), 336–8.

83. Steindór J. Erlingsson, 'Institutions and Innovation: Experimental Zoology and the Creation of the *British Journal of Experimental Biology* and the Society for Experimental Biology', *British Journal for the History of Science*, 46, no. 1 (2013): 73–95; Steindór J. Erlingsson, 'The Costs of Being a Restless Intellect: Julian Huxley's Popular and Scientific Career in the 1920s', *Studies in the History and Philosophy of Biology and the Biomedical Sciences*, 40, no. 2 (2009): 101–8.

84. Erlingsson, 'Institutions and Innovation', 78.

85. G. P. Bidder to THH, 4 August 1925, G. P. Bidder Papers, Archive of the Marine Biological Association, National Marine Biological Library, Plymouth, PBD 13.

86. Ronald A. Fisher, *The Genetical Theory of Natural Selection* (Oxford: Clarendon, 1930).

87. JSH, typescript, 'Darwin and the Idea of Evolution', May 1858 (1958), JSH Papers, RU, MS 50, Box 74: 5.

88. JSH, 'Mendelism', *Morning Post*, 1 June 1922.

89. JSH, 'The Darwinian Theory of Evolution: I – The Problem', *Manchester Guardian*, 13 August 1925; 'II – The Solution', *Manchester Guardian*, 14 August 1925; 'III – Corollaries and Consequences', *Manchester Guardian*, 15 August 1925.

90. These talks were published as JSH, *The Stream of Life* (1926).

91. W. B. Provine, *The Origins of Theoretical Population Genetics, with a New Afterword* (Chicago: University of Chicago Press, 2001). There are revisionist accounts of this twentieth-century history. See Vassiliki Betty Smocovitis, 'Unifying Biology: The Evolutionary Synthesis and Evolutionary Biology', *Journal of the History of Biology*, 25, no. 1 (1992), 1–65.

92. JSH, *Evolution: The Modern Synthesis*, 8.

93. JSH, *Evolution: The Modern Synthesis*, 24–5, 27.

94. JSH, *Evolution: The Modern Synthesis*, 28.

95. JSH, typescript preface to revised edition of *Living Thoughts of Darwin*, January 1958, JSH Papers, RU, MS 50, Box 74: 5.

96. Francis Huxley, 'Charles Darwin: Life and Habit', *The American Scholar*, 28, no. 4 (1959): 489–99.

97. Notes on Darwin Centennial, Chicago, 26 November 1959, JSH Papers, RU, MS 50, Box 108: 1.

98. News clipping, 5 December 1959, JSH Papers, RU, MS 50, Box 108: 4.

CHAPTER 3: MALADY OF THOUGHT

1. *L&L*, I, 8.

2. These folios have 'By LH?' handwritten, with a question mark. However, the handwriting is Julian's, JH Papers, RU, MS 474, Box 10: 7.

3. Adrian Desmond, *Huxley: From Devil's Disciple to Evolution's High Priest* (Harmondsworth: Penguin, 1998), 476.

4. JSH, *Memories* I, 15.

5. THH, 'notes on the unused pages at the end of his *Rattlesnake* diary', quoted in Ronald W. Clark, *The Huxleys* (New York: McGraw-Hill, 1968), 109.

6. THH, *Rattlesnake Diary*, 4 May 1847, 38.

7. THH, *Rattlesnake Diary*, 22 June 1847, 44.

8. *L&L*, I, 7.

9. Joseph Priestley, *Disquisitions Relating to Matter and Spirit* (London: J. Johnson, 1777).

10. THH, *Rattlesnake Diary*, 25 November 1847, 84–6.

11. JSH, comment in THH, *Rattlesnake Diary*, 120.

12. THH, *Rattlesnake Diary*, 120; 24 December 1847, 93; 5 December 1847, 88.

13. JSH, comment in THH, *Rattlesnake Diary*, 112.

14. THH, 'Science at Sea', *Westminster Review* (January 1854), quoted in *L&L*, I, 49.

15. JSH, comment in THH, *Rattlesnake Diary*, 124.

16. THH, *Rattlesnake Diary*, 6 April 1850, 265–6.

17. Henrietta Heathorn to THH, 22 and 31 May 1851, THHC, ICL, Series 22, Letter 148; and Henrietta Heathorn to THH, 30 April and 9 May 1852, THHC, ICL, Series 22, Letter 200.

18. Henrietta Heathorn to THH, 23 December 1847, THHC, ICL, Series 22, Letter 6.

19. THH to Charles Darwin, 24 April 1873, DCP, Letter 8873.

20. Charles Darwin to THH, 23 April 1873, DCP, Letter 8872.

21. JSH, *Memories* I, 54.

22. JSH to Kathleen Fordham, 30 June 1917, J&JH Papers, MS512, Box 37: 11.

23. JSH Journal, 4 October 1908, J&JH Papers, RU, MS512, Box 1.

24. Thomas K. Scott to E. O. Lovett, 31 December 1912, Rice Institute President Edgar Odell Lovett's Papers, Woodson Research Center, Fondren Library, Rice University, UA 014, Box 3: 8.

25. JSH to Henrietta Huxley, 30 July 1911, JH Papers, RU, MS474, Box 9: 11.

26. Letters between JSH, Trev and Kathleen, J&JH Papers, RU, MS512, Box 37: 11.

27. Mary Ward to JSH, 3 March 1912, MAWL, BL RP 8974.

28. JSH, *Memories* I, 97.

29. Kathleen Fordham, letters of thanks to President Lovett, Rice Institute President Edgar Odell Lovett's Papers, RU, UA 014, Box 3: 8.

30. Leonard Huxley to E. O. Lovett, 22 June 1913, Lovett's Papers, RU, UA 014, Box 43: 12.

31. JSH personal note, 29 November 1936, J&JH Papers, RU, MS512, Box 37: 4.

32. JSH, 'The Courtship Habits of the Great Crested Grebe', 491.

33. JSH, 'The Great Crested Grebe and the Idea of Secondary Sexual Characters', *Science*, 36, no. 931 (1912): 601–2.

34. JSH, 'The Courtship Habits of the Great Crested Grebe', 492.

35. Aldous Huxley to JSH, 3 July 1913, JH Papers, RU, MS474, Box 11: 2.

36. Trev Huxley to Kathleen Fordham, 24 February 1914, J&JH Papers MS512, Box 37: 11.

37. JSH, *Memories* I, 102.

38. Leonard to JSH, 19 August 1914, JH Papers, RU, MS474, Box 10: 2.

39. Julia Huxley to JSH, 21 June 1904, JH Papers, RU, MS474, Box 9: 13.

40. Personal communication with Angela Darwin, granddaughter of Leonard and Rosalind Huxley.

41. Juliette Huxley to May Sarton, 19 January 1976, in Susan Sherman (ed.), *Dear Juliette: Letters of May Sarton to Juliette Huxley* (New York: Norton, 1999), 353.

42. Sándor Forbát (ed.), *Love, Marriage, Jealousy* (London: Pallas, 1938).

43. C. J. H. van der Brock to THH, 29 December 1879, THHC, ICL, Series 1B, 11: 77.

44. Havelock Ellis and J. A. Symonds, *Das Konträre Geschlechtsgefühl* (Leipzig: Georg H. Wigand, 1896); Havelock Ellis and J. A. Symonds, *Studies in the Psychology of Sex: Sexual Inversion* (London and Leipzig: The University Press, 1900).

45. JSH, *If I Were a Dictator* (London: Methuen, 1934), 5. He discussed with Ellis a manuscript, 'very vivid & outspoken', of the 'making of an invert out of a neurotic ... personality', JSH to Havelock Ellis, 21 October 1932, Havelock Ellis Letters, Columbia University, Rare Book & Manuscript Library, Box 1.

46. JSH, 'Bird-Watching and Biological Science: Some Observations on the Study of Courtship in Birds', *The Auk*, 33, no. 3 (1916): 256–70.

47. Norman Haire, *Rejuvenation: The Work of Steinach, Voronoff and Others* (London: George Allen & Unwin, 1924).

48. Viola Ilma, credits page, table of contents, and 'We Take the Baton', editor's page, *Modern Youth*, 1, no. 1 (February 1933), 1–3; Julian Huxley, 'Hansen's Tomb', *Modern Youth*, 1, no. 1 (1933): 28–30.

49. Theodore Draper, 'American Youth Rejects Fascism', *New Masses*, 12, no. 9 (28 August 1934): 11–12.

50. Viola Ilma, *And Now, Youth!* (New York: Robert O. Ballou, 1934).

51. JSH and A. C. Haddon, *We Europeans: A Survey of 'Racial' Problems* (London: Jonathan Cape, 1935).

52. Viola Ilma to JSH, 29 October 1932, J&JH Papers, RU, MS512, Box 37: 5.

53. Juliette Huxley, typescript, n.d., J&JH Papers, RU, MS512, Box 37: 5.

54. Juliette Huxley to Dr Riddoch, typescript, n.d., J&JH Papers, RU, MS512, Box 37: 5.

55. Juliette Huxley to JSH, 31 August 1931, J&JH Papers, RU, MS512, Box 37: 4.

56. JSH to Juliette Huxley, 9 August 1932, J&JH Papers, RU, MS512, Box 37: 5.

57. Aldous Huxley to JSH, 9 January 1932, J&JH Papers, RU, MS512, Box 37: 5.

58. Juliette Huxley to Kot (Samuel Solomonovich Koteliansky), 23 October 1931, J&JH Papers, RU, MS512, Box 37: 9.

59. Aldous Huxley to JSH, 21 February 1931, and Leonard Huxley to JSH, 8 January 1933, J&JH Papers, RU, MS512, Box 37: 5.

60. Ethel Collier to JSH, 20 April 1931, J&JH Papers, RU, MS512, Box 37: 5.

61. Aldous Huxley to JSH, 10 May 1933, J&JH Papers, RU, MS512, Box 37: 5.

62. Aldous Huxley to JSH, early February 1932, J&JH Papers, RU, MS512, Box 37: 5.

63. JSH to Juliette Huxley, 1931, J&JH Papers, RU, MS 512, Box 37: 5.

64. 'Look Ahead 100 Years, Through the Eyes of Julian Huxley', *Tit-Bits*, 24 January 1931. In JSH Papers, MS 50, RU, Box 97: 11.

65. JSH, *If I Were a Dictator*, 93; JSH, 'Bird-Watching and Biological Science', 146.

66. JSH to Juliette Huxley, 22 July 1965, 11 p.m., J&JH Papers, RU, MS 512, Box 37: 4.

67. Julian's Fugue, J&JH Papers, RU, MS 512, Box 37: 5.

68. Viola Ilma to JSH, 9 March 1933 and 6 September 1934, J&JH Papers, RU, MS 512, Box 37: 13.

69. These ages according to May Sarton, 12 January 1976, in Sherman (ed.), *Dear Juliette*, 292–3.

70. May Sarton, 'The Leaves', included in May Sarton to Juliette Huxley, 22 June 1947, in Sherman (ed.), *Dear Juliette*, 156.

71. May Sarton to Juliette Huxley, 8 August 1987, in Sherman (ed.), *Dear Juliette*, 331.

72. Juliette Huxley to JSH, 10 November, 1933, J&JH Papers, RU, MS 512, Box 37: 5.

73. May Sarton to Juliette Huxley, 6 October 1977, in Sherman (ed.), *Dear Juliette*: 299.

74. Francis Huxley, Foreword, in Sherman (ed.), *Dear Juliette*, 15.

75. Juliette Huxley to May Sarton, 19 January 1976, in Sherman (ed.), *Dear Juliette*, 353.

76. May Sarton to Juliette Huxley, 25 January 1978, in Sherman (ed.), *Dear Juliette*, 302.

77. Juliette Huxley, *Leaves of the Tulip Tree*.

78. May Sarton to Juliette Huxley, 12 January 1976, in Sherman (ed.), *Dear Juliette*, 292.

79. Aldous Huxley, *On the Margin: Notes and Essays* (London: Chatto & Windus, 1923), 19–20.

80. Barming Asylum, Register of Private Patients 1850–1895, Kent Library and Archives Centre, MH/Md2/Ap3/1. George Huxley is not in any of the half-yearly returns, 1849–1855, for pauper and private patients.

81. Visiting Committee Report Book, 1833–1864, Kent Lunatic Asylum, Kent Library and Archives Centre, MH/Md2/Am1/1.

82. Visiting Committee Report Book, 1833–1864, Kent Lunatic Asylum, Kent Library and Archives Centre MH/Md2/Am1/1, entry for 15 April 1852.

83. J. E. Huxley and C. L. Roberts, 'On the Existing Relation between the Lunacy Commission and Medical Superintendents of Public Asylums', *Journal of Mental Science*, 5, no. 27 (1858), 95–102.

84. Visiting Committee Report Book, Kent Lunatic Asylum, 1833–1864, MH/Md2/Am1/1, 16 December 1861, 27 June 1856, 12 December 1861.

85. James Huxley, 'Remarks on the Tables', *Kent County Asylum, Sixteenth Annual Report, 1861–2*, 20.

86. James Huxley, 'Asylum Topics of More General Interest', *Annual Report, 1861–62*, 33. J. E. Huxley, 'History and Description of the Kent Asylum', *The Asylum Journal*, 1, no. 3 (1854): 39–45; J. E. Huxley, 'The Treatment of Melancholia with Refusal of Food', *The Asylum Journal*, 1, no. 13 (1855): 198–200.

87. List of Private Lunatics in the Asylum at Barming Heath, 1850–1855, Kent Library and Archives Centre, Q/GLP/2/21–34.

88. *L&L*, I, 248.

89. Kent County Lunatic Asylums, *Annual Report, 1907* (Maidstone: W. P. Dickinson, 1908), Kent County Library and Archives Centre, MH/T3/Aa5; Ernest W. White, 'James Edmund Huxley M.D.St.And.', *Journal of Mental Science*, 53, no. 221 (April 1907): 419.

90. Her obituary indicates a first Royal Academy exhibition in 1879, 'The Sins of the Fathers', 'favourably placed'. She also exhibited in 1881, including a full-length portrait of her sister Henrietta. At the Grosvenor Gallery, she exhibited child portraits and *By the Tideless Midland Sea*. Marian Collier, obituary, *The Times*, 23 November 1887.

91. THH to Michael Foster, [20 November 1887], Michael Foster and THH, Correspondence, Wellcome Trust Library, Letter 244.

92. Leonard Huxley to E. O. Lovett, 28 September 1913, President Edgar Odell Lovett's Papers, RU, UA 014, Box 43: 12.

93. JSH to Juliette Huxley, nd, J&JH Papers, RU, MS512, Box 37: 4.

94. JSH to Juliette Huxley, nd, J&JH Papers, RU, MS512, Box 37: 7.

95. And there was a physiology linked to his mind: 'Actions also want to disobey – result, conflict & tremor when not rigidly controlled; when relax, thoughts out of control again. When try to feel alive, nonsense words frame themselves. Haywire – needs typing and & can't find ends. A lot of regression.' JSH to Juliette Huxley, n.d., J&JH Papers, RU, MS512, Box 37: 7. Original emphasis.

96. JSH to Juliette Huxley, 29 June 1966, J&JH Papers, RU, MS512, Box 37: 7. Original emphasis.

97. JSH, *Memories* I, 98; JSH to Juliette Huxley, 20 April 1966, J&JH Papers, RU, MS512, Box 37: 7.

98. JSH notes, 25 November 1957, J&JH Papers, RU, MS512, Box 37: 7.

99. Juliette Huxley, 2 February, 6 March, 17 March, 24 March 1966, Journal (1966), J&JH Papers, RU, MS512, Box 2.

100. JSH to Hubert (Winthrop Young), 29 January 1932, J&JH Papers, RU, MS512, Box 37: 3.

101. Julian Huxley, *Education and the Humanist Revolution* (Southampton: University of Southampton, 1962), 8.

102. Notebooks of THH, 30 January 1842, THHC, ICL, Series 3, 31: 179.

103. Notebooks of THH, THHC, ICL, Series 3, 31: 178–9.

104. THH, 'Thoughts and Doings', 22 November 1840, THHC, ICL, Series 3, 31: 169.

105. THH, January–April 1856, printed syllabus of twelve lectures to the Royal Institution of Great Britain, 1856, on Physiology and Comparative Anatomy, THHC, ICL, Series 20, vol. 122: 8.

106. Thomas Capern, *Pain of Body and Mind Relieved by Mesmerism: A Record of Mesmeric Facts* (London: Bailliere, 1861).

107. Desmond, *Huxley*, 557.

108. Royal Commission on Vivisection [1875], *Report . . . with Minutes of Evidence* (London: George Eyre and William Spottiswoode, 1876), question 3209–11.

109. Henrietta Huxley, n.d., Reminiscences of early married life, THHC, ICL, Series 3, Box 31: 92, 95.

110. THH, 'On Descartes' "Discourse Touching the Method of Using One's Reason Rightly and of Seeking Scientific Truth"' (1870), in THH, *Method and Results*, 191.

111. THH, 'On the Physical Basis of Life' (1868), in THH, *Method and Results*, 154.

112. 'Andrew F. Huxley – Physiology or Medicine', 1963. NobelPrize.org.

113. Robert J. Richards, *Darwin and the Emergence of Evolutionary Theories of Mind and Behavior* (Chicago: University of Chicago Press, 1987).

114. Darwin, *On the Origin of Species*, 488.

115. Herbert Spencer, 'Morals and Moral Sentiments', *Fortnightly Review*, 9 n.s., no. 52 (1871): 419–32.

116. See Contents, Charles Darwin, *The Expression of the Emotions in Man and Animals* (London: John Murray, 1872), iv.

117. Ernest Jones, *Papers on Psychoanalysis* (London: Balliere, Tyndall and Fox, 1913), xii.

118. JSH, *Religion without Revelation*, 107–8, 104.

119. JSH to Kathleen [Fordham], 30 June 1917, J&JH Papers, RU, MS512, Box 37: 11.

120. JSH personal note, 29 November 1936, on Zoological Gardens letter-head, J&JH Papers, RU, MS512, Box 37: 4.

121. JSH, 'Evolutionary Ethics' in THH and JSH, *Evolution and Ethics*, 107, 110.

122. Marked copy of Marjorie Brierley, 'Present Tendencies in Psycho-analysis', *British Journal for Medical Psychology*, 14, no. 3 (1932): 216, JSH Papers, RU, MS50, Box 121: 5.

123. JSH, *Memories*, vol. II (New York: Harper & Row, 1973) [hereafter *Memories* II], 107.

124. JSH, *Unesco: Its Purpose and Philosophy* (Paris: Unesco, 1946), 36.

125. Julian Huxley, Ernst Mayer, Humphry Osmond and Abram Hoffer, 'Schizophrenia as a Genetic Morphism', *Nature*, 204, no. 4955 (1964): 220–21.

126. Francis Huxley, 'Marginal Lands of the Mind', in JSH (ed.), *The Humanist Frame* (London: George Allen & Unwin, 1961), 169–79. See also Aldous Huxley's chapter, 'Human Potentialities', 417–32.

127. Sumita Roy, Annie Pothen and K.S. Sunita (eds.), *Aldous Huxley and Indian Thought* (New Delhi: Sterling, 2003).

128. A. David Napier, 'Francis Huxley obituary', *Guardian*, 20 December 2016; Francis Huxley, *Affable Savages: An Anthropologist Among the Urubu Indians of Brazil* (New York: Viking, 1957); Francis Huxley, *The Invisibles: Voodoo Gods in Haiti* (London: Rupert Hart-Davis, 1966); Jeremy Narby and Francis Huxley, *Shamans Through Time: 500 Years on the Path to Knowledge* (New York: Penguin, 2001).

129. Francis Huxley, 'The Ritual of Voodoo and the Symbolism of the Body', *Philosophical Transactions of the Royal Society of London*, B, 251, no. 772 (1966): 427.

130. Francis Huxley, 'The Body and the Mind', 17 February 1966, R. D. Laing Collection, Special Collections Department, University of Glasgow Library, MS Laing L 226. See also Francis Huxley, 'Anthropology and ESP', in J. R. Smythies (ed.), *Science and ESP* (London: Routledge and Kegan Paul, 1967), 281–302; 'Psychoanalysis and Anthropology', in Peregrine Horden (ed.), *Freud and the Humanities* (London: Duckworth, 1985), 130–51; 'The Liberating Shaman of Kingsley Hall', obituary for R. D. Laing, *Guardian*, 25 August 1989.

131. JSH, typescript of preface to revised edition of *Living Thoughts of Darwin*, January 1958, JSH Papers, RU, MS50, Box 74: 5.

132. Cynthia Carson Bisbee et al. (eds.), *Psychedelic Prophets: The Letters of Aldous Huxley and Humphry Osmond* (Montreal: McGill-Queen's University Press, 2018).

133. For Osmond's research, see Matthew Oram, *The Trials of Psychedelic Therapy: LSD Psychotherapy in America* (Baltimore: Johns Hopkins University Press, 2018).

134. Laura Archera Huxley, *This Timeless Moment: A Personal View of Aldous Huxley* (London: Chatto & Windus, 1969), 295.

135. Francis Huxley, *The Eye – Seer and the Seen* (London: Thames and Hudson, 1990); Ron Roberts and Theodor Itten, *Francis Huxley and the Human Condition: Anthropology, Ancestry, Knowledge* (London: Routledge, 2020), 7, 198.

CHAPTER 4: CREATURES OF THE SEA AND SKY

1. Owen Stanley, *Whales Amusing Themselves*, watercolour from voyage of the H.M.S. *Rattlesnake*, 1846–1849, Mitchell Library, State Library of New South Wales, SAFE/PXC 281, f. 70.

2. THH, *The Oceanic Hydrozoa* (London: Ray Society, 1859), vii.

3. THH, *Rattlesnake Diary*, 113, 118.

4. Carl von Linné, trans. William Turton, *A General System of Nature*, vol. IV: *Worms* (1758) (London: Lackington, Allen & Co., 1806), 3–4.

5. Charles Bell, *The Hand; Its Mechanism and Vital Endowments as Evincing Design* (London: William Pickering, 1833), 309.

6. THH, *Rattlesnake Diary*, 15 June 1850, 329.

7. THH, 'Zoological Notes and Observations Made on Board H.M.S. *Rattlesnake*, Pt. 3: Upon *Thalassicolla*, a New Zoophyte', *Annals and Magazine of Natural History*, 2nd series, 8, no. 48 (1851): 433–42.

8. THH, *Rattlesnake Diary*, February 1848, 103.

9. THH, 'On the Anatomy and the Affinities of the Family of the Medusae', *Philosophical Transactions of the Royal Society of London*, 139 (1849): 413–34.

10. JSH comments in THH, *Rattlesnake Diary*, 66.

11. George J. Allman, testimonial for THH, 17 October 1851, THHC, ICL, Series 3, 31.69.

12. Richard Owen, testimonial for THH, 17 October 1851, THHC, ICL, Series 3, 31.72.

13. THH to Professor Leuckart, 30 January 1859, in *L&L*, I, 162–3; THH, 'A Lobster; Or The Study of Zoology' (1861), in *Discourses: Biological and Geological. Collected Essays* (London: Macmillan, 1894), 209.

14. Marsha L. Richmond, 'T. H. Huxley's Criticism of German Cell Theory', *Journal of the History of Biology*, 33, no. 2 (2000): 247–89.

15. JSH and Douglas Cleverdon, *Julian Huxley on T. H. Huxley: A New Judgment* (London, Watts & Co., 1945), 13.

16. Mary P. Winsor, *Starfish, Jellyfish, and the Order of Life: Issues in Nineteenth Century Science* (New Haven: Yale University Press, 1976).

17. THH, *The Oceanic Hydrozoa*, 22.

18. THH, *The Oceanic Hydrozoa*, 2.

19. THH, *The Oceanic Hydrozoa*, x.

20. THH, *The Oceanic Hydrozoa*, v.

21. Richard Owen, *Memoir on the Pearly Nautilus* (London: Richard Taylor 1832); THH, 'On Some Points in the Anatomy of *Nautilus pompilius*', *Journal of the Proceedings of the Linnaean Society*, August 1858, offprint, THHC, ICL, Series 5, Biology, Miscellaneous Papers, 1878–1893, 40. 1.

22. THH to Henrietta Heathorn, 23 September 1851, THHC, ICL, Series 22, Letter 166.

23. THH, Paper relating to Fisheries Commissions and Scottish Fishery Board, 11 May 1860, 20–22 August 1862, THHC, ICL, Series 7, 43.1, 3–4, 6; THH, Fisheries, questionnaire and report, 5 January 1858, THHC, ICL, Series 7, 43.76.

24. THH, deep sea soundings notebook, 1857, notes on *Globigerina*, 1857, THHC, ICL, Series 19, 116.1; THH, 'On Some Organisms Living at Great Depths in the North Atlantic Ocean', *Quarterly Journal of Microscopical Science*, 8 (1868): 203–12.

25. THH, 'On the Recent Work of the "Challenger" Expedition, and Its Bearing on Geological Problems', *Proceedings of the Royal Institution*, 7 (1875): 354–7; THH, 'On Some of the Results of the Expedition of H.M.S. "Challenger"', *The Contemporary Review*, 25 (March 1875): 639–60; THH, 'Notes from the "Challenger"', *Nature*, 12, no. 303 (19 August 1875): 315–16.

26. THH, 'The Problems of the Deep Sea' (1873), in THH, *Discourses: Biological and Geological: Collected Essays* (London: Macmillan, 1894), 51.

27. THH, *Rattlesnake Diary*, 16 June 1849, 186.

28. THH, 'Remarks upon *Archaeopteryx lithographica*', *Proceedings of the Royal Society*, 16 (30 January 1868): 248.

29. Brian Switek, 'Thomas Henry Huxley and the Reptile to Bird Transition', in Richard T. J. Moody et al. (eds.), *Dinosaurs and Other Extinct*

Saurians: A Historical Perspective (London: Geological Society, 2010), 251–63.

30. THH, *Lectures on the Elements of Comparative Anatomy*, 69.

31. THH to Ernst Haeckel, 21 January 1868, in *L&L*, I, 302–3.

32. Henrietta Huxley to JSH, 12 February 1903, JH Papers, RU, MS474, Box 9: 11.

33. JSH, *Memories* I, 66.

34. JSH, 'Metamorphosis of Axolotl Caused by Thyroid-feeding', *Nature*, 104 (1920): 435.

35. JSH, 'Metamorphosis of Axolotl', 435.

36. Juvenilia, JH Papers, RU, MS474, Box 23: 3; JSH, 'The Courtship Habits of the Great Crested Grebe', 516.

37. JSH to Julia Huxley, 29 April 1901, JH Papers, RU, MS474, Box 9: 13.

38. Leonard Huxley to JSH, 17 May 1896, JH Papers, RU, MS474, Box 10: 1.

39. Julia Huxley to JSH, 11 May 1903, JH Papers, RU, MS474, Box 9: 13.

40. Aldous Huxley to JSH, 7 December 1902, JH Papers, RU, MS474, Box 11: 2.

41. JSH, '1901–1903, Surrey and Hertfordshire', notebook, Edward Grey Institute of Field Ornithology, Oxford, Julian Sorell Huxley collection, Box 1; Juvenilia, JH Papers, RU, MS474, Box 23: 3.

42. JSH, 'Courtship Activities in the Red-throated Diver (*Colymbus stellatus* Pontopp.); Together with a Discussion of the Evolution of Courtship in Birds', *Zoological Journal of the Linnean Society, Zoology*, 35, no. 234 (1923): 253–92; JSH, 'A Natural Experiment on the Territorial Instinct', *British Birds*, 27 (1934): 270–77.

43. Niko Tinbergen, 'On Aims and Methods of Ethology', *Zeitschrift für Tierpsychologie*, 20 (1963): 410–33.

44. Richard J. Burkhardt Jnr, 'Huxley and the Rise of Ethology', in Waters and van Helden (eds.), *Julian Huxley*, 127–49; see JSH reference to Edmund Selous, 'Observations Tending to Throw Light on the Question of Sexual Selection in Birds, Part 1', *The Zoologist*, 10 (1906), 201–19, in JSH, 'A First Account of the Courtship of the Redshank', *Proceedings of the Zoological Society of London*, 82, no. 3 (1912): 648.

45. Richards, *Darwin and the Making of Sexual Selection*; Erika L. Milam, *Looking for a Few Good Males: Female Choice in Evolutionary Biology* (Baltimore: Johns Hopkins University Press, 2010).

46. JSH, 'A First Account of the Courtship of the Redshank', 647–55; JSH to Kathleen Fordham, 19 April 1911, JSH Papers, RU, MS 50, Box 162: 1.

47. JSH to Kathleen Fordham. See Burkhardt, 'Huxley and the Rise of Ethology', 130–31.

48. See John R. Durant, 'The Tension at the Heart of Huxley's Evolutionary Ethology', in Waters and Van Helden (eds.), *Julian Huxley*, 159; Richards, *Darwin and the Making of Sexual Selection*, 530–31.

49. Mary M. Bartley, 'Courtship and Continued Progress: Julian Huxley's Studies on Bird Behavior', *Journal of the History of Biology*, 28 (1995): 91–108; Burkhardt, 'Huxley and the Rise of Ethology', 127–49; JSH, 'Darwin's Theory of Sexual Selection and the Data Subsumed by It, in the Light of Recent Research', *American Naturalist*, 72 (1938): 416–33.

50. JSH, 'The Courtship Habits of the Great Crested Grebe', 516.

51. JSH, 'Courtship Activities in the Red-throated Diver', 286.

52. JSH, 'The Courtship Habits of the Great Crested Grebe', 517, 520, 544.

53. JSH, 'The Courtship Habits of the Great Crested Grebe', 507.

54. JSH, Birdwatching notes, 'April', JSH Papers, RU, MS 50, Box 56: 2.

55. Nigel Sitwell to JSH, 14 August 1964, JSH Papers, RU, MS 50, Box 37: 2.

56. JSH, 'The Courtship Habits of the Great Crested Grebe', 495, 510.

57. JSH, 'The Courtship Habits of the Great Crested Grebe', 559, 510.

58. JSH, *Memories* I, 103–05.

59. Lantern slides: two boxes on Spitzbergen, one box on egrets and herons, others on courtship and behaviour generally and various birds, Edward Grey Institute of Field Ornithology, Oxford, Julian Sorell Huxley collection, Box 8.

60. Charles Elton to Juliette Huxley, 22 March 1975, JH Papers, RU MS 474, Box 1: 26.

61. JSH comments in THH, *Rattlesnake Diary*, 53–5.

62. JSH, *Memories* II, 92, 178–9.

63. THH, Lectures on Natural History at the School of Mines. Manuscripts, Lectures 1866–8, vol. 59.

64. THH, 'A Lobster', 199, 202.

65. THH, 'A Lobster', 204–206.

66. THH, *The Crayfish. An Introduction to the Study of Zoology* (New York: D. Appleton & Co., 1880), 17–31, 48, 205, 221.

67. THH, *Crayfish*, 284–6, 318.

68. THH, *Crayfish*, 286, 317.

69. THH, *Crayfish*, 338.

70. Bernard Lightman, 'Darwin and the Popularization of Evolution', *Notes and Records of the Royal Society*, 64, no. 1 (2010): 5–24.

71. Charles Darwin to THH, 5 November 1864, DCP, Letter 4661.

72. JSH, 'Growing up with a legend', handwritten notes, no date, JSH Papers, RU, MS 50, Box 63: 3.

73. JSH to H. G. Wells, 24 July 1925, H. G. Wells Papers, Rare Book & Manuscript Library, University of Illinois at Urbana-Champaign, 01/01/MSS00071.

74. JSH, *Memories* I, 165.

75. H. G. Wells to JSH, 11 April 1926, JSH Papers, RU, MS 50, Box 9: 1.

76. JSH, *Memories* I, 159–60.

77. H.G. Wells to JSH, 17 November 1928, JSH Papers, RU, MS 50, Box 9: 8.

78. H.G. Wells to JSH, 3 October 1928, JSH Papers, RU, MS 50, Box 9: 8.

79. H.G. Wells to JSH, 22 December 1927, in JSH, *Memories* I, 158.

80. H.G. Wells to JSH, 4 July 1928, JSH Papers, RU, MS 50, Box 9: 8.

81. 'Thought in the Night', November 1928, JSH Papers, RU, MS 50, Box 9: 8.

82. H. G. Wells, JSH and G. P. Wells, *The Science of Life*, vol. III (London: Waverley, 1931), 976.

83. Wells, JSH and Wells, *Science of Life*, III, 641.

84. H. G. Wells, JSH and G. P. Wells, *The Science of Life*, vol. I (London: Waverley, 1931), 275. Original emphasis.

85. Wells, JSH and Wells, *Science of Life*, I, 151.

86. Wells, JSH and Wells, *Science of Life*, I, 152.

87. Wells, JSH and Wells, *Science of Life*, I, 145–6, 150.

88. THH, 'The Herring', a lecture delivered at the National Fishery Exhibition, Norwich, 21 April 1881, *The Popular Science Monthly*, 19 (August 1881): 433–50.

89. Wells, JSH and Wells, *Science of Life*, III, 663–4.

90. Juliette Huxley to JSH, 13 November, no year, J&JH Papers, RU, MS 512, Box 37: 5. Original emphasis.

91. JSH, 'Egrets in Louisiana', *Country Life*, 18 June 1921; JSH, 'Bird's Play and Pleasure', *New Statesman*, 19 January 1924.

92. JSH to Nigel Sitwell, 18 August 1964, JSH Papers, RU, MS 50, Box 37: 2.

93. Walt Disney to JSH, 12 November 1964, JSH Papers, RU, MS 50, Box 37: 5. See Marianne Sommer, 'Animal Sounds against the Noise of Modernity and War: Julian Huxley (1887–1975) and the Preservation of the Sonic World Heritage', *Journal of Sonic Studies*, 13 (2017): 10.

94. David Attenborough to Juliette Huxley, 4 September 1986, JH Papers, RU, MS474, Box 1: 3.

95. *Zoo Babies* (Strand Film Company, 1938), director Evelyn Spice, supervised by Julian Huxley, British Film Institute National Archives, 12443; *Monkey into Man* (Strand Film Company, 1938), director Stanley Hawes, supervised by Julian Huxley, British Film Institute National Archives, 13048. See also Marianne Sommer, *History Within: The Science, Culture and Politics of Bones, Organisms and Molecules* (Chicago: University of Chicago Press, 2016), 174.

96. *Free to Roam* (Strand Film Company, 1938), director Paul Burnford, supervised by Julian Huxley, British Film Institute National Archives, 12469.

97. *The Private Life of the Gannets* (London Film Productions, 1934), executive producer Alexander Korda, produced and script by Julian Huxley, British Film Institute National Archives, 11147. 'His great contribution to natural history film making was undoubtedly the gannet film – and that certainly must consciously or unconsciously have influenced anyone such as myself who subsequently tried to do the same sort of thing.' David Attenborough to Alison Bashford, 27 December 2021. See also Gregg Mitman, *Reel Nature: America's Romance with Wildlife on Film* (Cambridge, MA: Harvard University Press, 1999).

98. *Animals of the Rocky Shore* (Gaumont-British Instructional, 1937), director J. V. Durden, supervisor Julian Huxley, British Film Institute National Archives, 11269; *The Earthworm* (Gaumont-British Instructional, 1936), director H. R. Hewer, supervisor Julian Huxley, British Film Institute National Archives, 12046.

99. David Attenborough, *Life on Air: Memoirs of a Broadcaster* (London: BBC Books, 2002), 15, 31–2.

100. David Attenborough to Juliette Huxley, 19 February 1975 and 4 September 1986, JH Papers, RU, MS474, Box 1: 3.

CHAPTER 5: ANIMAL POLITICS

1. THH, 'Notes on biology by F. Field', May 1876, THHC, ICL, Series 5, 40.239. Original emphasis.

2. Harriet Ritvo, *The Animal Estate: The English and Other Creatures in the Victorian Age* (Cambridge, MA: Harvard University Press, 1987); Angelique Richardson (ed.), *After Darwin: Animals, Emotions, and the Mind* (Amsterdam: Rodopi, 2013); Rob Boddice, *The Science of*

Sympathy: Morality, Evolution, and Victorian Civilization (Urbana: University of Illinois Press, 2016).

3. Jeremy Bentham, letter to Editor, *Morning Chronicle*, 4 March 1825; Peter Singer, *Animal Liberation: A New Ethics for Our Treatment of Animals* (New York: Random House, 1975).

4. THH, 'Autobiography', 7.

5. THH, *Crayfish*, 87–8.

6. THH, 'On the Hypothesis That Animals Are Automata, and Its History' (1874), in THH, *Method and Results*, 199–250.

7. Herbert Spencer, *Justice: Being Part IV of the Principles of Ethics*, vol. II (London: Williams & Norgate, 1891), 1–24.

8. Herbert Spencer, 'On Justice', *Nineteenth Century*, March 1890 clipping, THHC, ICL, Series 9, 45.22–5.

9. Charles Darwin, *The Expression of the Emotions in Man and Animals* (London: John Murray, 1872).

10. Catherine Marshall, Bernard Lightman and Richard England (eds.), *The Metaphysical Society (1869–1880): Intellectual Life in Mid-Victorian England* (Oxford: Oxford University Press, 2019).

11. William Connor Magee to his wife, 13 February 1873, in MacDonnell John Cotter, *The Life and Correspondence of William Connor Magee*, vol. I (London: Isbister, 1896) 284.

12. THH, 'On the Hypothesis That Animals Are Automata'; THH, 'Has a Frog a Soul?' (1870), in Catherine Marshall, Bernard Lightman and Richard England (eds.), *The Papers of the Metaphysical Society 1869–1880*, vol. I (Oxford: Oxford University Press, 2015), 174–84.

13. THH, 'On the Hypothesis That Animals Are Automata', 201.

14. Bentham, letter to Editor, *Morning Chronicle*, 4 March 1825.

15. Frances Power Cobbe, 'Darwinism in Morals', *Theological Review*, 33 (April 1871): 167–92.

16. Frances Power Cobbe to Charles Darwin, 26 November 1872, DCP, Letter 8649; Frances Power Cobbe, 'The Consciousness of Dogs', *Quarterly Review*, 133 (October 1872): 419–51.

17. Charles Darwin to THH, 14 January 1875, DCP, Letter no. 9817.

18. THH to Charles Darwin, 19 May 1875, DCP, Letter no. 9985.

19. David Feller, 'Dog Fight: Darwin as Animal Advocate in the Antivivisection Controversy of 1875', *Studies in History and Philosophy of Biological and Biomedical Sciences*, 40, no. 4 (2009): 265–71.

20. Evidence of Arthur de Noé Walker, *Report of the Royal Commission on the Practice of Subjecting Live Animals to Experiments for*

Scientific Purposes (London: George Edward Eyre and William Spottiswoode, 1876), Question 1768, 93 [hereafter *Royal Commission on Vivisection*].

21. Evidence of John Anthony, *Royal Commission on Vivisection*, Q2537–48, 135–6.

22. Evidence of John Anthony, *Royal Commission on Vivisection*, Q2596, 137.

23. THH, *Lessons in Elementary Physiology*, 8th edn (London: Macmillan, 1874), 252.

24. 'Report on the Debate on Vivisection in the House of Lords on Moving Lord Carnarvon's Cruelty to Animals Bill', *The Times*, 23 May 1876, 6; THH, Professor Huxley on Lord Shaftesbury, letter to Editor, *The Times*, 26 May 1876, 10; Evidence of Mr G. R. Jesse, *Royal Commission on Vivisection*, Q6507, 322.

25. Friedrich Leopold Göltz, *Wider die Humanaster. Rechtfertigung eines Vivisektors* (Strasbourg: Karl J. Trübner, 1883).

26. THH, 'On the Hypothesis That Animals Are Automata', 225.

27. Evidence of Charles Darwin, *Royal Commission on Vivisection*, Q4668–72, 234.

28. Evidence of John Anthony, *Royal Commission on Vivisection*, Q2602–03, 138.

29. Evidence of Sir William Withey Gull, *Royal Commission on Vivisection*, Q5522–38, 269.

30. Evidence of John Colam, *Royal Commission on Vivisection*, Q1581–89, 82.

31. Evidence of David Ferrier, *Royal Commission on Vivisection*, Q3283–89, 172.

32. Carmela Morabito, 'David Ferrier's Experimental Localization of Cerebral Functions and the Anti-Vivisection Debate', *Nuncius,* 32 (2017): 146–65.

33. Charles Brown-Séquard to THH, 12 February 1874, THHC, ICL, Series 20, 121/28.

34. For example, Evidence of John Colam, *Royal Commission on Vivisection*, Q1698–1700, 88.

35. Evidence of Arthur de Noé Walker, *Royal Commission on Vivisection*, Q1741–46, 92.

36. Evidence of Alfred Henry Garrod, *Royal Commission on Vivisection*, Q1997–99, Q2002, 107.

37. Evidence of Frederick William Pavy and Phillip H. Pye-Smith, *Royal Commission on Vivisection*, Q2113, 111.

38. Evidence of Arthur de Noé Walker, *Royal Commission on Vivisection*, Q1736, 91.

39. 'The Sin of Vivisection', n.d. leaflet, THHC, ICL, Series 11, 49.145–6.

40. 'Thoughts on Vivisection: Vivisection in Demoniacal Physiology', THHC, ICL, 22 March 1875, Series 11, 49.152.

41. *Final Report of the Royal Commission on Vivisection* (London: His Majesty's Stationery Office, 1912); *Protection of Animals Act 1911* (1911 Chapter 27 1 & 2 Geo 5).

42. B. E. Kidd, 'Anti-vivisectors and the Zoo', letter to the Editor, *Spectator*, 7 September 1934; Joe Cain, 'Julian Huxley, General Biology and the London Zoo, 1935–42', *Royal Society Journal of the History of Science*, 64, no. 4 (2010): 359–78.

43. THH to J. D. Hooker, 6 July 1855, in *L&L*, I, 185, see also 65, 127, 128.

44. William Hamilton Drummond, *The Rights of Animals, and Man's Obligation to Treat Them with Humanity* (London: John Mardon, 1838).

45. *Blackwood's* magazine, June 1837, quoted in John Styles, *The Animal Creation: Its Claims on Our Humanity. Stated and Enforced* (London: Thomas Ward & Co, 1839), 346.

46. Henry Scherren, *The Zoological Society of London* (London: Cassell, 1905), 85.

47. Charles Darwin, Notebook C, Transmutation of species [1838.02–1838.07], CUL-DAR122, Paragraphs 79 and 196–7, Darwin Online.

48. [Thomas Boreman], *A Description of Some Curious and Uncommon Creatures* (London: Printed for Richard Ware and Thomas Boreman, 1739), 24.

49. Wilfrid J. W. Blunt, *The Ark in the Park: The Zoo in the Nineteenth Century* (London: Hamish Hamilton, 1976).

50. John S. Allen, et al., 'The Chimpanzee Tea Party: Anthropomorphism, Orientalism and Colonialism', *Visual Anthropology Review*, 10, no. 2 (1994): 48.

51. Cain, 'Julian Huxley, General Biology, and the London Zoo', 363.

52. Receipt from Gerrard's, 8 July 1935, JSH Papers, RU, MS50, Box 120: 6.

53. JSH, *Memories* I, 230, 235.

54. JSH, *Memories* I, 231.

55. Cain, 'Julian Huxley, General Biology and the London Zoo', 359–78.

56. JSH, *At the Zoo* (London: G. Allen & Unwin, 1936).

57. JSH, *Memories* I, 231–3.

58. Sommer, 'Animal Sounds', 3.

59. JSH, *Memories* I, 249–50.

60. JSH, *Memories* I, 253–6.

61. Charles Elton to Julian and Juliette Huxley, 21 March 1969, JH Papers, RU, MS474, Box 1: 26.

62. *Kingdom of the Beasts*, text by JSH, photographs by W. Suschitzky (London: Thames & Hudson, 1956).

63. [J. B. Morton and T. H. White], 'Man Versus the Rest', review of *Kingdom of the Beasts*, *Times Literary Supplement*, 13 July 1956.

64. JSH to Solly Zuckerman [hereafter SZ], 29 June 1957; and JSH to SZ, 15 May 1959, SZ Papers, UEA Archives, SZ/GEN/HUX.

65. JSH to Hugh Casson, 26 April 1960, SZ Papers, UEA Archives, SZ/GEN/HUX.

66. JSH to SZ, 20 March 1967, SZ Papers, UEA Archives, SZ/GEN/HUX.

67. Robin W. Doughty, *Feather Fashions and Bird Preservation: A Study in Nature Protection* (Berkeley: University of California Press, 1975), 5–6.

68. JSH correspondence with the Royal Society for the Protection of Birds, 1920–1972, JSH Papers, MS50, RU, Box 116: 7.

69. Peter Marren, *The New Naturalists* (London: Harper Collins, 1995).

70. *Conservation of Nature in England and Wales. Report*, National Parks Committee, Wild Life Conservation Special Committee (England and Wales) (London: His Majesty's Stationery Office, 1947).

71. JSH, *Memories* I, 179.

72. JSH, *Memories* I, 188. Original emphasis.

73. This trip is summarized in JSH, *Memories* I, 179–96.

74. Colonial Office: Colonial Higher Education Commission (Asquith Commission, 1943–4), The National Archives, Kew, CO 958.

75. Tim Livsey, *Nigeria's University Age: Reframing Decolonisation and Development* (London: Palgrave Macmillan, 2017), 19. Thanks to Helen Tilley for this reference.

76. JSH, 'Poaching: The Shocking Slaughter of Africa's Wild Life', *UNESCO Courier*, September 1961, 8.

77. The Covenant of the League of Nations, Article 22.

78. Address to the Third Session of the United Nations General Assembly, 3 November 1948, by Pandit Jawaharlal Nehru.

79. JSH, 'Africa's Wild Life in Peril', *UNESCO Courier*, September 1961, 4.

80. JSH, 'Poaching', 8.

81. Ramachandra Guha, 'The Authoritarian Biologist and the Arrogance of Anti-humanism: Wildlife Conservation in the Third World', in Vasant

K. Saberwal and Mahesh Rangarajan (eds.), *Battles over Nature: Science and the Politics of Conservation* (Delhi: Permanent Black, 2003), 139–57.

82. JSH, 'Introduction', Juliette Huxley, *Wild Lives of Africa* (London: Collins, 1963), 5.

83. Juliette Huxley, *Wild Lives*, quoted in JSH, *Memories* I, 192.

84. Juliette Huxley, *Wild Lives*, 194.

85. JSH, 'Africa's Wild Life in Peril', 4.

86. JSH, *Memories* II, 210–19, 252.

87. JSH, 'Poaching', 10, 13.

88. Alison Bashford, *Global Population: History, Geopolitics and Life on Earth* (New York: Columbia University Press, 2014).

89. JSH, 'Africa's Wild Life in Peril', 4.

90. William M. Adams and Martin Mulligan (eds.), *Decolonizing Nature: Strategies for Conservation in a Post-Colonial Era* (London: Earthscan, 2003).

91. Juliette Huxley, *Wild Lives*, 192.

92. JSH, *Memories* I, 106.

93. Juliette Huxley, *Wild Lives*, 191.

94. Juliette Huxley, *Wild Lives*, 196–7; JSH, 'Introduction', Joy Adamson, *Living Free: The Story of Elsa and Her Cubs* (London: Collins & Harvill, 1961), xii; Joy Adamson to JSH, 5 May 1961, JH Papers, RU, MS474, Box 1: 2.

95. Juliette Huxley, *Wild Lives*, 195.

96. Gerald Durrell, *The Ark's Anniversary* (London: Fontana, 1991), 24–5.

97. JSH, *Memories* II, 101.

98. African photos, Album, J&JH Papers, RU, MS512, Box 23: 4; JSH, 'Introduction', to *Living Free*, ix–xiii.

CHAPTER 6: PRIMATES

1. Andrew Pulver, interview with Wolfgang Suschitzky, 'My Best Shot', *Guardian*, 16 July 2009.

2. Adrian Desmond, *The Ape's Reflexion* (London: Blond and Briggs, 1979).

3. THH, *Evidence as to Man's Place in Nature*; Charles Darwin to THH, 18 [February 1863], DCP, Letter 3996

4. Jane Goodall to JSH, 16 September 1971, JSH Papers, RU, MS50, Box 37: 1; see also Shirley C. Strum and Linda Marie Fedigan (eds.), *Primate*

Encounters: Models of Science, Gender and Society (Chicago: University of Chicago Press, 2000).

5. *L&L*, I, 194.

6. See also Erika Lorraine Milam and Robert A. Nye (eds.), *Scientific Masculinities* (Chicago: University of Chicago Press, 2016).

7. Christopher Cosans, 'Anatomy, Metaphysics and Values: The Ape Brain Debate Reconsidered', *Biology and Philosophy*, 9 (1994): 129–65.

8. Richard Owen, 'Osteological Contributions to the Natural History of Chimpanzees (*Troglodytes*) and Orangs (*Pithecus*). No. IV.', *Transactions of the Zoological Society*, 4 (1853): 75–88.

9. Nicolaas A. Rupke, *Richard Owen: Biology without Darwin* (Chicago: University of Chicago Press, 2009).

10. [Richard Owen], Review of *Origin* and other works, *Edinburgh Review*, 111 (April 1860): 487–532.

11. Primates notebook, *c.* 1859–64, THHC, ICL, Series 18, vol. 14, 27–8, 81, 123, 139.

12. THH, 'On the Zoological Relations of Man with the Lower Animals', *Natural History Review*, 1 (n.s.) (1861): 67–84.

13. Allen Thomson to THH, 24 May 1860, THHC, ICL, Series 19, Primates notebook, 1860–61, vol. 118, 99–100.

14. THH, 'On the Zoological Relations of Man with the Lower Animals'; THH, 'Man and the Apes', letters to the *Athenaeum*, 30 March and 21 September 1861, 433 and 498; THH, 'The Brain of Man and Apes', letter to *Medical Times & Gazette*, 25 October 1862, 449; THH, 'On Some Fossil Remains of Man', *Proceedings of the Royal Institution of Great Britain*, 3 (1858–62): 420–22.

15. Filippo Pigafetta, *A Report of the Kingdome of Congo, a Region of Africa* (London: John Wolfe, 1597), 89.

16. *Purchas His Pilgrimes*, quoted in THH, *Evidence as to Man's Place in Nature*, 3.

17. *Purchas His Pilgrimes*, quoted in THH, *Evidence as to Man's Place in Nature*, 4–6.

18. Joanna Bourke, *What It Means to Be Human: Reflections from 1791 to the Present* (London: Virago, 2011), 31–43.

19. Samuel Pepys, *The Diary of Samuel Pepys*, vol. II, ed. R. Latham and W. Matthews (London: Bell & Hyman, 1985), 160. See Caroline Grigson, *Menagerie: The History of Exotic Animals in England* (Oxford: Oxford University Press, 2016), 38.

20. Gregory Radick, *The Simian Tongue: The Long Debate about Animal Language* (Chicago: University of Chicago Press, 2008).

21. THH, *Evidence as to Man's Place in Nature*, 8.

22. Edward Tyson, *Orang-outang, sive Homo Sylvestris; or the Anatomy of a Pygmie compared with that of a* Monkey, *an* Ape, *and a* Man (London: Thomas Benett, 1699), dedicatory preface, no page.

23. Tyson, *Orang-outang*, preface, no page.

24. Peter Bowler, *Progress Unchained: Ideas of Evolution, Human History and the Future* (Cambridge: Cambridge University Press, 2021).

25. THH, *Evidence as to Man's Place in Nature*, 13.

26. Robert Hartmann, *Anthropoid Apes* (London: Kegan Paul, Trench, 1885), 267.

27. THH, *Evidence as to Man's Place in Nature*, 13.

28. Georges Louis Leclerc de Buffon, *Buffon's Natural History*, vol. IX (London: Symonds, 1797), vol. IX, 186.

29. THH, *Evidence as to Man's Place in Nature*, 17.

30. Richard Owen, 'On the Osteology of the *Chimpanzee* and *Orang Outan*', *Transactions of the Zoological Society of London*, 1, no. 4 (December 1835): 343–79.

31. Leclerc de Buffon, *Buffon's Natural History*, IX, 109.

32. Londa Schiebinger, *Nature's Body: Gender in the Making of Modern Science* (Boston: Beacon Press, 1993), 94–8.

33. THH, *Evidence as to Man's Place in Nature*, 49.

34. William Smith, 'A New Voyage to Guinea' (1744), quoted in THH, *Evidence as to Man's Place in Nature*, 11–12.

35. Donna Haraway, *Primate Visions: Gender, Race, and Nature in the World of Modern Science* (New York: Routledge, 1989).

36. Charles Waterton, 'The Monkey Family', in *Essays on Natural History* (London: Frederick Warne & Co., 1871), 170.

37. *Proceedings of the Zoological Society of London*, 17 April 1877, 303–4; see also Holly Dugan, 'Renaissance Gorillas', *Criticism*, 62, no. 3 (2020): 387–410.

38. Richard Owen, 'Osteological Contributions to the Natural History of the Chimpanzee (*Troglodytes*, Geoffroy), Including the Description of the Skull of a Large Species (*Troglodytes gorilla*, Savage) Discovered by Thomas S. Savage, M.D., in the Gaboon Country, West Africa', *Transactions of the Zoological Society of London*, 6 (1849): 381–422.

39. T. S. Savage and J. Wyman, 'Notice of the External Characters and Habits of Troglodytes Gorilla', *Boston Journal of Natural History*, 5, no. 4 (December 1847): 417–43; Geoffroy St Hilaire, 'Sur les rapports naturels du Gorille: Remarques faites à la suite de la lecture de M. Duvernoy', *Comptes rendus hebdomadaires des séances de l'Académie des*

Sciences, 36 (Paris, 1853), 933–6; Colin P. Groves, 'A History of Gorilla Taxonomy', in Andrea B. Taylor and Michele L. Goldsmith (eds.), *Gorilla Biology* (Cambridge: Cambridge University Press, 2003), 16–17.

40. Richard Owen, 'Description of the Cranium of an Adult Male Gorilla from the River Danger, West Coast of Africa . . . ', *Transactions of the Zoological Society of London*, 4, no. 3 (April 1853): 75–88.

41. Richard Owen, cited in Paul B. Du Chaillu, *Explorations and Adventures in Equatorial Africa* (London: John Murray, 1861), 366.

42. THH, *Evidence as to Man's Place in Nature*, 22.

43. Ted Gott and Kathryn Weir, *Gorilla* (London: Reaktion Books, 2013), 44.

44. Dr Burt and W. Turner, 'Exhibition of Three Skulls of the Gorilla, Received from M. Du Chaillu, with Observations Relative to Their Anatomical Features', *Proceedings of the Royal Society of Edinburgh*, 5 (1866): 341–50.

45. 'Gorillas at the National Museum', Frederick McCoy to Editor, *The Argus*, 20 June 1865.

46. Du Chaillu, *Explorations and Adventures*, iii.

47. Du Chaillu, *Explorations and Adventures*, 60.

48. Du Chaillu, *Explorations and Adventures*, 61–2, 434–5, 366–70.

49. THH, *Evidence as to Man's Place in Nature*, 24–5.

50. THH, *Evidence as to Man's Place in Nature*, 53–4; Stuart McCook, '"It May Be Truth, but It Is Not Evidence": Paul du Chaillu and the Legitimation of Evidence in the Field Sciences', *Osiris*, 11 (1996): 177–97.

51. [Philip Egerton, 'Gorilla'], 'Monkeyana: Am I a Man and a Brother?', *Punch*, 18 May 1861.

52. James L. Newman, *Encountering Gorillas: A Chronicle of Discovery, Exploitation, Understanding, and Survival* (Lanham, MD: Rowman & Littlefield, 2013), 33.

53. Hartmann, *Anthropoid Apes*, 220.

54. T. Alexander Barns, *The Wonderland of the Eastern Congo* (London and New York: G. P. Putnam's Sons, 1922).

55. JSH, Guy the Gorilla, typescript, 1958, JSH Papers, RU, MS 50, Box 74: 12.

56. JSH, 'Paintings by Chimpanzees', typescript article for *The New York Times*, 20 September 1957, JSH Papers, RU, MS 50, Box 74: 5.

57. C. R. Darwin, [Orang utans at] Zoological Gardens, 2 September 1838, CUL-DAR 191.1–2, in John van Wyhe (ed.), *The Complete Work of*

Charles Darwin Online, 2002– (http://darwin-online.org.uk/); see also John van Wyhe and Peter C. Kjærgaard, 'Going the Whole Orang: Darwin, Wallace and the Natural History of Orangutans', *Studies in History and Philosophy of Biological and Biomedical Sciences*, 51 (2015): 53–63.

58. Charles Darwin, *The Expression of the Emotions in Man and Animals* (London: John Murray, 1872), 95, 135, 138.

59. SZ, *Functional Affinities of Man, Monkeys and Apes* (London: Kegan Paul, Trench, Trubner, 1933).

60. SZ, *The Social Life of Monkeys and Apes* (London: Kegan Paul, 1932).

61. For the legacy of this kind of research, a generation later, see Erika L. Milam, *Creatures of Cain: The Hunt for Human Nature in Cold War America* (Princeton: Princeton University Press, 2019).

62. JSH to SZ, 19 August 1936, and JSH to SZ, 8 June 1936, SZ Papers, UEA Archives, SZ/GEN/HUX.

63. JSH, Guy the Gorilla typescript, 1958, 3.

64. Alyse Cunningham, 'A Gorilla's Life in Civilization', *Zoological Society Bulletin* [New York], 24 (1921): 118–24.

65. A photo courtesy of E. Reichenow, of a gorilla who died at ten months. The Yerkes do not make anything of this photo, or comment on it. Robert M. Yerkes and Ada W. Yerkes, *The Great Apes: A Study of Anthropoid Life* (New Haven: Yale University Press, 1929), 443, 448–9.

66. Julia Macdonald to JSH, 18 August 1964, JSH Papers, RU, MS 50, Box 37: 2.

67. JSH, Guy the Gorilla typescript.

68. Juliette Huxley, *Wild Lives*, 210–13.

69. Haraway, *Primate Visions*.

70. JSH in *Journal of the Royal College of Surgeons of Edinburgh*, cited in George B. Schaller, *The Year of the Gorilla* (Chicago: University of Chicago Press, 1964), 176.

71. Jane Goodall to JSH, 10 July 1964, JSH Papers, RU, MS 50, Box 37: 1.

72. JSH to Robert Hinde, 30 October 1964, JSH Papers, RU, MS 50 Box 37: 4.

73. Jane van Lawick Goodall, *My Friends the Wild Chimpanzees* (Washington, DC: National Geographic Society, 1967), 21.

74. Jane Goodall to JSH, 2 September 1964, JSH Papers, RU, MS 50, Box 37: 3.

75. Jane Goodall, *Through a Window: My Thirty Years with the Chimpanzees of Gombe* (Boston: Houghton Mifflin Company, 1990); Jane

Goodall, 'Unusual Violence in the Overthrow of an Alpha Male Chimpanzee at Gombe', in *Topics in Primatology 1: Human Origins*, (1992): 131–42. See also J. M. Williams et al., 'Causes of Death in the Kasekela Chimpanzees of Gombe National Park, Tanzania', *American Journal of Primatology*, 70, no. 8 (2008): 766–77.

76. JSH to SZ, 20 March 1962, SZ Papers, UEA, SZ/GEN/HUX.

77. SZ to JSH, 14 April 1962, JSH to SZ, 2 May 1962, JSH to SZ, 9 June 1962, SZ Papers, UEA, SZ/GEN/HUX.

78. JSH to SZ, 2 May 1962, JSH to SZ, 31 January 1963, SZ Papers, UEA, SZ/GEN/HUX.

79. SZ to JSH, 2 February 1963, SZ Papers, UEA, SZ/GEN/HUX.

80. Dr Jane Goodall, CBE, 'The Scientific Study of Primates and Its Impact on Contemporary World-Views', Royal Anthropological Institute of Great Britain, Huxley Memorial Medal and Lecture, 2002.

81. Charles Bell, *The Hand; Its Mechanism and Vital Endowments as Evincing Design* (London, William Pickering, 1833), iii, 45, 309, 209.

82. THH, *Evidence as to Man's Place in Nature*, 86.

83. JSH, 'From Fin to Fingers: The Evolution of Man's Hand', *Illustrated London News*, December 1930, 1138–9; 'Fingers and Thumbs', Series 2, *The Animal Kingdom*, Strand Film Zoological Productions, 1938.

84. Cast of a hand, said to be Thomas Henry Huxley's, THHC, ICL, Series 13.

85. Charlotte Wolff, with preface by Aldous Huxley, *Studies in Hand-Reading* (London: Chatto & Windus, 1936); Charlotte Wolff to 'Mr Chatto and Mr Windus', n.d., Charlotte Wolff Papers [hereafter CWP], Wellcome Library, London, PSY/WOL/2/27; Aldous Huxley, typescript preface, CWP, Wellcome Library, London, PSY/WOL/6/3/2.

86. Wolff, *Studies in Hand-Reading*, 75–80.

87. Handprints I, Aldous Huxley Family, CWP, Wellcome Library, London, PSY/WOL/2/1.

88. Charlotte Wolff, *The Human Hand* (London: Methuen, 1942), 68.

89. Charlotte Wolff, 'A Comparative Study of the Form and Dermatoglyphs of the Extremities of Primates', *Proceedings of the Zoological Society of London*, 108, no. 1 (April 1938): 143–61.

90. Superintendent, Regent's Park to Charlotte Wolff, 27 January 1939, CWP, Wellcome Library, London, PSY/WOL/6/3/2.

91. Album of prints of chimpanzees from the Zoological Gardens of London, CWP, Wellcome Library, London, PSY/WOL/3/3.

92. Charlotte Wolff, 'The Form and Dermatoglyphs of the Hands and Feet of Certain Anthropoid Apes', *Proceedings of the Zoological Society of London*, A107, no. 3 (September 1937): 348.

93. Wolff, 'The Form and Dermatoglyphs of the Hands and Feet of Certain Anthropoid Apes'.

94. Wolff, typescript, 'La Forme et les dermatoglyphes des mains et pieds d'un gorilla', CWP, Wellcome Library, London, PSY/WOL/2/27.

95. Wolff, 'A Comparative Study of the Form and Dermatoglyphs of the Extremities of Primates', 157. Figure VII is Mok's hand. Other prints from Mok are in the CWP, Wellcome Library, London, PSY/WOL/3/7.

96. Wolff, 'A Comparative Study of the Form and Dermatoglyphs of the Extremities of Primates', 157.

97. JSH, Guy the Gorilla typescript, 1.

98. See, for example, portraits by Wolfgang Suschitzky of Anthony and Francis Huxley, 1940, in JSH, *Memories* I, 193, and of Juliette Huxley, 1964, in JSH, *Memories* II, 128. Portraits of the Huxleys are in the National Portrait Gallery, London.

99. 'The airing courts were no longer surrounded by a high wall but by a sunk fence', describing changes in 1845–6. Kent County Lunatic Asylums, *Annual Report*, 1907 (Maidstone: W. P. Dickinson, 1908), 13, Kent County Library and Archives, MH/T3/Aa5.

100. John Berger, *Why Look at Animals?* (1977) (London: Penguin, 2009), 37.

CHAPTER 7: THE MAN FAMILY

1. Martin J. S. Rudwick, *Worlds Before Adam: The Reconstruction of Geohistory in the Age of Reform* (Chicago: University of Chicago Press, 2008), 48. For palaeontology, see Gowan Dawson, *Show Me the Bone: Reconstructing Prehistoric Monsters in Nineteenth-Century Britain and America* (Chicago: University of Chicago Press, 2016).

2. Martin J. S. Rudwick, *Bursting the Limits of Time: The Reconstruction of Geohistory in the Age of Revolution* (Chicago: University of Chicago Press, 2005). For history and science in this period, see Ian Hesketh, *The Science of History in Victorian Britain: Making the Past Speak* (London: Pickering & Chatto, 2011).

3. Historians often seek this of Huxley, reading him backwards through Darwin's evolution and *Descent of Man*, and fail to appreciate how minimally Darwinian *Evidence as to Man's Place in Nature* is. See, for example, Tom Gundling, *First in Line: Tracing Our Ape Ancestry* (New

Haven, CT: Yale University Press, 2005), 36. Gundling regrets the absence of a family tree of primates in Huxley's book.

4. For the journey from bones to molecules, see Marianne Sommer, *History Within: The Science, Culture, and Politics of Bones, Organisms, and Molecules* (Chicago: University of Chicago Press, 2016), 20.

5. THH, *Evidence as to Man's Place in Nature*, 57.

6. THH, *Evidence as to Man's Place in Nature*, 60–67.

7. THH, *Evidence as to Man's Place in Nature*, 110, 67–8.

8. THH, *Evidence as to Man's Place in Nature*, 90–91.

9. THH, *Evidence as to Man's Place in Nature*, 59, 104, 119, 78.

10. THH, *Evidence as to Man's Place in Nature*, 109.

11. THH, *Evidence as to Man's Place in Nature*, 104.

12. THH, *Evidence as to Man's Place in Nature*, 106.

13. [THH], 'Review of *On the Origin of Species by Means of Natural Selection*, by Charles Darwin', *The Times*, 26 December 1859, 8–9. Republished as 'The Darwinian Hypothesis' in THH, *Darwiniana*, 1–21.

14. THH, *Evidence as to Man's Place in Nature*, 106.

15. 'Note on the Resemblances and Differences in the Structure and the Development of the Brain in Man and Apes' by Professor Huxley, FRS, in Charles Darwin, *The Descent of Man and Selection in Relation to Sex*, 2nd edn (London: John Murray, 1877), 199–206.

16. THH to Charles Darwin, 16 April 1874, *L&L* I, 419.

17. Darwin, *Descent of Man*, 609.

18. Gowan Dawson, '"A Monkey into a Man": Thomas Henry Huxley, Benjamin Waterhouse Hawkins, and the Making of an Evolutionary Icon', in Ian Hesketh (ed.), *Imagining the Darwinian Revolution: Historical Narratives of Evolution from the Nineteenth Century to the Present* (Pittsburgh: University of Pittsburgh Press, 2022), ch. 4.

19. THH, *Evidence as to Man's Place in Nature*, 76.

20. THH, *Evidence as to Man's Place in Nature*, 78–9, 81.

21. Arthur O. Lovejoy, *The Great Chain of Being* (Cambridge, MA: Harvard University Press, 1936), 80, 90, 190–91, 252, 265; Peter Bowler, *Progress Unchained: Ideas of Evolution, Human History and the Future* (Cambridge: Cambridge University Press, 2021), Part 1, 'The Ladder of Progress and the End of History', 27–49.

22. Aaron Garrett, 'Anthropology: the "Original" of Human Nature', in Alexander Broadie (ed.), *The Cambridge Companion to the Scottish Enlightenment* (Cambridge: Cambridge University Press, 2003), 79–93; Paul Wood, 'The Natural History of Man in the Scottish Enlightenment', *History of Science*, 28, no. 1 (1990): 89–123.

23. THH, 'On the Methods and Results of Ethnology' [1865], in *Man's Place in Nature and Other Anthropological Essays: Collected Essays* (London: Macmillan, 1894), 212. Hereafter *Anthropological Essays*.

24. THH to Charles Kingsley, 5 May 1863, Huxley Papers, THHC, ICL, Series 1K, 19.217.

25. THH, *Evidence as to Man's Place in Nature*, 144 (my italics).

26. THH, 'On the Relations of Man to the Lower Animals' (1863), *Anthropological Essays*, 153, 154.

27. Darwin, *Descent of Man*, 618–619.

28. Darwin, *Descent of Man*, 619. See also Richards, *Darwin and the Making of Sexual Selection*, 3–34.

29. Darwin, *Descent of Man*, 619.

30. THH, *Evidence as to Man's Place in Nature*, 144, 153, 71.

31. THH, *Evidence as to Man's Place in Nature*, 145, 146, 149–51, 153.

32. Londa Schiebinger, *Nature's Body: Gender in the Making of Modern Science* (Boston: Beacon Press, 1993).

33. THH, *Evidence as to Man's Place in Nature*, 143, 152, 144, 77. Rudolph Wagner, *Vorstudien zu einer wissenschaftlichen Morphologie und Physiologie des menschlichen Gehirns als Seelenorgan* [Preliminary Studies on a Scientific Morphology of the Human Brain as a Soul Organ] (Göttingen: Dieterichschen, 1860).

34. THH, *Evidence as to Man's Place in Nature*, 153.

35. THH, *Evidence as to Man's Place in Nature*, 119, 143.

36. THH, *Evidence as to Man's Place in Nature*, 156.

37. Professor D. Schaaffhausen, 'On the Crania of the Most Ancient Races of Man' (1858), trans. George Busk, quoted in THH, *Evidence as to Man's Place in Nature*, 128–9.

38. THH, *Evidence as to Man's Place in Nature*, 156.

39. THH, 'Further Remarks Upon the Human Remains from the Neanderthal', *Natural History Review*, n.s. 4, no. 1 (1864): 429.

40. THH, *Evidence as to Man's Place in Nature*, 157–8. For prehistory, see Matthew R. Goodrum, 'The Idea of Human Prehistory: The Natural Sciences, the Human Sciences, and the Problem of Human Origins in Victorian Britain', *History and Philosophy of the Life Sciences*, 34, no. 1–2 (2012): 117–45; Donald R. Kelley, 'The Rise of Prehistory', *Journal of World History*, 14, no. 1 (2003): 17–36; Hesketh, *The Science of History in Victorian Britain*.

41. THH, *Evidence as to Man's Place in Nature*, 159.

42. Daniel Burr to Sir Roderick Murchison, 19 September 1866, THHC, ICL, Series 20, vol. 121/29.

43. 'The Australian Museum', *Sydney Morning Herald*, 13 May 1864, 2.

44. Henry Morton Stanley, *Through the Dark Continent: Or, The Sources of the Nile, Around the Great Lakes of Equatorial Africa, and Down the Livingstone River to the Atlantic Ocean*, vol. II (London: Sampson, Low, Marston, Searle, & Rivington, 1878), 143–4.

45. Fleetwood Isham Edwards to THH, 18 February 1880, THHC, ICL, Series 1E, 15.162.

46. Austin N. Cooper, FRCS to THH, 10 October 1894, THHC, ICL, Series 20, vol. 121/45.

47. Herbert H. Gower to THH, 24 September 1894, THHC, ICL, Series 20, vol. 121/78.

48. Charles Lyell, *The Geological Evidences of the Antiquity of Man* (London: John Murray, 1863), 476–8.

49. Ten years later in the fourth edition, the subtitle had expanded. *The Geological Evidences of the Antiquity of Man; With an Outline of Glacial and Post-Tertiary Geology, and Remarks on the Origin of Species, with Special Reference to Man's First Appearance on the Earth* (London: John Murray, 1873).

50. 'To man alone is given this belief, so consonant to his reason, and so congenial to the religious sentiments implanted by nature in his soul, a doctrine which tends to raise him morally and intellectually in the scale of being, and the fruits of which are, therefore, most opposite in character to those which grow out of error and delusion.' Lyell, *The Antiquity of Man*, 4th edn, 537. Browne, *Charles Darwin: The Power of Place*, 218.

51. Charles Lyell, *Principles of Geology, Being an Attempt to Explain the Former Changes of the Earth's Surface, by Reference to Causes Now in Operation* (London: John Murray, 1833), III, 137.

52. James A. Secord, 'Introduction' to Charles Lyell, *Principles of Geology* (Harmondsworth: Penguin, 1997), xxxiv.

53. THH, 'On Species and Races and Their Origin', fragment of a lecture at the Royal Institution, 10 February 1860, in Lectures (Species and Races etc.) notebook, 1860–83, THHC, ICL, Series 19, vol. 82.10.

54. Robert King to THH, 8 July 1863, THHC, ICL, Series 1K, Scientific and General Correspondence, 19.155–7.

55. David N. Livingstone, *Adam's Ancestors: Race, Religion and the Politics of Human Origins* (Baltimore: Johns Hopkins University Press, 2008), 138; W. B. Clarke to THH, 6 May 1864, THHC, ICL, Series 18, Anthropology, vol. 2, c. 1865–1875, vol. XVI:181.

56. THH, 'On the Ethnology of Britain', 10 May 1870, THHC, ICL, Series 4, Anthropology and ethnology, Presidential address by THH, 33.62.

57. Samuel Laing, *Pre-historic Remains of Caithness, with Notes on the Human Remains by T. H. Huxley* (London: Williams & Norgate, 1866); Edwin Brown to THH, 10 April 1865, THHC, ICL, Series 18, Anthropology, vol. 2, c. 1865–1875, 188; Photographs of crania, 1857–71, Series 21, Box H, 139–54.

58. Karl Penka, *Origines Ariacae. Linguistisch-ethnologische Untersuchungen zur ältesten Geschichte der arischen Völker und Sprachen* (Vienna: Karl Prochaska, 1883).

59. Isaac Taylor, 'The Origin and Primitive Seat of the Aryans', *Journal of the Anthropological Institute of Great Britain and Ireland*, 17 (1888): 238–75.

60. F. Max Müller, *Biographies of Words and the Home of the Aryas* (London: Longmans, Green, and Co., 1888), 120; Dorothy M. Figueira, *Aryans, Jews, Brahmins: Theorizing Authority through Myths of Identity* (Albany: State University of New York Press, 2002), 45.

61. Isaac Taylor, *The Origin of the Aryans* (London: Walter Scott, 1889), 81, 122.

62. Joan Leopold, 'British Applications of the Aryan Theory of Race to India, 1850–1870', *English Historical Review*, 89, no. 352 (July 1974): 578–603.

63. THH, 'The Aryan Question and Prehistoric Man', *Nineteenth Century*, 165 (November 1890): 753, 756, 766, 776.

64. H. G. Wells, *The Outline of History: Being a Plain History of Life and Mankind*, 3rd edn (New York: Macmillan, 1920), 167–82.

65. Taylor, *The Origin of the Aryans*, 332.

66. For inter-war literature reviews of the idea, see Frank H. Hankins, *The Racial Basis of Civilization: A Critique of the Nordic Doctrine* (New York: Knopf, 1931), ch. 1; V. Gordon Childe, *The Aryans: A Story of Indo-European Origins* (London: K. Paul, Trench, Trubner & Co., 1926).

67. Adolf Hitler, *Mein Kampf* (New York: Stackpole Sons, 1939), 288.

68. Christopher M. Hutton, *Race and the Third Reich: Linguistics, Racial Anthropology and Genetics in the Dialectic of Volk* (Cambridge: Polity Press, 2005), 80–81, 93.

69. Elazar Barkan, *The Retreat of Scientific Racism* (Cambridge: Cambridge University Press, 1992).

70. JSH and Haddon, *We Europeans*, 7, 21.

71. JSH and Haddon, *We Europeans*, 145–50.

72. JSH and Haddon, *We Europeans*, 150–51. Parts of *We Europeans* were abridged and republished within the Oxford Pamphlets on World Affairs series, as JSH, *Race in Europe* (Oxford: Clarendon Press, 1939).

73. 'I have declared again and again that if I say Aryans, I mean neither blood nor bones, nor hair, nor skull: I mean simply those who speak an Aryan language.' Müller, *Biographies of Words*, 245, quoted in JSH and Haddon, *We Europeans*, 151.

74. JSH and Haddon, *We Europeans*, 163.

75. Emily Kern, 'Out of Asia: A Global History of the Scientific Search for the Origins of Humankind, 1800–1965', PhD Thesis, Princeton University, 2018.

76. JSH, *Africa View* (London: Chatto & Windus, 1931), 251.

77. R. A. Dart, 'Australopithecus africanus: The Man-Ape of South Africa', *Nature*, 115, no. 2884 (1925): 195–9.

78. Raymond Dart, *Adventures with the Missing Link* (London: Hamish Hamilton, 1959), 6.

79. THH, *Elementary Lessons in Physiology* (London: Macmillan, 1871) was Elliot Smith's foundational anatomical book.

80. H. G. Wells, JSH and G. P. Wells, *The Science of Life*, vol. II (London: Waverley, 1931), 533.

81. Wells, JSH and Wells, *The Science of Life*, II, 537.

82. John Evangelist Walsh, *Unravelling Piltdown: The Science Fraud of the Century and Its Solution* (New York: Random House, 1996).

83. Wells, JSH and Wells, *The Science of Life*, I, 266.

84. JSH, *We Europeans*, 57–9.

85. Diagram of Climate and Prehistory in Western Europe, JSH and Haddon, *We Europeans*, 55.

86. Samuel Taylor Coleridge, 'Kubla Khan: or, A Vision in a Dream', quoted in JSH, *Africa View*, 252. See also L. S. B. Leakey, *The Stone Age Cultures of Kenya Colony* (London: Cambridge University Press, 1931).

87. JSH, *Africa View*, 255.

88. Juliette Huxley, *Wild Lives*, 135–6.

89. Juliette Huxley, *Wild Lives*, 137–9.

90. JSH, *Memories II* (New York: Harper & Row, 1973), 212.

91. JSH to SZ, 3 March 1961, SZ Papers, UEA, SZ/GEN/HUX. See also SZ to JSH, 20 March 1962, JSH to Zuckerman, 13 May 1964, and JSH to SZ, 13 April 1972, SZ Papers, UEA, SZ/GEN/HUX, in which he is still wondering about Leakey being elected to the Royal Society. Louis Leakey was never a fellow, though his son, Richard Leakey, also a palaeo-anthropologist, was.

92. SZ, 'Taxonomy and Human Evolution', *Biological Reviews*, 25, no. 4 (October 1950), 435–85.

93. SZ, 'Correlation of Change in the Evolution of Higher Primates', 303, 325–6, 347–9.

94. JSH, *We Europeans*, 48.

95. Darwin, *Descent of Man*, 11–25.

96. Wells, JSH and Wells, *The Science of Life*, I, 267–8.

97. Wells, JSH and Wells, *The Science of Life*, I, 269.

98. Soraya de Chadarevian, 'Chromosome Photography and the Human Karyotype', *Historical Studies in the Natural Sciences* 45, no. 1 (February 2015), 115–46.

99. JSH, 'The Uniqueness of Man', *Man in the Modern World* (London: Chatto & Windus, 1947), 4–5.

100. JSH, *Evolution in Action*, 136–7.

101. JSH, *We Europeans*, 135–6.

102. JSH, *Evolution in Action*, 137–8.

103. JSH and Haddon, *We Europeans*, 111, 134, 201.

104. JSH and Haddon, *We Europeans*, 266, 270; Sommer, *History Within*, 20.

105. JSH, 'The Uniqueness of Man', 11.

106. JSH and Haddon, *We Europeans*, 146.

107. JSH, *Evolution in Action*, 135.

108. JSH, 'Notes on the Scientific and Cultural History of Mankind', May 1948, JSH Papers, RU, MS 50, Box 118.

109. JSH, *Memories* I, 270–71; JSH, *Memories* II, 207.

110. Victoria Huxley, Julian's granddaughter, remembers this cast. It is not clear whether it was original or a copy, or what happened to it after Julian and Juliette's death. Personal communication with Victoria Huxley, 29 October 2020. See also photograph of Huxley's library in JSH, *Memories* II, 129, while other photos are clearly Clarke's sculpture. For Dora Clarke's sculpture, see JSH, *Memories* II, 89–90. Dora Clarke, 'Negro Art: Sculpture from West Africa', *African Affairs*, 34, no. 135 (1935): 129–37.

111. JSH, *Memories* I, 270, n. 1.

CHAPTER 8: HUMAN DIFFERENCE

1. JSH, 'The Family of Man', *Unesco Courier*, February 1956, 3.

2. THH, *Rattlesnake Diary*, 19 July 1849, 208.

3. Armlet, British Museum, donated by Thomas Henry Huxley, Oc.5395.

4. David Hume, *A Treatise of Human Nature* (1739), vol. III (Oxford: Clarendon, 1896): ''tis utterly impossible for men to remain any considerable time in that savage condition, which precedes society; but that his very first state and situation may justly be esteem'd social', 493. T. H. Huxley, *Hume* (London: Macmillan, 1879).

5. THH, *Rattlesnake Diary*, 8 July 1849, 206.

6. John MacGillivray, *Narrative of the Voyage of H.M.S. Rattlesnake*, vol. I (London: T. & W. Boone, 1852), ch. 1.6.

7. THH, 'Autobiography', 12–13.

8. THH to Lizzie Salt [sister], 14 March 1849, in *L&L*, I, 44–5.

9. THH, *Rattlesnake Diary*, 21 May 1848, 127.

10. THH, 'Huxley Being Painted by a Native Acquaintance', *Rattlesnake Diary*, Plate 6, 126.

11. Anthropological Photographs, THHC, ICL, Series 21, Box G and Box H; see Philippa Levine, 'States of Undress: Nakedness and the Colonial Imagination', *Victorian Studies*, 50, no. 2 (2008): 189–219; Philippa Levine, 'Naked Truths: Bodies, Knowledge, and the Erotics of Colonial Power', *Journal of British Studies*, 52, no. 1 (2013): 5–25; James R. Ryan, *Picturing Empire: Photography and the Visualization of the British Empire* (London: Reaktion Books, 1997).

12. Anthropological Photographs, THHC, ICL, Series 21, Box G, Photograph 4a.

13. 'Bushman Genealogy', Anthropological Photographs, THHC, ICL, Series 21, Box G.

14. Alfred C. Haddon, 'Imperial Institute of Ethnology'; THH to Haddon, 1 January 1892, Haddon Papers, Cambridge University Library, Folder 5061.

15. Edward Clodd, preface to 1908 edition of *Man's Place in Nature* (London: Watts, 1908), 3–4.

16. A. C. Haddon, *The Study of Man* (London: John Murray, 1898).

17. JSH, 'Huxley and the Savages', in THH, *Rattlesnake Diary*, 170.

18. JSH, *Memories* II, 111–12.

19. JSH, *Memories* II, 119.

20. JSH, *Evolution in Action* (London: Chatto & Windus, 1953), 139–40.

21. E. E. Evans-Pritchard, *Social Anthropology* (London: Cohen & West, 1951), 78–9.

22. Francis Huxley, chart 'showing how the members of Pari's and Antonio-hu's villages were related', *Affable Savages: An Anthropologist Among the Urubu Indians of Brazil* (London: Travel Book Club, 1956), 287.

23. Francis Huxley, *Affable Savages*, 12–13.

24. Ron Roberts and Theodor Itten, *Francis Huxley and the Human Condition: Anthropology, Ancestry and Knowledge* (Abingdon: Routledge, 2020), 188–94; Edwin Brooks et al., *Tribes of the Amazon Basin in Brazil 1972: Report for the Aborigines Protection Society* (London: Charles Knight & Co., 1973).

25. Survival International, https://www.survivalinternational.org/, July 2021.

26. Alison Bashford and Joyce E. Chaplin, *The New Worlds of Thomas Robert Malthus: Re-reading the Principle of Population* (Princeton: Princeton University Press, 2016).

27. JSH, 'Introduction: The Ritualization of Behaviour in Animals and Man', *Philosophical Transactions of the Royal Society of London*, Series B, Biological Sciences, 251, 772 (1966): 263–4.

28. Francis Huxley, 'The Marginal Lands of the Mind', in JSH (ed.), *The Humanist Frame* (London: Allen & Unwin, 1961), 167–80.

29. JSH, 'The Ritualization of Behaviour in Animals and Man', 265, citing Francis Huxley, *The Invisibles: Voodoo Gods in Haiti* (London: Rupert Hart-Davis, 1966).

30. Arthur Keith, 'Huxley as Anthropologist', *Nature*, 115 (9 May 1925): 719–23.

31. THH, 'On the Methods and Results of Ethnology', *Fortnightly Review*, 1 (1865): 257, 258.

32. James Hunt, 'On the Negro's Place in Nature', *Memoirs Read Before the Anthropological Society of London, 1863–64* (London: Trübner, 1865), 3; see Efram Sera-Shriar, *The Making of British Anthropology, 1813–1871* (London: Pickering and Chatto, 2013), 53–79.

33. Dr Seeman, response to James Hunt, 'On the Negro's Place in Nature', *Journal of the Anthropological Society of London*, 2 (1864): xviii.

34. James Hunt, 'On the Physical and Mental Characters of the Negro', *Report of the British Association for the Advancement of Science, Newcastle, 1863* (London: John Murray, 1864): 140.

35. Hunt, 'On the Negro's Place in Nature', 2; James Hunt, *The Negro's Place in Nature* (New York: Van Evrie, Horton & Co., 1864), i.

36. Berthold Seemann, response to W. Winwood Reade, 'On the Bush Tribes of Equatorial Africa', *Transactions of the Anthropological Society of London*, 1 (1863), xxiii; George Busk, 'Observations on a Systematic Mode of Craniometry', *Transactions of the Ethnological Society of London*, 1(n.s.) (1861): 341–8. See also Paul Turnbull, 'British Anthropological Thought in Colonial Practice', in Bronwen Douglas and Chris Ballard (eds.), *Foreign Bodies: Oceania and the Science of Race, 1750–1940* (Canberra: ANU ePress, 2008), 205–28.

37. It was that organization into which his great-grandson Francis was elected fellow, nominated in 1948 by Evans-Pritchard.

38. [THH], 'Darwin On the Origin of Species' (1859), *The Times*, republished as 'Darwinian Hypothesis' in THH, *Darwiniana*, 2.

39. THH, 'The Structure and Classification of the Mammalia,' Lecture 9, *Medical Times and Gazette*, 2 April 1864, 369–70.

40. THH, 'Emancipation – Black and White' (1865), in *Science and Education: Collected Essays* (London: Macmillan, 1893), 66–75.

41. Mrs P. A. Taylor, *Professor Huxley on the Negro Question* (London: For the Ladies' London Emancipation Society [E. Faithfull], 1864).

42. THH, 'Emancipation – Black and White', 67–8.

43. THH to John Tyndall, 9 November 1866, in *L&L*, I, 283.

44. [Thomas Carlyle], 'Occasional Discourse on the Negro Question', *Fraser's* magazine, 40 (1849), 670–79; John Stuart Mill, 'The Negro Question', *Fraser's* magazine, 41 (1850): 25–31.

45. Editor, *Pall Mall Gazette*, 29 October 1866, 9; THH, letter to Editor, *Pall Mall Gazettte*, 31 October 1866, 3.

46. It appeared in the *Reader* in May, and the Jamaican march to the courthouse in Morant Bay, led by Paul Bogle, in which twenty-five protesters were killed by a militia, was not until October that year.

47. THH to Lizzie Salt (later Scott), 4 May 1864, in *L&L*, I, 251.

48. Adrian Desmond and James Moore, *Darwin's Sacred Cause: Race, Slavery and the Quest for Human Origins* (Chicago: University of Chicago Press, 2009).

49. THH to Lizzie Salt (later Scott), 4 May 1864, in *L&L*, I, 251–2.

50. This was a reprint of his lecture on human types – 'The Structure and Classification of Mammalia' – delivered at the Royal College of Surgeons and printed in the short-lived weekly the *Reader* (which Huxley and his X Club friends were about to take over in December 1864 until July 1865), along with an abolitionist editorial summary, far more abolitionist than Huxley and his lecture warranted. 'Professor Huxley's Lectures on "The Structure and Classification of the Mammalia"', *Reader*, 3 (1864): 266–8, also printed in the *Medical Times and Gazette*.

51. Taylor, *Professor Huxley on the Negro Question*, 13–14.

52. Jane Rendall, 'Langham Place group (1857–1866)', *Oxford Dictionary of National Biography*, online, 2005.

53. Eliza Lynn Linton to THH, 11 November 1868, THHC, ICL, Series 1L, 21: 223. See also Evelleen Richards, 'Huxley and Woman's Place in Science: The "Woman Question" and the Control of Victorian Anthropology',

in *Ideology and Evolution in Nineteenth Century Britain* (London: Routledge, 2020), 240–67.

54. Maria Grey, 'Education of Women', *The Times*, 23 May 1872, 5.

55. Maria Grey to Henrietta Huxley, 13 July 1871, THHC, ICL, Series 1G, 17: 148. See 'Education of Women', *The Times*, 14 November 1871, 4.

56. THH to Sophia Jex-Blake, 28 October 1872, in *L&L*, I, 387.

57. THH to Michael Foster, 8 January 1872, THHC, ICL, Series 1F, 4: 35.

58. Emily A. E. Shirreff, *The Work of the National Union* (London: William Ridgway, 1873). See also James Elwick, 'Economies of Scales: Evolutionary Naturalists and the Victorian Examination System', in Bernard Lightman and Gowan Dawson (eds.), *Victorian Scientific Naturalism: Community, Identity, Continuity* (Chicago: University of Chicago Press, 2014), 131–56.

59. Sarah Hackett Stevenson, *Boys and Girls in Biology; or, Simple Studies of the Lower Forms of Life, Based Upon the Lectures of Prof. T. H. Huxley and Published by His Permission. Illustrated by Miss M. A. I. Macomish* (New York: D. Appleton, 1875).

60. F. M. Sperry, *A Group of Distinguished Physicians and Surgeons of Chicago* (Chicago: J. H. Beers, 1904).

61. Mlle M. Halboister to THH, 18 February 1894, THHC, ICL, Series 1H, 17: 229.

62. THH to Anton Dohrn, 3 January 1872, in *L&L*, I, 368.

63. Immediately after Julia's death in 1908, Ethel Arnold toured the United States, and again over 1910 and 1912, lecturing on women's civil rights. Phyllis E. Wachter, 'Ethel M. Arnold (1865–1930)', *Victorian Periodicals Review*, 20, no. 3 (1987): 107–11.

64. 'Women's National Anti-Suffrage League Manifesto' (1910), Records of the Women's National Anti-Suffrage League, Women's Library Archives, London School of Economics, GB 106 2WNA/D.

65. JSH to Mary Ward, n.d. [1908], MAWL, BL RP8974.

66. Mary Ward to JSH, 3 March 1912, MAWL, BL RP8974.

67. Mary Ward to JSH, 26 September 1916, MAWL, BL RP8974.

68. Mary Ward to JSH, 3 March 1912, MAWL, BL RP8974.

69. JSH, *Memories* I, 197. Huxley names this the Association of Scientific Workers.

70. Max Nicholson and Julian Huxley Papers, Woodson Research Center, Fondren Library, Rice University, MS 54, Box 3: 1.

71. Conrad Hal Waddington, *The Scientific Attitude* (Harmondsworth: Penguin, 1948); Thomas Mougey, 'Needham at the Crossroads: History,

Politics and International Science in Wartime China, 1942–1946', *British Journal for the History of Science*, 50, no. 1 (2017): 83–109.

72. JSH, *Soviet Genetics and World Science: Lysenko and the Meaning of Heredity* (London: Chatto & Windus, 1949), 24–5; Hermann Muller, 'The Destruction of Science in the U.S.S.R', *Saturday Review of Literature*, 4 December 1948, 13–15, 63–5; Hermann Muller, 'Back to Barbarism Scientifically', *Saturday Review of Literature*, 11 December 1948, 8–10.

73. F. A. E. Crew et al., 'Social Biology and Population Improvement', *Nature*, 144, no. 3646 (16 September 1939): 521–2.

74. Crew et al., 'Social Biology and Population Improvement', 521.

75. Kirrill O. Rossianov, 'Editing Nature: Joseph Stalin and the "New" Soviet Biology', *Isis*, 84, no. 4 (1993): 728–45; Valery N. Soyfer, *Lysenko and the Tragedy of Soviet Science* (New Brunswick, NJ: Rutgers University Press, 1994).

76. Mark B. Adams, 'The Politics of Human Heredity in the USSR, 1920–1940', *Genome*, 31, no. 2 (1989): 879–84.

77. JSH, *Soviet Genetics*, 23, vii, viii–ix, ch. 5.

78. JSH, *Soviet Genetics*, 1.

79. *A Month in Russia* (1931). Julian Huxley, director of photography. British Film Institute National Archives, 21542.

80. JSH, *Memories* I, 287.

81. JSH, *Memories* I, 284.

82. JSH, *Soviet Genetics*, 34, 35; JSH, *Memories*, I, 282.

83. JSH, *Soviet Genetics*, 153; JSH, *Memories* I, 284–5. See also Diane B. Paul, 'A War on Two Fronts: J. B. S. Haldane and the Response to Lysenkoism in Britain', *Journal of the History of Biology*, 16, no. 1 (1983): 1–37.

84. Crew et al., 'Social Biology and Population Improvement', 521.

85. JSH, *Memories* II, 13–14.

86. JSH, *Memories* II, 23–4.

87. JSH, *Memories* II, 40–43.

88. World Wildlife Fund documents, JSH Papers, RU, MS50, Box 120: 3.

89. JSH, *Memories* II, 68; JSH, *From an Antique Land: Ancient and Modern in the Middle East* (London: Max Parrish, 1954).

90. JSH, *Unesco: Its Purpose and Its Philosophy* (Paris: Unesco, 1946), 45; Glenda Sluga, 'UNESCO and the (One) World of Julian Huxley', *Journal of World History*, 21, no. 3 (2010), 393–418.

91. John S. Partington, 'H. G. Wells and the World State: A Liberal Cosmopolitan in a Totalitarian Age', *International Relations*, 17, no. 2 (2003), 233–46.

92. JSH, Preface, THH and JSH, *Evolution and Ethics*, vii.

93. JSH, *Unesco*, 61.

94. JSH, *Unesco*, 41, 61.

95. JSH, *Unesco*, 7–8, 35.

96. René Maheu to JSH, 1 July 1964, JSH Papers, RU, MS 50, Box 37: 1.

97. The Constitution of UNESCO, signed on 16 November 1945, ratified 4 November 1946 by twenty countries.

98. JSH, *Unesco*, 6–7.

99. Unesco, 'Statement on Race, Paris, July 1950', in *Four Statements on the Race Question* (Paris: Unesco, 1969), 35; Michele Brattain, 'Race, Racism and Antiracism: Unesco and the Politics of Presenting Science to the Post-war Public', *American Historical Review*, 112, no. 5 (2007): 1386–1413; Elazar Barkan, *The Retreat of Scientific Racism* (Cambridge: Cambridge University Press, 1992), 341–6.

100. Unesco, 'Statement on Race', 33; Charles Darwin, *The Descent of Man*, quoted in Unesco, 'Statement on Race', 33.

101. JSH, Introduction, in THH and JSH, *Evolution and Ethics*, 9.

102. Montagu, *Man – One Family* (1946).

103. JSH, 'America Revisited: The Negro Problem', *The Spectator*, 29 November 1924, 821–2. Barkan, *The Retreat of Scientific Racism*, 178–89.

104. Henry Louis Gates Jr, 'W. E. B. Du Bois and the Encyclopedia Africana, 1909–63', *Annals of the American Academy of Political and Social Science*, 568 (March 2000): 203–19.

105. W. E. B. Du Bois to JSH, 7 October 1937, JSH to W. E. B. Du Bois, 22 October 1927, W. E. B. Du Bois Papers, Special Collections and University Archives, University of Massachusetts Amherst Libraries, MS 312.

106. Notes on the French Revolution and *Déclaration des droits de l'homme et du citoyen* etc., n.d., THHC, ICL, Series 20, 122: 142–50.

107. THH, 'Natural Rights and Political Rights' (1890), in THH, *Method and Results*, 336–82; see also Kevin Currie-Knight, 'Rival Visions: J. J. Rousseau and T. H. Huxley on the Nature (or Nurture) of Inequality and What It Means for Education', *Philosophical Studies in Education*, 42 (2011): 25–35.

108. THH, 'On the Natural Inequality of Men', in THH, *Methods and Results*, 313, 322.

109. JSH, 'Evolutionary Ethics', in THH and JSH, *Evolution and Ethics*, 116–17.

110. JSH, *Unesco*, 19.

111. JSH, *Unesco*, 15, 18–21.

112. JSH (ed.), *The Humanist Frame* (London: George Allen & Unwin, 1961), 24.

CHAPTER 9: TRANSHUMANS

1. 'AEM' to Julian Huxley, 3 January 1937, Wellcome Library, Eugenics Society Papers, SA/EUG/C.186.

2. Angus McLaren, *Reproduction by Design: Sex, Robots, Trees, and Test-Tube Babies in Inter-war Britain* (Chicago: University of Chicago Press, 2012).

3. Artificial Insemination by Donor, Wellcome Library, Eugenics Society Papers, SA/EUG/D.6: Box 30.

4. Susan Squier, *Babies in Bottles: Twentieth-Century Visions of Reproductive Technology* (New Brunswick: Rutgers University Press, 1994); Scott Gelfand (ed.), *Ectogenesis: Artificial Womb Technology and the Future of Human Reproduction* (Amsterdam: Rodopi, 2006).

5. JSH to THH, and THH to JSH, 24 March 1892, in *L&L*, II, 436–9; Piers J. Hale, 'Monkeys into Men and Men into Monkeys: Chance and Contingency in the Evolution of Man, Mind and Morals in Charles Kingsley's Water Babies', *Journal of the History of Biology*, 46, no. 4 (2013): 551–97.

6. Bashford, *Global Population*.

7. Aldous Huxley, *Brave New World Revisited* (London: Chatto & Windus, 1959), 18. First US edition, 1958.

8. Aaron O'Neill, 'Child Mortality Rate (Under Five Years Old) in the United Kingdom from 1800 to 2020', statista.com, 9 September 2019.

9. THH, 'The Struggle for Existence in Human Society' (1888), in *Evolution and Ethics: Collected Essays* (London: Macmillan, 1894), 210.

10. Simon Szreter (ed.), *The Hidden Affliction: Sexually Transmitted Infections and Infertility in History* (Rochester, NY: University of Rochester Press, 2019).

11. JSH, *Memories* I, 18.

12. Charles Darwin, *The Descent of Man*, 2nd edn (London: John Murray, 1877), 607, 618.

13. THH, 'Government: Anarchy or Regimentation' (1890), in *Method and Results*, 428.

14. THH, 'The Struggle for Existence', 209, 216.

15. THH, 'The Struggle for Existence', 203–6.

16. Laura Schwartz, *Infidel Feminism: Secularism, Religion, and Women's Emancipation, England 1830–1914* (Manchester: Manchester University Press, 2013).

17. THH to Michael Foster, 18 July 1883, in 'Michael Foster and Thomas Henry Huxley, Correspondence, Letters 100 through 134, 1865–1895', *Medical History Supplement*, 28 (2009): 74–104.

18. JSH, 'Birth Control and Hypocrisy', *The Spectator*, 19 April 1924, 630–31.

19. JSH, 'The Problem of Birth Control: A Vigorous Viewpoint', *The Humanist*, October 1925, offprint, JSH Papers, RU, MS 50, Box 97: 4.

20. JSH, 'Birth Control and Hypocrisy', 630–31.

21. JSH, 'The Ethics of Birth Control', *Nature*, 116 (26 September 1925): 455–7.

22. JSH, 'Are We Over-populated?', *Evening Standard*, 25 October 1927, offprint, JSH Papers, RU, MS 50, Box 97: 7.

23. JSH, 'World Population', in *The Human Sum*, ed. C. H. Rolph (New York: Macmillan, 1957), 14–15; Huxley, *Brave New World Revisited*, 15.

24. Juliette Huxley, Journal 1947–8, 26 October 1947, J&JH Papers, RU, MS 512, RU, Box 1.

25. Minutes of the First Meeting of Population Ltd, 12 January 1957. JSH Papers, MS 50, RU, Box 115: 6.

26. Aldous Huxley, interview on *The Mike Wallace Interview*, 18 May 1958, Openculture.com.

27. 'Plenary Session on the Control of Global Population', *British Medical Journal*, 2, no. 5455 (24 July 1965): 224; this included a session on eugenics and population control; see also Claude Davies, General Secretary, BMA to JSH, 15 July 1964, JSH Papers, RU, MS 50 Box 37: 1.

28. C. H. Sharpe to the Manager, BBC, 18 November 1926, JSH Papers, RU, MS 50, Box 9: 2.

29. Carrie Pitzulo, *Bachelors and Bunnies: The Sexual Politics of Playboy* (Chicago: University of Chicago Press, 2011).

30. Murray Fisher to JSH, 3 December 1964, JSH Papers, RU, MS 50, Box 37: 6.

31. Unknown to JSH, n.d. [1965], JSH Papers, RU, MS 50, Box 37: 7.

32. Unknown to JSH, n.d. [1965], JSH Papers, RU, MS 50, Box 37: 7.

33. THH to Charles Kingsley, 5 May 1863, in *L&L*, I, 240.

34. Francis Galton, *Hereditary Genius* (London: Macmillan, 1869), 64.

35. Francis Galton, *Inquiries into Human Faculty and Its Development* (London: Macmillan, 1883), 25.

36. Charles Darwin to Francis Galton, 4 January 1873, DCP, Letter 8724.

37. Darwin, *The Descent of Man*, 617.

38. Darwin, *The Descent of Man*, 617–18.

39. JSH, 'Eugenics in Evolutionary Perspective', Second Galton Lecture Draft, 1962, JSH Papers, RU, MS50, Box 79: 4, ff. 19–20. See also Paul Weindling, 'Julian Huxley and the Continuity of Eugenics in Twentieth Century Britain', *Journal of Modern European History*, 10, no. 4 (2012): 480–99.

40. Darwin, *The Descent of Man*, 618.

41. Henry Fairfield Osborn, 'Birth Selection versus Birth Control', in Henry F. Perkins and Harry H. Laughlin (eds.), *A Decade of Progress in Eugenics. Scientific Papers of the Third International Congress of Eugenics, 1932* (Baltimore: William and Wilkins, 1934), 29–41.

42. JSH, *Evolution: The Modern Synthesis*, 398.

43. Henry H. Goddard, *The Kallikak Family: A Study in the Heredity of Feeble-Mindedness* (New York: Macmillan, 1912).

44. Leonard Huxley, *Progress and the Unfit* (London: Watts & Co., 1926), 30–31.

45. E. J. Lidbetter and E. Nettleship, 'On a Pedigree Showing Both Insanity and Complicated Eye Disease', *Brain*, 35, no. 3 (1913): 195–221.

46. JSH, 'Are We Over-populated', *Evening Standard*, 25 October 1927, JSH Papers, RU, MS50, Box 97: 7.

47. 'Sterilization of Defectives: Departmental Committee's Report', *The British Medical Journal*, 1, no. 3812 (27 January 1934), 161–5; John Macnicol, 'The Voluntary Sterilization Campaign in Britain, 1918–1939', *Journal of the History of Sexuality*, 2, no. 3 (January 1992): 422–38.

48. 'Unfit Parents', *Daily Mail*, 29 May 1928, Lady Askwith files, Eugenics Society Papers, Wellcome Library, SAEUG/C/7.

49. JSH, 'Eugenics in Evolutionary Perspective', f. 34.

50. JSH, 'Address on Voluntary Sterilization to the Public Health Congress and Exhibition, 1934', JSH Papers, RU, MS50, Box 106: 1, ff. 305, 307, 309.

51. 'Look Ahead 100 Years Through the Eyes of Julian Huxley', *Tit-Bits*, 24 January 1931, offprint, JSH Papers, RU, MS50, Box 97: 11.

52. List of Galton Lectures, JSH Papers, RU, MS50, Box 112: 5.

53. Leonard Darwin to JSH, 31 October 1928, JSH Papers, RU, MS50, Box 9: 8.

54. List of Galton Lectures, JSH Papers, RU, MS50, Box 112: 5.

55. JSH to SZ, 21 April 1960, SZ Papers, UEA Archives, SZ/GEN/HUX.

56. For example, Paul Glumaz, 'T. H. Huxley's Hideous Revolution in Science', *Executive Intelligence Review*, 42 (2015), 18–29.

57. JSH, 'Eugenics in Evolutionary Perspective'.

58. JSH, 'Eugenics in Evolutionary Perspective'.

59. JSH, *Memories*, I, 228–9.

60. JSH, 'The Tissue-Culture King: A Biological Fantasy', *Cornhill* magazine, 60 (n.s.), no. 358 (April 1926), 422–58, reprinted in *Amazing Stories*, 1927.

61. JSH, 'The Tissue-Culture King', in *Amazing Stories*, 451–9.

62. Groff Conklin (ed.), *Great Science Fiction by Scientists* (New York: Collier, 1962); JSH, *Memories* I, 229.

63. For example, JSH and Lancelot T. Hogben, 'Experiments on Amphibian Metamorphosis and Pigment Responses in Relation to Internal Secretions', *Proceedings of the Royal Society of London B: Biological Sciences*, 93, no. 649 (1922): 36–53.

64. JSH, 'The World with Mechanism Enslaved', *Evening Standard*, 12 October 1927, offprint, JSH Papers, RU, MS50, Box 97: 7.

65. Newsclipping, 1921, JSH Papers, RU, MS50, Box 6: 4.

66. JSH, 'Man's Place in the Universe' (1933), typescript, JSH Papers, RU, MS50, Box 63: 7.

67. JSH, 'Glands, Growth, and Race', *Everyman*, 20 March 1920, offprint, JSH Papers, RU, MS50, Box 97: 1.

68. Unsigned letter to JSH, Karachi, 30 December 1920, JSH Papers, RU, MS50, Box 9: 2.

69. For a summary of Steinach's procedures in English, see Norman Haire, *Rejuvenation: The Work of Steinach, Voronoff, and Others* (London: G. Allen & Unwin, 1924).

70. JSH, 'Man's Place in the Universe'.

71. JSH, *Problems of Relative Growth* (New York: Dial Press, 1932), 229.

72. JSH, 'Man's Place in the Universe', handwritten note on f. 56.

73. Aldous Huxley, *After Many a Summer* (London: Chatto & Windus, 1939).

74. JSH, 'Evolution, Introduction 1958–59', JSH Papers, RU, MS50, Box 75: 2; untitled typescript, JSH Papers, RU, MS50, Box 63: 7; JSH, 'Eugenics in Evolutionary Perspective'.

75. JSH, *Evolution in Action* (London: Chatto & Windus, 1953), 153.

76. JSH, 'Transhumanism', in *New Bottles for New Wine* (London: Chatto & Windus, 1957), 13–17.

77. Citing Aldous Huxley, *Ends and Means* (1937), in JSH, *Evolution: The Modern Synthesis*, 573.

78. JSH, *Evolution: The Modern Synthesis*, 573.

79. JSH, *Evolution: The Modern Synthesis*, 577.

80. William B. Provine, 'Progress in Evolution and Meaning in Life', in Waters and van Helden (eds.), *Julian Huxley*, 165–80.

81. JSH, *Evolution in Action*, 135.

82. JSH, *Evolution: The Modern Synthesis*, 572; JSH, 'Transhumanism', *New Bottles for New Wine*, 13–17.

83. JSH, 'The Future of Man – Evolutionary Aspects', in Gordon E. W. Wolstenholme (ed.), *Man and His Future* (London: J. & A. Churchill, 1963), 2.

84. Wolstenholme (ed.), *Man and His Future*, v.

85. JSH, 'The Future of Man – Evolutionary Aspects', in Wolstenholme (ed.), *Man and His Future*, 8–10.

86. JSH, 'The Future of Man – Evolutionary Aspects', in Wolstenholme (ed.), *Man and His Future*, 12, 16.

87. Gregory Pincus, 'Control of Reproduction in Mammals', in Wolstenholme (ed.), *Man and His Future*, 79, 89.

88. Lara V. Marks, *Sexual Chemistry: A History of the Contraceptive Pill* (New Haven: Yale University Press, 2001); Laura Briggs, *Reproducing Empire: Race, Sex, Science and U.S. Imperialism in Puerto Rico* (Berkeley: University of California Press, 2002).

89. Comment by Klein, in Wolstenholme (ed.), *Man and His Future*, 100–101.

90. Betty Friedan, *The Feminine Mystique* (London: Gollancz, 1963).

91. Hermann J. Muller, 'Genetic Progress by Voluntarily Conducted Germinal Choice', in Wolsteholme (ed.), *Man and His Future*, 258–9.

92. Martin Richards, 'Artificial Insemination and Eugenics: Celibate Motherhood, Eutelegenesis and Germinal Choice', *Studies in History and Philosophy of Biological and Biomedical Sciences*, 39, no. 2 (2008): 211–21.

93. Hermann Muller to JSH, 17 September 1964, JSH Papers, RU, MS 50, Box 37: 3.

94. JSH, 'Eugenics in Evolutionary Perspective', f. 37.

95. JSH, 'The Future of Man – Evolutionary Aspects', in Wolstenholme (ed.), *Man and His Future*, 17.

96. THH to JSH, 24 March 1892, in *L&L*, II, 436–8.

97. Virginia Rozée and Sayeed Unisa (eds.), *Assisted Reproductive Technologies in the Global South and North: Issues, Challenges and the Future* (New York: Routledge, 2016).

CHAPTER 10: OTHER WORLDS

1. Max Weber, 'Science as a Vocation' (original German lecture, 1918). Weber's thesis has been discussed ever since. See Egil Asprem, *The Problem of Disenchantment: Scientific Naturalism and Esoteric Discourse 1900–1939* (Leiden: Brill, 2014); Jason Josephson-Storm, *The Myth of Disenchantment: Magic, Modernity, and the Birth of the Human Sciences* (Chicago: University of Chicago Press, 2017); Joshua Landy and Michael Saler (eds.), *The Re-Enchantment of the World: Secular Magic in a Rational Age* (Stanford: Stanford University Press, 2009).

2. Bernard Lightman, *The Origins of Agnosticism: Victorian Unbelief and the Limits of Knowledge* (Baltimore: Johns Hopkins University Press, 1987).

3. Aldous Huxley, *The Doors of Perception* (London: Chatto & Windus, 1954).

4. Henrietta Heathorn to THH, quoted in Martin Huxley Cooke, *The Evolution of Nettie Huxley* (Chichester: Phillimore, 2008), 36.

5. Henrietta Heathorn, Diary, 3 June 1849, in THH, *Rattlesnake Diary*, 290.

6. Henrietta Heathorn, Diary, 20 January 1850, in THH, *Rattlesnake Diary*, 290.

7. From Henrietta A. Huxley, 'An Agnostic Hymn', in *Poems* (1913), 74–5.

8. Julian Huxley commentary on 'Miss Heathorn and her Journal', *Rattlesnake Diary*, 291.

9. Browne, *Charles Darwin: The Power of Place*, 151–2.

10. THH to Charles Kingsley, 23 September 1860, in *L&L*, I, 217.

11. Peter J. Bowler, *Reconciling Science and Religion: The Debate in Early Twentieth-Century Britain* (Chicago: University of Chicago Press, 2001), 16.

12. THH to Charles Kingsley, 23 September 1860, in *L&L*, I, 217–21.

13. Leonard Huxley, *L&L*, I, 216.

14. THH to Charles Kingsley, 23 September 1860, in *L&L*, I, 217, 219.

15. THH, 'Agnosticism' (1889), in THH, *Science and Christian Tradition: Collected Essays* (London: Macmillan, 1894), 245.

16. THH, *Crayfish*, 318.

17. THH quoted in JSH and Douglas Cleverdon, *Julian Huxley on T. H. Huxley: A New Judgment* (London: Watts & Co, 1945), 17.

18. Leonard Huxley, *Thomas Henry Huxley: A Character Sketch* (London: Watts, 1920), 93.

19. 'Professor Huxley on Agnosticism', *Agnostic Journal*, 27 April 1889, newscutting in THHC, ICL, Series 10, 47.62.

20. G. W. Foote, *Freethinker*, 1 (May 1881): 1.

21. Bowler, *Reconciling Science and Religion*, 18.

22. THH to Charles Kingsley, 5 May 1863, *L&L*, I, 241; THH, 'Agnosticism', 237–8.

23. THH, 'On the Methods and Results of Ethnology' (1865), in *Anthropological Essays*, 243.

24. THH, 'Biogenesis and Abiogenesis' (1870), in *Discourses: Biological and Geological*, 231, 235.

25. THH, 'Biogenesis', 255–6.

26. Alfred Pagan to THH, 5 February 1892, THHC, ICL, Series 20, 121: 97–8.

27. THH, *Science and Christian Tradition* (1894); THH, *Science and Hebrew Tradition: Collected Essays* (London: Macmillan, 1895).

28. THH, *Rattlesnake Diary*, 18 October 1849, 246.

29. THH, 'Evolution and Ethics' (1893), 68, 62–3. See Vijitha Rajapakse, 'Buddhism in Huxley's "Evolution and Ethics": A Note on a Victorian Evaluation and Its Comparativist Dimension', *Philosophy East and West*, 35, no. 3 (1985): 295–304.

30. Swami Ranganathananda, *Vedanta and Modern Science, Being Correspondence between Sir Julian Huxley and Swami Ranganathananda on 'The Message of the Upsanisads'* (Bombay: Bharatiya Vidya Bhavan, 1971); Sumita Roy, Annie Pothen and K. S. Sunita (eds.), *Aldous Huxley and Indian Thought* (New Delhi: Sterling, 2003).

31. T. W. Rhys Davids, *On Nirvāna, and on the Buddhist Doctrines of the 'Groups' the Sanskāras, Karma, and the 'Paths'* (London, 1877); T. W. Rhys Davids, *Buddhism: Being a Sketch of the Life and Teachings of Gautama, the Buddha* (London: Society for Promoting Christian Knowledge, 1880).

32. Hesketh, *Of Apes and Ancestors*.

33. THH to Mary Ward, 1892, quoted in Mrs Humphry Ward, *A Writer's Recollections*, II, 173, 175, 121.

34. John William Parker (ed.), *Essays and Reviews* (London: John W. Parker, 1860).

35. Julia Huxley to JSH, 21 March 1906, JH Papers, RU, MS474, Box 9: 13.

36. Matthew Arnold, *God and the Bible* (1875) (London: John Murray, 1906), xiv.

37. THH notes, 'The Ark of the Covenant and the Tabernacle', and 'Further notes and materials on Old Testament subjects'. Stone Tables, THHC, ICL, Series 10, 46. 1, 59–68.

38. THH, 'Further notes and materials on Old Testament subjects', THHC, ICL, Series 10, 46.68.

39. THH to Mary Ward, 1892, quoted in Ward, *A Writer's Recollections*, II, 173–4.

40. THH to Mary Ward, 1892, quoted in Ward, *A Writer's Recollections*, II, 172–5.

41. *Christianity and Agnosticism: A Controversy* (New York: Humboldt Publishing, 1889). Papers by Henry Mace, Thomas Henry Huxley, the Bishop of Peterborough, W. H. Mallock and Mrs Humphry Ward.

42. Gail Hamilton, 'A New Analysis of the Agnosticism of Robert Elsmere', *Chicago Daily Tribune*, 17 March 1889, 25.

43. Ward, *A Writer's Recollections*, II, 172–3.

44. Ward, *A Writer's Recollections*, II, 97.

45. Mrs Humphry Ward, *Robert Elsmere* (1888) (London: Smith, Elder, 1914), 328.

46. Hamilton, 'A New Analysis of the Agnosticism of Robert Elsmere', 25.

47. Michael R. Watts, 'The Seal and Servant of Christianity: The Spiritualist Alternative', in *The Dissenters: The Crisis and Conscience of Nonconformity* (Oxford: Oxford University Press, 2015), 41; Gowan Dawson, '"The Cross-Examination of the Physiologist": T. H. Huxley and the Resurrection', in Catherine Marshall, Bernard Lightman and Richard England. (eds.), *The Metaphysical Society (1869–1880): Intellectual Life in Mid-Victorian England* (Oxford: Oxford University Press, 2019), 91–118.

48. THH to Mary Ward, 15 March 1888, in *L&L*, II, 193–4.

49. William S. Peterson, 'Gladstone's Review of Robert Elsmere: Some Unpublished Correspondence', *Review of English Studies*, 21, no. 84 (1970): 442–61.

50. Edward Everett Hale, Marion Harland, Joseph Cook and Julia Ward Howe, 'Robert Elsmere's Mental Struggles', *North American Review*, 1 January 1889, 119–27. Gladstone's review was reprinted and included in this symposium.

51. W. E. Gladstone, *Robert Elsmere and the Battle of Belief* (New York: Anson D. F. Randolph, 1888). Bernard Lightman, '*Robert Elsmere* and the Agnostic Crisis of Faith', in Richard J. Helmstadter and Bernard Lightman (eds.), *Victorian Faith in Crisis* (London: Palgrave Macmillan, 1990), 283–311.

52. Esther Grace Wiseman, 'Mrs Humphry Ward's *Robert Elsmere* and Its Effect upon the Press', MA thesis, University of Illinois, 1915, 2.

53. Thomas Arnold, *Passages in a Wandering Life* (London: Edward Arnold, 1900).

54. Mary Ward to JSH and Trevenen Huxley, 8 November 1908, JH Papers, RU, MS474, Box 10: 19.

55. Julia Huxley to Mary Ward, 4 November 1908, MAWL, BL RP8974.

56. THH, 'Spiritualism Unmasked', *Pall Mall Gazette*, 49, no. 7423 (1 January 1889), 1–2.

57. THH, Report of a Séance, handwritten note, 27 January 1874, THH ICL, Series 11, Box 49: 121–6.

58. George Howard Darwin to THH, 22 January 1874 and 28 January 1874, THH ICL, Series 1d, vol. 13: 87–9.

59. THH to Charles Kingsley, 22 May 1863, in *L&L*, I, 242–3.

60. THH, 'Spiritualism Unmasked', 2.

61. Herbert Junius Hardwicke to THH, 24 February 1893, THHC, ICL, Series 1h, 18: 28–9.

62. Elizabeth Blackwell to THH 4 July 1892, THHC, ICL, Series 1b, 11: 11.

63. See Ann Braude, *Radical Spirits: Spiritualism and Women's Rights in Nineteenth Century America* (Bloomington: Indiana University Press, 2001).

64. Anna Blackwell, *The Philosophy of Existence* (London: J. Burns, 1871); Anna Blackwell, *Spiritualism and Spiritism* (Glasgow: Printed for the Writer by H. Nisbet, 1873).

65. Elizabeth Blackwell to THH, 21 July 1892, THHC, ICL, Series 1b, 11: 12.

66. Elizabeth Blackwell to THH, 5 September 1892, THHC, ICL, Series 1b, 11: 14.

67. A. R. Wallace, 'Spiritualism and Science', *The Times*, 4 January 1873, 10. See also Ian Hesketh, 'The Future Evolution of "Man"', in Efram Sera-Shriar, *Historicizing Humans: Deep Time, Evolution, and Race in Nineteenth-Century British Sciences* (Pittsburgh: University of Pittsburgh Press, 2018), 193–281.

68. THH to Wallace, 1 November 1886, in J. Marchant (ed.), *Alfred Russel Wallace: Letters and Reminiscences*, vol. II (London: Cassell and Co., 1916), 187–8.

69. THH, 'On the Classification and Distribution of the *Alectoromorphae* and *Heteromophae*', *Proceedings of the Zoological Society of London*, 14 May 1868, 313.

70. Gerald Massey, *Concerning Spiritualism* (London: James Burns, 1871), 55.

71. Wells, JSH and Wells, *The Science of Life*, III, 933–4, 938.

72. Wells, JSH and Wells, *The Science of Life*, III, 887, 935.
73. Wells, JSH and Wells, *The Science of Life*, III, 941–2, photograph on 934.
74. Wells, JSH and Wells, *The Science of Life*, III, 885–6, 895–6.
75. Wells, JSH and Wells, *The Science of Life*, III, 896–7, 929.
76. Wells, JSH and Wells, *The Science of Life*, III, 931–32.
77. Wells, JSH and Wells, *The Science of Life*, III, 940.
78. JSH to Havelock Ellis, 21 October 1932, Havelock Ellis Letters, Columbia University, Box 1.
79. JSH, *Memories* I, 174–5.
80. JSH, *Evolution: The Modern Synthesis,* 578.
81. JSH, Darwin and the Idea of Evolution, typescript, May 1958, JSH Papers, RU, MS50, Box 74: 5
82. Julia Huxley to JSH, 31 March 1908, JH Papers, RU, MS474, Box 9: 13.
83. JSH, *Religion without Revelation*, 112.
84. Obituary, 'Death of Mrs Leonard Huxley', newsclipping, n.d., JH Papers, RU, MS474, Box 9: 13.
85. JSH, *Religion without Revelation*, 118, 119–20.
86. John Collier, *The Religion of an Artist* (London: Watts, 1926).
87. John Collier to JSH, 19 May 1926, JH Papers, RU, MS474, Box 9: 2.
88. Wells, Huxley, and Wells, *The Science of Life*, I, 3–4.
89. H. G. Wells, *God the Invisible King* (London: Cassell, 1917), preface.
90. Quote from jacket of the book series. Agnes Maude Roydon, *I Believe in God* (New York: Harper, 1927); Ronald Knox, *The Belief of Catholics* (London: E. Benn, 1927).
91. JSH, *Religion without Revelation*, 29–31; Peter J. Bowler, 'Julian Huxley (1887–1975): Religion without Revelation', in Nicolaas Rupke (ed.), *Eminent Lives in Twentieth Century Science and Religion* (Frankfurt am Main: Peter Lang, 2009), 215–32.
92. JSH, *Religion without Revelation*, 30.
93. JSH, *Religion without Revelation*, 127–8.
94. JSH, 'What I Believe', draft BBC talk, January 1957, JH Papers, RU, MS474, Box 23: 2.
95. JSH, *Memories* I, 205, 287.
96. JSH to John Craven Pritchard, 11 December 1968, Papers of John Craven Pritchard (1899–1992), Political and Economic Planning Correspondence, UEA Archives, PP/4/1/11/10.
97. JSH, 'Introduction', Pierre Teilhard de Chardin, *The Phenomenon of Man* (1959) (New York: Harper, 2008), 11.
98. JSH, *Memories* II, 27–9.
99. JSH, 'Introduction', *The Phenomenon of Man*, 11–30.

100. JSH, 'Introduction', *The Phenomenon of Man*, 13–18.

101. JSH, 'Introduction', *The Phenomenon of Man*, 26.

102. JSH to John Craven Pritchard, 11 December 1968, Papers of John Craven Pritchard (1899–1992), Political and Economic Planning Correspondence, UEA Archives, PP/4/1/11/10.

103. T. Phillips and H. Aarons, 'Looking "East": An Exploratory Analysis of Western Disenchantment', *International Sociology*, 22, no. 3 (2007): 325–41.

104. Aldous Huxley, *Island* (1962) (London: Chatto & Windus, 1975), ch. 8.

105. Aldous Huxley, 'Human Potentialities', in JSH (ed.), *The Humanist Frame* (London: George Allen & Unwin, 1961), 424–5.

106. THH, notebook, 22 November 1840, THHC, ICL, Series 3, 31: 172–3. See also Todd H. Weir (ed.), *Monism: Science, Philosophy, Religion and the History of a Worldview* (Basingstoke: Palgrave Macmillan, 2012).

107. Ernst Haeckel, *Monism as Connecting Religion and Science: The Confession of Faith of a Man of Science*, trans. J. Gilchrist (London: A. & C. Black, 1894).

108. Evelleen Richards, '"Metaphorical Mystifications": The Romantic Gestation of Nature in British Biology', in *Ideology and Evolution in Nineteenth Century Britain*, 27–39; Nicholas Jardine, '*Naturphilosophie* and the Kingdoms of Nature', in Nicholas Jardine, James A. Secord and Emma C. Spary (eds.), *Cultures of Natural History* (Cambridge: Cambridge University Press, 1996), 230–45.

109. Julian Huxley, *Education and the Humanist Revolution* (Southampton: Southampton University Press, 1962), 8.

110. Aldous Huxley to JSH, summer 1908, Prior's Field, JH Papers, RU, MS474, Box 11: 2.

111. Aldous Huxley, 'Human Potentialities', 424. R. S. Deese, *We Are Amphibians: Julian and Aldous Huxley on the Future of Our Species* (Berkeley: University of California Press, 2014).

112. William Blake, 'Auguries of Innocence' (1803), published 1863.

113. JSH, *Religion without Revelation*, 368.

114. JSH, *Religion without Revelation*, 344.

115. Laura Archera Huxley, *This Timeless Moment: A Personal View of Aldous Huxley* (London: Chatto & Windus, 1969), 295.

116. JSH, *Religion without Revelation*, 123–5.

117. JSH, *Religion without Revelation*, 114.

118. JSH, *Religion without Revelation*, 106.

119. JSH, *Religion without Revelation*, 134–6.

120. Roberta Russell and Ronald David Laing, *R. D. Laing and Me: Lessons in Love* (New York: Hillgarth Press, 1992), 63.

121. JSH, *Religion without Revelation*, 139–40.

122. Francis Huxley, *The Way of the Sacred* (London: Aldus Books, 1974), 209.

123. Francis Huxley, *The Way of the Sacred*, 278–9.

124. Francis Huxley, *The Way of the Sacred*, 6, 244.

125. Wells, JSH and Wells, *The Science of Life*, I, 274.

126. Francis Huxley, *The Way of the Sacred*, 6.

127. JSH, *Religion without Revelation*, 368.

128. JSH, *Religion without Revelation*, 356.

129. JSH to Juliette Huxley, 29 June 1966, J&JH Papers, RU, MS512, Box 37: 7.

130. JSH, 'Beethoven's Missa Solemnis in Ely Cathedral', attached to JSH to SZ, 20 February 1974, SZ Papers, UEA Archives, SZ/GEN/HUX.

131. Professor David Napier, personal communication with the author, 1 July 2021.

132. THH, *Crayfish*, 3.

EPILOGUE

1. From Henrietta A. Huxley, 'Browning's Funeral', *Poems* (1913), 52. The final three lines, slightly altered, are Thomas Henry Huxley's epitaph.

2. JSH to SZ, 23 August 1973, JSH to SZ, 29 October 1974, ZS to JSH, 2 November 1974, memorial meeting sheet for Julian Sorell Huxley, SZ Papers, UEA Archives, SZ/GEN/HUX.

3. Joseph Needham, 'Huxley Remembered', *Nature*, 254 (6 March 1975): 2; Peter Medawar, René Maheu, E. B. Ford, G. P. Wells, 'Huxley: The Presence of the "S" in UNESCO is Largely Due to Him', *Nature*, 254 (6 March 1975): 4–5; Milo Keynes and G. Ainsworth Harrison (eds.), *Evolutionary Studies: A Centenary Celebration of the Life of Julian Huxley* (London: Macmillan, 1989).

4. THH, 'Abstract of a Friday Evening Discourse', 30 April 1852, *Proceedings of the Royal Institution*, 1 (1851–4): 1849.

5. JSH, 'The Meaning of Death', *Cornhill* magazine (April 1911): 492–4; JSH, *The Individual in the Animal Kingdom* (Cambridge: The University Press, 1912), 19.

6. Wells, JSH and Wells, *The Science of Life*, III, 942.

7. JSH, *The Stream of Life*, 5.

8. JSH, *The Individual in the Animal Kingdom*, 26.

9. Julian Huxley, 'Thomas Henry Huxley and Julian Huxley: An Imaginary Interview (1942)', in *On Living in a Revolution* (London: Chatto & Windus, 1944), 83–9.

10. Andrew Huxley, 'The Galton Lecture for 1987: Julian Huxley – A Family View', in Keynes and Harrison (eds.), *Evolutionary Studies*, 9–25.

11. Juliette Huxley to May Sarton, 22 June 1985, in Susan Sherman (ed.), *Dear Juliette: Letters of May Sarton to Juliette Huxley* (New York: Norton, 1999), 364.

12. JSH, *The Individual in the Animal Kingdom*, 19–20.

13. From Henrietta A. Huxley, 'Thoughts on a Birth', in Henrietta A. Huxley, *Poems* (1913), 154.

14. From THH, 'Westminster Abbey', in Henrietta A. Huxley, *Poems* (1913), 3–4.

15. JSH, 'The Meaning of Death', 507.

16. J. Charles Deduchson (Editor of *Toronto Mail*) to THH, 24 January 1887, THHC, ICL, Series ID, 13: 136–8.

Index